NAVAIR 01-F14AAA-1

F-14 TOMCAT
PILOT'S FLIGHT OPERATING MANUAL

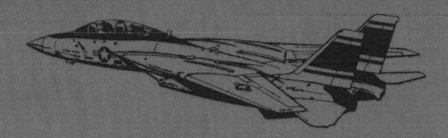

ISSUED BY AUTHORITY OF THE CHIEF OF NAVAL OPERATIONS
AND UNDER THE DIRECTION OF THE COMMANDER,
NAVAL AIR SYSTEMS COMMAND

Reprint ©2009 Periscope Film LLC
All Rights Reserved
www.PeriscopeFilm.com

ISBN # 978-1-935327-71-4   1-935327-71-2

THE AIRCRAFT	1
INDOCTRINATION	2
NORMAL PROCEDURES	3
FLIGHT CHARAC	4
EMER PROCEDURES	5
ALL-WEA OPERATION	6
COMM PROCEDURES	7
WEAPONS SYSTEMS	8
FLT CREW COORD	9
NATOPS EVAL	10
PERFORM DATA	11
INDEX & FOLDOUTS	

NAVAIR 01-F14AAA-1

NATOPS FLIGHT MANUAL
NAVY MODEL

F-14A
AIRCRAFT

1 August 1981

THIS MANUAL SUPERSEDES NAVAIR 01-F14AAA-1
DATED 1 JANUARY 1980, CHANGED 15 JULY 1981
AND NAVAIR 01-F14AAA-1P DATED 1 JANUARY 1980,
CHANGED 15 MARCH 1981, WHICH SHOULD BE DESTROYED

ISSUED BY AUTHORITY OF THE CHIEF OF NAVAL OPERATIONS
AND UNDER THE DIRECTION OF THE COMMANDER,
NAVAL AIR SYSTEMS COMMAND

Reprint ©2009 Periscope Film LLC
All Rights Reserved
www.PeriscopeFilm.com

ISBN # 978-1-935327-71-4   1-935327-71-2

THE AIRCRAFT	1
INDOCTRINATION	2
NORMAL PROCEDURES	3
FLIGHT CHARAC	4
EMER PROCEDURES	5
ALL-WEA OPERATION	6
COMM PROCEDURES	7
WEAPONS SYSTEMS	8
FLT CREW COORD	9
NATOPS EVAL	10
PERFORM DATA	11
INDEX & FOLDOUTS	

CHANGE NOTICE

THESE ARE SUPERSEDING OR SUPPLEMENTARY PAGES TO SAME PUBLICATION OF PREVIOUS DATE
Insert these pages into basic publication. Destroy superseded pages.

NAVAIR 01-F14AAA-1

NATOPS FLIGHT MANUAL
NAVY MODEL
F-14A
AIRCRAFT

GRUMMAN AEROSPACE CORPORATION

TACTICAL SOFTWARE EFFECTIVITY: 111D/P7D TAPE

This publication is required for official use or for administrative or operational purposes only. Distribution is limited to U.S. Government agencies. Other requests for this document must be referred to Commanding Officer, Naval Air Technical Services Facility, 700 Robbins Avenue, Philadelphia, PA 19111.

ISSUED BY AUTHORITY OF THE CHIEF OF NAVAL OPERATIONS
AND UNDER THE DIRECTION OF THE COMMANDER,
NAVAL AIR SYSTEMS COMMAND

THE AIRCRAFT	1
INDOCTRINATION	2
NORMAL PROCEDURES	3
FLIGHT CHARAC	4
EMER PROCEDURES	5
ALL-WEA OPERATION	6
COMM PROCEDURES	7
WEAPONS SYSTEMS	8
FLT CREW COORD	9
NATOPS EVAL	10
PERFORM DATA	11
INDEX & FOLDOUTS	

1 August 1981
Change 1 — 1 June 1982

Reproduction for non-military use of the information or illustrations contained in this publication is not permitted without specific approval of the Commander, Naval Air Systems Command. The policy for military use reproduction is established for the Air Force in AFR 205-1 and for Navy and Marine Corps in OPNAVINST 5510.1 Series.

LIST OF EFFECTIVE PAGES

Dates of issue for original and change pages are:

Original 0 . . . . 1 August 1981
Change 1 . . . . 1 June 1982

TOTAL NUMBER OF PAGES IN THIS PUBLICATION IS 681 CONSISTING OF THE FOLLOWING:

Page No.	Issue
*Title	1
*A–B	1
C–G	0
Letter of Promulgation (Reverse blank)	0
i–viii	0
(ix blank)/x	0
1-1–1-4	0
*1-5–1-13	1
1-14–1-207	0
*1-208	1
*1-208A/(1-208B blank)	1
*1-209	1
1-210–1-213	0
*1-214	1
1-215/(1-216 blank)	0
2-1–2-4	0
3-1–3-26	0
*3-27	1
3-28–3-43	0
*3-44	1
*3-44A/(3-44B blank)	1
3-45–3-80	0
4-1–4-16	0
4-17/(4-18 blank)	0
*5-1–5-2	1
5-3–5-5	0
*5-6	1
*5-6A/(5-6B blank)	1
*5-7	1
5-8–5-23	0
*5-24	1
*5-24A/(5-24B blank)	1
*5-25–5-27	1
5-28–5-36	0
*5-37	1
5-38–5-64	0
6-1–6-20	0

Page No.	Issue
6-21/(6-22 blank)	0
7-1–7-82	0
*8-1–8-40	0
9-1–9-8	0
*9-8A/(9-8B blank)	1
9-10	0
*9-11	1
9-12–9-13	0
*9-14	1
9-15–9-16	0
*9-17	1
9-18	deleted
9-20–9-60	0
10-1–10-20	0
10-21/(10-22 blank)	0
11-1/11-2 blank)	0
FO-0 (Reverse blank)	0
FO-1 (Reverse blank)	0
FO-2 (Reverse blank)	0
FO-3 (Reverse blank)	0
FO-4 (Reverse blank)	0
FO-5 (Reverse blank)	0
FO-6 (Reverse blank)	0
FO-7 (Reverse blank)	0
FO-8 (Reverse blank)	0
FO-9 (Reverse blank)	0
FO-10 (Reverse blank)	0
FO-11 (Reverse blank)	0
FO-12 (Reverse blank)	0
FO-13 (Reverse blank)	0
FO-14 (Reverse blank)	0
FO-15 (Reverse blank)	0
FO-16 (Reverse blank)	0
FO-17 (Reverse blank)	0
FO-18 (Reverse blank)	0
FO-19 (Reverse blank)	0
FO-20 (Reverse blank)	0
*Index 1 – Index 24	1
*Index 25/(Index 26 blank)	1

*The asterisk indicates pages changed, added, or deleted by the current change

NAVAIR 01-F14AAA-1

INTERIM CHANGE SUMMARY

The following Interim Changes have been canceled or previously incorporated in this manual:

INTERIM CHANGE NUMBER(S)	REMARKS/PURPOSE
1 through 79	Incorporated or canceled.

The following Interim Changes have been incorporated in this Change/Revision:

INTERIM CHANGE NUMBER(S)	REMARKS/PURPOSE
72	Held in abeyance (fuel migration).
80	Revised ECS emergency procedures.
81	Revised maneuvering limits.
82	Revised upright departure/flap spin procedures.

NAVAIR 01-F14AAA-1

Interim Changes Outstanding - To be maintained by the custodian of this manual:

INTERIM CHANGE NUMBER	ORIGINATOR/DATE (or DATE/TIME GROUP)	PAGES AFFECTED	REMARKS/PURPOSE

PART 2 — SYSTEMS

AIR INLET CONTROL SYSTEM

The purpose of the air inlet control system (AICS) is to decelerate supersonic air and to provide even, subsonic airflow to the engine throughout the aircraft flight envelope. The AICS consists of two variable geometry intakes one on each side of the fuselage at the intersection of the wing glove and fuselage. Intake inlet geometry is varied by three automatically controlled hinged ramps on the upper side of the intakes. The ramps are positioned to decelerate supersonic air by creating a compression field outside the inlet and to regulate the amount and quality of air going to the engine. The rectangular intakes are spaced away from the fuselage to minimize boundary layer ingestion and are highly raked to optimize operation at high angle-of-attack.

Inlet ramps are positioned by electrohydraulic actuators, which respond to fixed schedules in the AICS programmers. Separate programmers, probes, sensors, actuators and hydraulic power systems provide completely independent operation of the left and right air inlet control systems. Figure 1-3 shows the basic elements of AICS mechanization. Electrical power is from the essential no. 2 ac and dc buses. Hydraulic power is supplied individually to the left AICS from the combined hydraulic system and to the right AICS from the flight hydraulic system.

Normal AICS Operations

No pilot control is required during the normal (AUTO) mode of operation. Electronic monitoring in the AICS detects failures that would degrade system operation and performance (refer to AICS BIT). AICS caution lights (L and R INLET, L and R RAMPS), and INLET RAMPS switches are shown in figure 1-4.

Sectional side views of representative variable geometry inlet configurations scheduled by AICS programmers and descriptive nomenclature are shown in figure 1-5.

During ground static and low-speed (Mach < 0.35) operation, the inlet ramps are mechanically restrained in the stowed (retracted) position. The predominant airflow is concentrated about the lower lip of the inlet duct and is supplemented by reverse airflow through the bleed door, around the forward lip of the third ramp. As flight speed is increased to 0.35 Mach, hydraulic power is ported to the ramp actuators, but the ramps are not scheduled out of the stowed position until Mach 0.5. (See figure 1-6.) The bleed slot bleeds low-energy, boundary-layer air from the movable ramps.

At airspeeds greater than 0.5 Mach, the ramps program primarily as a function of Mach for optimum AICS performance. At transonic speeds, a normal shock wave attaches to the second movable ramp. The third ramp deflects above 0.9 Mach to maintain proper bleed slot height (Δh) for transonic and low supersonic flight.

At supersonic speeds, four shock waves compress and decelerate the inlet air. The bleed slot removes boundary layer air and stabilizes the shock waves. This design results in substantially higher performance above Mach 2 than simpler inlet designs.

AICS Test

Two types of AICS tests are provided to check the general condition of the AICS and to pinpoint system components causing detected failures; AICS Built-in Test (BIT) and On-board Check (OBC).

AICS BUILT-IN-TEST (BIT)

BIT in the AICS computer programmer is automatically and continually initiated within the programmer to check AICS components when the programmer is energized.

The operational status of the AICS depends on BIT detected failures in AICS components. Failures of static or total pressure sensors, ramp no. 1, 2, or 3 positioning, programmer continuous end-to-end BIT, or hydraulic pressure to any of the ramp actuators, would seriously degrade AICS performance. Detected failures of these items cause the AICS to automatically transfer to a significantly degraded fail-safe mode of operation, indicated by illumination of an INLET caution light.

Detected failures of angle-of-attack, total temperature, engine fan speed, mid-compression by-pass (MCB) discrete, or out-of calibration detection of the difference between P1 and P2 (ΔP), Ps or Pt sensors will cause the AICS to revert to the slightly degraded fail-operational mode of operation. Nominal values of angle-of-attack, total temperature, or engine fan speed are substituted for the failed values in the AICS programmer, without illumination of an INLET caution light.

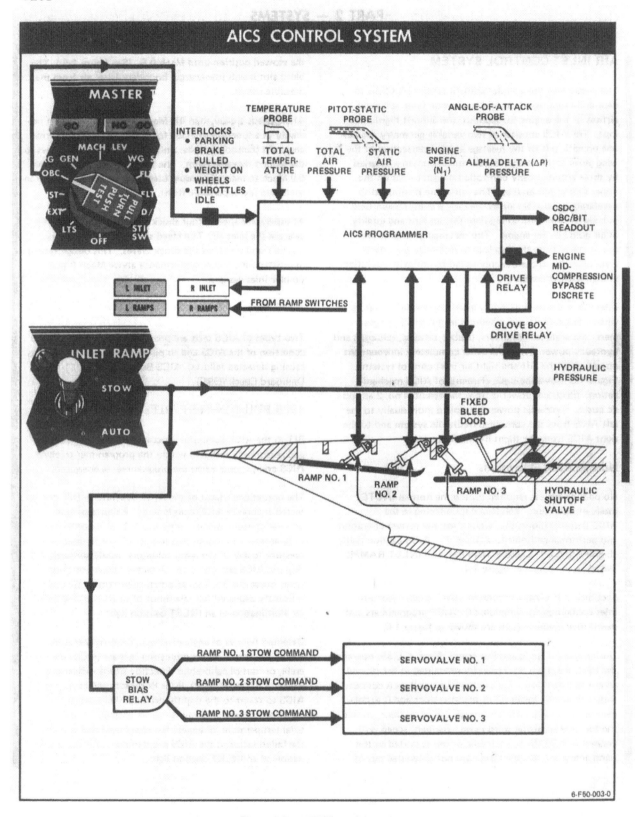

Figure 1-3. AICS Control System

NAVAIR 01-F14AAA-1

SUMMARY OF APPLICABLE TECHNICAL DIRECTIVES

INFORMATION RELATING TO THE FOLLOWING TECHNICAL DIRECTIVES HAS BEEN INCORPORATED IN THIS MANUAL

CHANGE NUMBER	DESCRIPTION	DATE INCORPORATED IN MANUAL	VISUAL IDENTIFICATION
AAC 645	AIM-54A ECS Heating	1 Nov 75	None
AAC 688	Sidewinder Cannot be Jettisoned	1 May 77	Decal is applied to LAU-7.
AAC 733	AIM-54 Coolant Pump Thermal Switch Interface	15 Mar 79	None
ACC 390	Modification of Emergency Harness Release Handle	1 July 78	None
ACC 404	Single Personnel Services Disconnect	1 June 74	Single connection.
AFB 198	Inspection of Flight Control Systems Angle-of-Attack Filter	15 Mar 79	None
AFC 166	Super Single CADC Processor	1 Nov 73	None
AFC 291	Preclude Inadvertent Firing Signals to GCU	1 June 74	None
AFC 304	Elimination of Static/Dyna Unbalance of Lateral Control System	1 May 77	None
AFC 338	Dual UHF	1 June 75	Radio Panels Changed
AFC 341	Negative g Vent Inlet Flapper	1 May 77	None
AFC 361	Modification of Seat Detonation Lines	1 June 75	None
AFC 364	Maneuver Slat	1 May 77	Barber poles on slat indicator with slats 7° or less
AFC 365	Integrated Trim	1 Nov 75	INTEG TRIM advisory light
AFC 371	AFCS Interim Autopilot Relief Mode	1 May 77	None
AFC 400	Aileron to Rudder Interconnect (ARI)	1 May 77	None
AFC 410	Wing Sweep Emergency Handle Locks	15 May 76	None
AFC 441	AICS Reprogramming (Rev 4)	1 Nov 76	None

D

NAVAIR 01-F14AAA-1

SUMMARY OF APPLICABLE TECHNICAL DIRECTIVES

INFORMATION RELATING TO THE FOLLOWING TECHNICAL DIRECTIVES HAS BEEN INCORPORATED IN THIS MANUAL

CHANGE NUMBER	DESCRIPTION	DATE INCORPORATED IN MANUAL	VISUAL IDENTIFICATION
AFC 451	Aircraft Defueling	15 May 76	None
AFC 454	Increase Strength of Longeron and Eliminate Fatigue Cracks on Upper Deck	1 May 77	None
AFC 461	Reinforce Upper Nacelle Longeron	1 May 77	None
AFC 467	Gun Bay General Cleanup	1 May 77	None
AFC 488	Activate AFCS, ACLS, PCD, and Vector Modes	15 May 76	Mach position deleted from AFCS control panel
AFC 494	Replace AN/ARC-51A with AN/ARC-159	15 May 76	AN/ARC 159 control panel
AFC 500	Flap/Slat Flex Drive Shaft	1 May 77	None
AFC 520	Incorporation of COATS	1 July 78	AWG-15 BIT failure time on TID
AFC 523	Replace Three Spoiler Module Circuit Breakers With One	1 Mar 78	Single SPOILER MODULE PUMP circuit breaker
AFC 531	Engine Fan Blade Integral Containment	1 May 77	None
AFC 536	Forward Fixed Cowl Skin Reinforcement	1 May 77	None
AFC 541	Fuel Vent Tank Modification	1 Mar 78	None
AFC 554	Revert From AUTO to BOOST Mode with CAGE button	15 Mar 79	None
AFC 556	AICS Dump Inhibit	1 Mar 78	None
AFC 561	Installation of ECM Equipment	1 Jun 74	ALR-45/50 and ALE-39 installed
AFC 564	Voltage Monitor Control Unit	1 Mar 78	None
AFC 566	Modify ARI	1 Mar 78	None
AFC 570	Arresting Hook Modification	1 Jan 79	Leaf spring wiper on uplock release mechanism

E

NAVAIR 01-F14AAA-1

SUMMARY OF APPLICABLE TECHNICAL DIRECTIVES

INFORMATION RELATING TO THE FOLLOWING TECHNICAL DIRECTIVES HAS BEEN INCORPORATED IN THIS MANUAL

CHANGE NUMBER	DESCRIPTION	DATE INCORPORATED IN MANUAL	VISUAL IDENTIFICATION
AFC 573	Spoiler Override Panel	15 Mar 79	Spoiler override panel on pilot's right console
AFC 577	IDG Scavenge Filter	1 Jan 80	None
AFC 578	Tactical Air Reconnaissance Pod System (TARPS)	1 Jan 80	Reconnaissance controller processor signal unit on RIO left console
AFC 579	HOT OIL Light Modification	1 Jan 79	Raised temperature setting: less illumination anomalies
AFC 580	AWG-9 Coolant Pump Thermal Switch	15 Mar 79	None
AFC 582	Forward Anticollision Light	1 Nov 79	None
AFC 587	Flight Control Binding on Caution Advisory Panel	15 Mar 79	Blank panel above caution advisory panel
AFC 592	Mid Compression Bypass Test Panel	15 Mar 79	Mid Compression Bypass Test panel on RIO's right console
AFC 595	APC Modification	1 Nov 79	Boost to manual mode in 0.25 second
AFC 597	AIM-54 Coolant Pump Thermal Switch	15 Mar 79	None
AFC 601	Nose Gear Strut Switch	1 Nov 79	None
AFC 610	Transfer Critical Items — ac/dc Essential No. 1 and No. 2 Circuit Breakers	1 May 80	None
AFC 618	HUD Camera Circuit Breaker Installation	1 Oct 80	HUD camera ac and dc circuit breakers
AFC 620	Turn and slip indicator lighting control	15 July 81	None
AFC 622	Brake Pressure Gage	1 Oct 80	Two needles on gage
AFC 628	EMERG STORES JETT switch modification	1 Oct 80	EMERG JETTISON caution light on pilot's caution/advisory panel

F

NAVAIR 01-F14AAA-1

SUMMARY OF APPLICABLE TECHNICAL DIRECTIVES

INFORMATION RELATING TO THE FOLLOWING TECHNICAL DIRECTIVES HAS BEEN INCORPORATED IN THIS MANUAL

CHANGE NUMBER	DESCRIPTION	DATE INCORPORATED IN MANUAL	VISUAL IDENTIFICATION
AFC 636	Postlight for BRAKE PRESSURE Gage	1 Oct 80	Postlight above brake pressure gage
AFC 655	Red stripes on nose gear drag brace oleo.	1 Aug 81	Red stripes painted on nose gear drag brace oleo.
AVC 1135	APC Modification	1 Nov 79	Boost to manual mode in 0.25 second.
AVC 1539	Modification of Control Stick Trigger Switch	1 Nov 75	None
AVC 1588	Approach Power Compensator	1 June 74	None
AVC 1709	Modification of Control Stick Trigger Switch	1 Nov 75	None
AVC 2102	AIM-7 Interface modification, AIM-7 Autotone/software timer	1 Jan 80	None
AVC 2212	Mid Compression Bypass System Modification	15 Mar 79	Mid compression bypass test panel on RIO right console.
AVC 2235	APC Modification	1 Nov 79	Boost to manual mode in 0.25 second.
AYC 598	267-Gallon External Fuel Tank Modification	1 Mar 78	Tank fin and tailcone removed.
AYC 633	APC Modification	1 Nov 79	Boost to manual mode in 0.25 second.
AYC 660P1	Elimination of Automatic Flap Retraction	1 July 80	None
IAFC 623	APC emergency disengage	1 Jan 80	None
IAFC 2342	APC emergency disengage	1 Jan 80	None

DEPARTMENT OF THE NAVY
OFFICE OF THE CHIEF OF NAVAL OPERATIONS
WASHINGTON, D.C. 20350

1 August 1981

LETTER OF PROMULGATION

1. The Naval Air Training and Operating Procedures Standardization Program (NATOPS) is a positive approach toward improving combat readiness and achieving a substantial reduction in the aircraft accident rate. Standardization, based on professional knowledge and experience, provides the basis for development of an efficient and sound operational procedure. The standardization program is not planned to stifle individual initiative, but rather to aid the Commanding Officer in increasing his unit's combat potential without reducing his command prestige or responsibility.

2. This manual standardizes ground and flight procedures but does not include tactical doctrine. Compliance with the stipulated manual procedure is mandatory except as authorized herein. In order to remain effective, NATOPS must be dynamic and stimulate rather than suppress individual thinking. Since aviation is a continuing, progressive profession, it is both desirable and necessary that new ideas and new techniques be expeditiously evaluated and incorporated if proven to be sound. To this end, Commanding Officers of aviation units are authorized to modify procedures contained herein, in accordance with the waiver provisions established by OPNAVINST 3510.9 series, for the purpose of assessing new ideas prior to initiating recommendations for permanent changes. This manual is prepared and kept current by the users in order to achieve maximum readiness and safety in the most efficient and economical manner. Should conflict exist between the training and operating procedures found in this manual and those found in other publications, this manual will govern.

3. Checklists and other pertinent extracts from this publication necessary to normal operations and training should be made and may be carried in Naval Aircraft for use therein. It is forbidden to make copies of this entire publication or major portions thereof without specific authority of the Chief of Naval Operations.

W. L. McDONALD
Vice Admiral, USN
Deputy Chief of Naval Operations
(Air Warfare)

THIS PAGE INTENTIONALY LEFT BLANK.

NAVAIR 01-F14AAA-1

TABLE OF CONTENTS

SECTION	I	THE AIRCRAFT	1-1
Part	1	Aircraft and Engine	1-2
Part	2	Systems	1-5
Part	3	Aircraft Servicing	1-180
Part	4	Operating Limitations	1-195
SECTION	II	INDOCTRINATION	2-1
SECTION	III	NORMAL PROCEDURES	3-1
Part	1	Flight Preparation	3-2
Part	2	Shore-Based Procedures	3-3
Part	3	Carrier-Based Procedures	3-48
Part	4	Special Procedures	3-60
Part	5	Functional Checkflight Procedures	3-66
SECTION	IV	FLIGHT CHARACTERISTICS	4-1
SECTION	V	EMERGENCY PROCEDURES	5-1
Part	1	Ground Emergencies	5-3
Part	2	Takeoff Emergencies	5-6
Part	3	In-Flight Emergencies	5-10
Part	4	Landing Emergencies	5-40
Part	5	Ejection and Bailout	5-48
SECTION	VI	ALL-WEATHER OPERATIONS	6-1
Part	1	Instrument Procedures	6-1
Part	2	Extreme Weather Operations	6-16
SECTION	VII	COMMUNICATIONS-NAVIGATION EQUIPMENT AND PROCEDURES	7-1
Part	1	Communications	7-1
Part	2	Navigation	7-31
Part	3	Identification	7-75
SECTION	VIII	WEAPONS SYSTEMS	8-1
Part	1	Vertical Display Indicator Group (AN/AVA-12)	8-1
Part	2	Multiple Display Indicator Group (AN/ASA-79)	8-21
Part	3	System Overview	*
Part	4	Computer Subsystem	*
Part	5	Radar Subsystem	*
Part	6	RIO's Controls and Displays	*
Part	7	Pilot's Controls and Displays	*
Part	8	Air-to-Air Tactical Functions	*

*REFER TO NAVAIR 01-F14AAA-1A.

Part	9	Armament Control System	*
Part	10	Air-to-Air Weapons Tactical Functions	*
Part	11	Air-to-Ground	*
Part	12	Data Link System	*
Part	13	In-Flight Training	*
Part	14	Defensive Electronic Countermeasures	**
Part	15	ECM/ECCM Systems	**
Part	16	TARPS Computer Subsystem and Displays	8-33

SECTION	IX	FLIGHTCREW COORDINATION	9-1
Part	1	Flightcrew Coordination	9-1
Part	2	Aircraft Self-Test	9-5

SECTION	X	NATOPS EVALUATION	10-1

SECTION	XI	PERFORMANCE DATA	11-1
Part	1	Standard Data	***
Part	2	Takeoff	***
Part	3	Climb	***
Part	4	Cruise	***
Part	5	Endurance	***
Part	6	Descent	***
Part	7	Landing	***
Part	8	Mission Planning	***

FOLDOUT ILLUSTRATIONS	FO-0

ALPHABETICAL INDEX	Index-1

Note

List of illustrations titles are included in alphabetical index.

*Refer to NAVAIR 01-F14AAA-1A
**Refer to NAVAIR 01-F14AAA-1A.1
***Refer to NAVAIR 01-F14AAA-1.1

FOREWORD

SCOPE

This NATOPS Flight Manual is issued by the authority of the Chief of Naval Operations and under the direction of Commander, Naval Air Systems Command in conjunction with the Naval Air Training and Operating Procedures Standardization (NATOPS) Program. This manual, together with the supplemental manuals listed below, contain information on all aircraft systems, performance data, and operating procedures required for safe and effective operations. Supplemental manuals have been separated from this manual for your convenience in handling classified information and performance charts. However, it is not a substitute for sound judgment. Compound emergencies, available facilities, adverse weather or terrain, or considerations affecting the lives and property of others may require modification of the procedures contained herein. Read this manual from cover to cover. It is your responsibility to have a complete knowledge of its contents.

APPLICABLE PUBLICATIONS

The following applicable publications complement this manual:
- NAVAIR 01-F14AAA-1.1 (Performance Charts)
- NAVAIR 01-F14AAA 1A (Supplemental)
- NAVAIR 01-F14AAA-1A.1 (Secret Supplemental Flight Manual)
- NAVAIR 01-F14AAA-1B (Pocket Checklist)
- NAVAIR 01-F14AAA-1F (Functional Checkflight Checklist)
- NWP 55-5-F14 (Tactical Manual, Volumes I through III)
- NWP 55-5F14(PG) Tactical Pocket Guide

HOW TO GET COPIES

Each flightcrew member is entitled to personal copies of the NATOPS Flight Manual and appropriate applicable publications.

Automatic Distribution

To receive future changes and revisions to this manual or any other NAVAIR aeronautical publication automatically, a unit must be established on an automatic distribution list maintained by the Naval Air Technical Services Facility (NATSF). To become established on the list or to change existing NAVAIR publication requirements, a unit must submit the appropriate tables from NAVAIR 00-25DRT-1 (Naval Aeronautic Publications Automatic Distribution Requirement Tables) to NATSF, Code 321, 700 Robbins Avenue, Philadelphia, Pa. 19111. Publication requirements should be reviewed periodically and each time requirements change, a new NAVAIR 00-25DRT-1 should be submitted. NAVAIR 00-25DRT-1 only provides for future issues of basic, changes, or revisions and will not generate supply action for the issuance of publications from stock. For additional instructions, refer to NAVAIRINST 5605.4 series and Introduction to Navy Stocklist of Publications and Forms NAVSUP Publication 2002 (S/N 0535-LP-004-0001).

Additional Copies

Additional copies of this manual and changes thereto may be procured by submitting DD Form 1348 to NAVPUBFORMCEN Philadelphia in accordance with Introduction to Navy Stocklist of Publications and Forms NAVSUP Publication 2002.

UPDATING THE MANUAL

To ensure that the manual contains the latest procedures and information, NATOPS review conferences are held in accordance with OPNAVINST 3510.9 series.

CHANGE RECOMMENDATIONS

Recommended changes to this manual or other NATOPS publications may be submitted by anyone in accordance with OPNAVINST 3510.9 series.

Routine change recommendations are submitted directly to the Model Manager in OPNAV Form 3500-22 shown on the next page. The address of the Model Manager of this aircraft is:

> Commanding Officer
> Fighter Squadron 124
> Naval Air Station
> Miramar, California 92145
>
> Attn: F-14 Model Manager

Change recommendations of an URGENT nature (safety of flight, etc.) should be submitted directly to the NATOPS Advisory Group Member in the chain of command by priority message.

YOUR RESPONSIBILITY

NATOPS Flight Manuals are kept current through an active manual change program. Any corrections, additions, or constructive suggestions for improvement of its content should be submitted by routine or urgent change recommendation, as appropriate, at once.

NATOPS FLIGHT MANUAL INTERIM CHANGES

Flight Manual Interim Changes are changes or corrections to the NATOPS Flight Manuals promulgated by CNO or NAVAIRSYSCOM. Interim Changes are issued either as printed pages, or as a naval message. The Interim Change Summary page is provided as a record of all interim changes. Upon receipt of a change or revision, the custodian of the manual should check the updated Interim Change Summary to ascertain that all outstanding interim changes have been either incorporated or cancelled; those not incorporated shall be recorded as outstanding in the section provided.

CHANGE SYMBOLS

Revised text is indicated by a black vertical line in either margin of the page, adjacent to the affected text, like the one printed next to this paragraph. The change symbol identifies the addition of either new information, a changed procedure, the correction of an error, or a rephrasing of the previous material.

WARNINGS, CAUTIONS, AND NOTES

The following defintions apply to "WARNINGS", "CAUTIONS", and "NOTES" found throughout this manual.

WARNING

An operating procedure, practice, or condition, etc., which may result in injury or death if not carefully observed or followed.

CAUTION

An operating procedure, practice, or condition, etc., which may result in damage to equipment if not carefully observed or followed.

Note

An operating procedure, practice, or condition, etc., which is essential to emphasize.

WORDING

The concept of word usage and intended meaning which has been adhered to in preparing this Manual is as follows:

"Shall" has been used only when application of a procedure is mandatory.

"Should" has been used only when application of a procedure is recommended.

"May" and "need not" have been used only when application of a procedure is optional.

"Will" has been used only to indicate futurity, never to indicate any degree of requirement for application of a procedure.

NATOPS/TACTICAL CHANGE RECOMMENDATION
OPNAV FORM 3500/22 (5-69) 0107-722-2002   DATE

TO BE FILLED IN BY ORIGINATOR AND FORWARDED TO MODEL MANAGER						
FROM (originator)			Unit			
TO (Model Manager)			Unit			
Complete Name of Manual Checklist		Revision Date	Change Date	Section/Chapter	Page	Paragraph

Recommendation (be specific)

☐ CHECK IF CONTINUED ON BACK

Justification

Signature		Rank	Title	
Address of Unit or Command				

TO BE FILLED IN BY MODEL MANAGER (Return to Originator)

FROM	DATE
TO	

REFERENCE
(a) Your Change Recommendation Dated ....

☐ Your change recommendation dated .... is acknowledged. It will be held for action of the review conference planned for .... to be held at ....

☐ Your change recommendation is reclassified URGENT and forwarded for approval to .... by my DTG ....

/S/ ....   MODEL MANAGER   AIRCRAFT

GLOSSARY

A

A/A	Air-to-air
AAC	Aviation Armament Change
AB	Afterburner
ac	Alternating current
ACL	Automatic Carrier landing
ACLS	Automatic carrier landing system
ACM	Air combat maneuver
A/D	Analog-to-digital
ADD	Angle of attack as measured by the alpha probe
ADF	Automatic direction finder
ADI	Attitude director indicator
AFB	Air Frame Bulletin
AFC	Air Frame Change
AFCS	Automatic flight control system
A/G	Air-to-ground
AGL	Above ground level
AHRS	Attitude heading reference system
AICS	Air inlet control system
AIM-54A	Phoenix missile
AM	Amplitude modulation
AOA	Angle-of-attack
APC	Approach power compensator
ARI	Aileron rudder interconnect
ASH	Automatic stored heading
ATDC	Airborne Tactical Data Control
AVC	Avionic Change
AWCS	Airborne Weapons Control System
AWL	All-weather landing

B

BARO	Barometric
BDHI	Bearing distance heading indicator
BINGO	Return fuel state
BIT	Built-in-test
BLS	Basic landing display
Bolter	Hook down, unintentional touch and go

C

CADC	Central air data computer
CAINS	Carrier Aircraft Inertial Navigation System
CAP	Combat air patrol/computer address panel
CARQUAL	Carrier qualifications
CAS	Calibrated air speed
CAT	Catapult

C (cont)

CATCC	Carrier Air Traffic Control Center
cb	Circuit breaker
CCA	Carrier controlled approach
cg	Center of gravity
CGTL	Command track line
Charlie Time	Expected time over ramp
CICU	Computer integrated converter unit
CIPDU	Control indicator power distrubution unit
CM	Continuous monitor
CMPTR	Computer
COATS	CICAS operational assessment
CPS	Controller processor signal unit
CSD	Constant speed drive
CSDC	Computer signal data converter
CSS	Control stick steering
CV	Aircraft carrier

D

D/A	Digital-to-analog
dB	Decibel
dc	Direct current
DDD	Detail data display
DDI	Digital data indicator
DDS	Data Display System
DECM	Defensive electronic countermeasures
DEST	Destination
DFM	Dog Fight mode
DG	Directional gyro
D/L	Data link
DLC	Direct lift control
DMA	Degraded mode assessment
DRO	Destructive Readout
DROT	Degraded range on target

E

EAC	Expected Approach Clearance time
EAS	Equivalent airspeed
ECM	Electronic countermeasures
ECMD	Electronic countermeasures display
ECS	Environmental control system
EGT	Exhaust gas temperature
EIF	Exposure interval factor
EMCON	Electronic radiation control
EPR	Engine pressure ratio

F

FAM	Familiarization
FCLP	Field carrier landing practice
FD	Fault direction
FF	Fuel flaw
FI	Fault isolation
FL	Flight level
FMLP	Field mirror landing practice
FOD	Foreign object damage
FOV	Field-of-view
FWD	Forward

G

G	Guard channel
g	Gravity
GCU	Generator control unit
GHz	Gigahertz

H

HCU	Hand control unit
HDG	Heading
HERO	Hazards of electromagnetic radiation to ordnance
Hot start	A start that exceeds normal starting temperatures
HSD	Horizontal situation display
HSI	Horizontal situation indicator
HUD	Heads-up display
Hung start	A start that results in a stagnated rpm and temperature

I

IAS	Indicated airspeed
ICAO	International Civil Aviation Organization
ICS	Intercommunications
IDG	Integrated-drive generator
IFF	Identification, friend or foe
IFR	Instrument flight rules
ILS	Instrument landing system
IMN	Indicated Mach number
IMU	Inertial measurement unit
INS	Inertial navigation system
IP	Initial point
IR	Infrared
IRLS	IR line scanner
IRNR	IR not ready
IRW	IR wider
ITS	Integrated trim system

K

KCAS	Knots calibrated airspeed
kHz	Kilohertz
KIAS	Knots indicated airspeed
KTS	Knots

L

LAOT	Low-prf antenna on target
LARI	Lateral automatic rudder interconnect
LBA	Limits of basic aircraft
LE	Leading edge
LOX	Liquid oxygen
LPA	Life Preserver Assembly
LS	Line scanner
LSO	Landing Signal Officer (Paddles)

M

M	Mach
MAC	Mean aerodynamic chord
MAG VAR	Magnetic variation
MAS	Missile auxiliary subsystem
MAX	Maximum
MCB	Mid compression bypass
MCM	Monitor control message
MDIG	Multipurpose display indicator group
Meatball	Glide slope image of mirror landing system
MHz	Megahertz
MIL	Military
MOAT	Missile on aircraft test
MR	Maintenance readout
MRT	Military rated thrust
MSL	Mean sea level
MTDS	Marine Tactical Data System
MTM	Magnetic tape memory

N

NAG	Air-to-ground mode
NATOPS	Naval Air Training and Operating Procedures Standardization
NDRO	Non-destructive readout
NFO	Naval Flight Officer
NFOV	Narrow field of view
nmi	Nautical miles
NOTAM	Notices to Airmen
NOZ	Nozzle
NPS	Navigation power supply
NR	Number
NTDS	Naval Tactical Data System

N (cont)

NWP	Naval Warfare Publications
N_1	Low-pressure compressor rotor speed
N_2	High-pressure compressor rotor speed

O

OAT	Outside air temperature
OBC	On-board check
OFT	Operational flight trainer
OWF	Overwing Fairing

P

Paddles	Landing Signal Officer
PAN	Panoramic
PCD	Precision course direction
PDCP	Pilot's display control panel
PDS	Pulse Doppler Search
PDSTT	Pulse Doppler Single Target Track
PH	Phoenix missile
PIO	Pilot induced oscillation
PP	Peak power
PPH	Pounds per hour
P_s	Static pressure
psi	Pounds per square inch
PSTT	Pulse Single Target Track
PT	Engine power trim
P_{T7}	Turbine exhaust pressure
P_t	Total pressure

Q

Q	Dynamic pressure

R

RARI	Rudder automatic rudder interconnect
RECON	Reconnaissance
r-f	Radio frequency
RIO	Radar Intercept Officer
ROT	Range on target
RPM	High-pressure compressor rotor speed (N_2)
RTGS	Real time gun sight
RWS	Range while search

S

SAM	Surface-to-air missile
SAR	Participating units and procedures

S (cont)

SAS	Stability augmentation system
SAT	Simultaneous alignment and test
SC	Sensor control
SIF	Selective identification feature
SINS	Ship's inertial navigation system
SMAL	Single mode alinement
STBY	Standby
STT	Single Target Track
SW	Sidewinder missile

T

TACAN	Tactical air navigation
TARPS	Tactical air reconnaissance pod system
TAS	True airspeed
TCR	Time code readout
TCS	Television camera set
TED	Trailing edge down
TEU	Trailing edge up
TID	Tactical information display
TIT	Turbine inlet temperature
T_S	Static temperature
T_{T2}	Compressor inlet temperature
T_{T4}	Compressor discharge temperature

U

UHF	Ultra-high frequency
UTM	Universal Test message

V

VERT	Vertical
VDI	Vertical display indicator
VDIG	Vertical display indicator group
VFR	Visual flight rules
V_g/H	Velocity divided by altitude
VID	Visual identification
VMCU	Voltage monitor control unit
V_R	Rotation speed
V_1	Critical engine failure speed

W

WCS	Weapons Control System
WFOV	Wide field of view
WOD	Wind over the deck
WRA	Weapons replaceable assembly
WST	Weapons system trainer

THIS PAGE INTENTIONALY LEFT BLANK.

NAVAIR 01-F14AAA-1

F-14A TOMCAT

Figure 1-0

SECTION I – THE AIRCRAFT

TABLE OF CONTENTS

PART 1 – AIRCRAFT AND ENGINE
- Aircraft 1-2

PART 2 – SYSTEMS
- Air Inlet Control System 1-5
- Engines 1-13
- Engine Bleed Air 1-18
- Engine Ignition System 1-25
- Engine Starting System 1-25
- Engine Compartment Ventilation 1-28
- Engine Oil System 1-29
- Engine Instruments 1-31
- Fire Detection System 1-33
- Fire Extinguishing System 1-34
- Engine Fuel System 1-35
- Afterburner Fuel System 1-37
- Aircraft Fuel System 1-40
- Precheck System 1-53
- Electrical Power Supply System 1-56
- Hydraulic Power Supply Systems 1-64
- Pneumatic Power Supply Systems ... 1-71
- Central Air Data Computer (CADC) .. 1-72
- Wing Sweep System 1-74
- Flaps and Slats 1-82
- Glove Vanes 1-88
- Speed Brakes 1-89
- Flight Control Systems 1-90
- Automatic Flight Control System (AFCS) 1-105
- Landing Gear Systems 1-113
- Wheel Brake System 1-116
- Nosewheel Steering System 1-120
- Nose Gear Catapult System 1-121
- Arresting Hook System 1-126
- Environmental Control System (ECS) 1-128
- Oxygen System 1-139
- Pitot-Static System 1-140
- Flight Instruments 1-142
- Angle-of-Attack System 1-144
- Canopy System 1-148
- Ejection System 1-151
- Lighting System 1-159
- Jettison System 1-165
- Miscellaneous Equipment 1-169
- Tactical Air Reconnaissance Pod System (TARPS) 1-171

PART 3 – AIRCRAFT SERVICING
- Servicing Data 1-180
- Ground Handling 1-187

PART 4 – OPERATING LIMITATIONS
- Limitations 1-195
- Airspeed Limitations 1-200
- Acceleration Limits 1-200
- Angle-of-Attack Limits 1-206
- Maneuvering Limits 1-208
- SAS Limits 1-208
- Takeoff and Landing Flap and Slat Transition Limits 1-211
- Approach Power Compensator (APC) System Limits 1-211
- Autopilot Limits 1-211
- Gross Weight Limits 1-211
- Barricade Engagement Limits 1-213
- Center of Gravity Position Limits .. 1-213
- Crosswind Limits 1-213
- Ground Operation Limits 1-213
- Ejection Seat Operation Limits 1-214
- Prohibited Maneuvers 1-214
- External Stores and Gun Limits 1-214
- Banner Towing Restrictions 1-215

PART 1 — AIRCRAFT AND ENGINE

AIRCRAFT

The F-14A aircraft is a supersonic, two-place, twin-engine, swing-wing, air-superiority fighter designed and manufactured by Grumman Aerospace Corporation. In addition to its primary fighter role, carrying missiles (Sparrow and/or Sidewinder) and an internal 20-millimeter gun, the aircraft is designed for fleet air defense (Phoenix missiles) and ground attack (conventional ordnance) missions. Armament and peculiar auxiliaries used only during secondary missions are installed in low-drag, external configurations. Mission versatility and tactical flexibility are enhanced through independent operational capability or integration under existing tactical data systems.

The forward fuselage, containing the flightcrew and electronic equipment, projects forward from the mid-fuselage and wing glove. Outboard pivots in the highly swept wing glove support the movable wing panels, which incorporate integral fuel cells and full-span leading edge slats and trailing edge flaps for supplemental lift control. In flight, the wings may be varied in sweep, area, camber, and aspect ratio by selection of any leading-edge sweep angle between 20° and 68°. Wing sweep can be automatically or manually controlled to optimize performance and thereby enhance aircraft versatility. Separate variable-geometry air inlets, offset from the fuselage in the glove, direct primary airflow to two TF30-P-414 dual axial-compressor, turbofan engines equipped with afterburners for thrust augmentation. The displaced engine nacelles extend rearward to the tail section, supporting the twin vertical tails, horizontal tails, and ventral fins. The middle and aft fuselage, which contains the main fuel cells, tapers off in depth to the rear where it accommodates the speed brake surfaces and arresting hook. Retractable vanes in the glove leading edges extend to supplement lift and compensate for changes in the aircraft aerodynamic center. All control surfaces are positioned by irreversible hydraulic actuators to provide desired control effectiveness throughout the flight envelope. Stability augmentation features in the flight control system enhance flight characteristics and thereby provide a more stable and maneuverable weapons delivery platform. The tricycle-type, forward-retracting landing gear is designed for nose gear capapult launch and carrier landings. Missiles and external stores are carried from eight hardpoint stations on the center fuselage between the nacelles and under the nacelles and wing glove; no stores are carried on the movable portion of the wing. The fuel system incorporates both inflight and single-point ground refueling capabilities. Aircraft general dimensions are shown in figure 1-1. Foldout pages FO-1 and FO-2 show the general placement of components within the aircraft.

Aircraft Weight

The zero fuel and stores gross weight of the aircraft is approximately 41,587 pounds, which includes trapped fuel, oil, gun, pilot, and RIO. Consult the applicable Handbook of Weight and Balance for the exact weight of any series aircraft.

Cockpit

The aircraft accommodates a two-man crew consisting of a pilot and RIO in a tandem seating arrangement. To maximize external field of view, stations within the tandem cockpit are prominently located atop the forward fuselage and enclosed by a single clamshell canopy. Integral boarding provisions to the cockpit and aircraft top deck are on the left side of the fuselage. Each crew station incorporates a rocket ejection seat that is vertically adjustable for crew accommodation. A single environmental control system provides conditioned air to the cockpit and electronics bays for pressurization and air conditioning. Oxygen for breathing is supplied to the crew under pressure from liquid storage bottles. The cockpit arrangement provides minimum duplication of control capability, which, of necessity, requires two crewmen for flight.

The forward station of the cockpit is arranged and equipped for the pilot. In addition to three multipurpose electronic displays for viewing tactical, flight, navigational, and ECM data, the pilot's instrument panel also contains armament controls, flight and engine instruments. Engine controls, fuel management, auxiliary devices, autopilot and communications control panels are on the left console. Display, power, lighting, and environmental controls are on the right console.

The aft station of the cockpit is equipped for the RIO. This instrument panel contains controls and displays for the AN/AWG-9 Airborne Weapon Control Systems and navigational flight instruments. Armament controls, sensor controls, keyboard panels, and communications panels are on the left console. The right console contains an ECM/navigational display, ECM controls, data link controls and lighting, and the IFF panel. Refer to pages FO-3 through FO-6 for illustrations of cockpit arrangements.

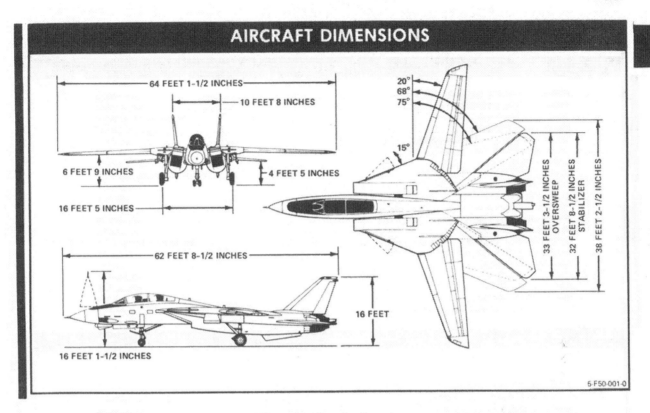

Figure 1-1. Aircraft Dimensions

Electronic Nomenclature

Figure 1-2 is an alphabetical listing of the tactical, communication, navigation, flight control, and instruments in the aircraft.

Technical Directives

As technical changes are made to the aircraft, those that affect aircraft operation or pilot and RIO need-to-know operation will be incorporated in the appropriate sections and listed in the Summary of Applicable Technical Directives in the front of this manual. In some instances, Technical Directives may be incorporated on the aircraft while it is still on the production line before delivery. Check the Technical Directives Section of the Aircraft Log Book for applicable modifications. The following are types of Technical Directives used in this manual:

AFC	Airframe Change
AFB	Airframe Bulletin
AVC	Avionics Change
ACC	Air Crew System Change
AAC	Aviation Armament Change
AYC	Accessories Change
IAVC	Interim Avionics Change

Block Numbers

The following production block numbers correspond to aircraft serial numbers (BUNO).

Block No.	Serial No. (BUNO)
65	158620 – 158637
70	158978 – 159006
75	159007 – 159429
80	159430 – 159468
85	159588 – 159637
90	159825 – 159874
95	160379 – 160414
100	160652 – 160696
105	160887 – 160930
110	161133 – 161168
115	161270 – 161293
120	161416 – 161445

ELECTRONICS NOMENCLATURE

TACTICAL

AIRBORNE WEAPON CONTROL SYSTEM	AN/AWG-9
AIRCRAFT FIRE CONTROL SET	AN/AWG-15
CENTRAL AIR DATA COMPUTER	CP-1066/A, CP-1166B/A
CHAFF DISPENSING SET	AN/ALE-29A/39
COMPUTER SIGNAL DATA CONVERTER	CP-1165A/A
DATA DISPLAY SYSTEM	AN/ASQ-172
DIGITAL DATA LINK	AN/ASW-27B
ELECTRONIC COUNTERMEASURES SET	AN/ALQ-100 OR AN/ALQ-126
FUZE FUNCTION CONTROL SET	AN/AWW-4
GUN CONTROL UNIT	C-8571/A
IFF INTERROGATOR SET	AN/APX-76(V)
IFF TRANSPONDER SET	AN/APX-72
IR RECONNAISSANCE SET	AN/AAD-5
INTERFERENCE BLANKER	MX-9467/A, MX-10161/A
MULTIPLE DISPLAY INDICATOR GROUP	AN/ASA-79
PANORAMIC CAMERA	KA-99A
RADAR RECEIVER SET	AN/ALR-50
RADAR WARNING SET	AN/ALR-45
SERIAL FRAME CAMERA	KS-87B
VERTICAL DISPLAY INDICATOR GROUP	AN/AVA-12

COMMUNICATION

CRYPTOGRAPHIC SYSTEM	KY-28
INTERCOMMUNICATIONS SYSTEM	LS-460/B
UHF AUXILIARY RECEIVER SET	AN/ARR-69
UHF COMMUNICATIONS SET	AN/ARC-51
UHF COMMUNICATIONS SET	AN/ARC-159

NAVIGATION

ATTITUDE HEADING REFERENCE SET	A/A24G-39
INERTIAL NAVIGATION SYSTEM	AN/ASN-92(V)
RADAR ALTIMETER	AN/APN-194(V)
RADAR BEACON AUGMENTOR SET	AN/APN-154B(V)
RECEIVER-DECODER GROUP	ARA-63A
TACTICAL NAVIGATION SET	AN/ARN-84
UHF-ADF DIRECTIONAL FINDER	AN/ARA-50

FLIGHT CONTROL AND INSTRUMENTS

AIR INLET CONTROL PROGRAMMER	C-8684A/A
AIRSPEED AND MACH NUMBER INDICATOR	AVU/-24/A
APPROACH POWER CONTROL SET	AN/ASW-105
AUTOMATIC FLIGHT CONTROL SET	AN/ASW-32, AN/ASW-43
BEARING DISTANCE HEADING INDICATOR	ID-633-C/U
COCKPIT ALTIMETERS	AAU-19/A
VERTICAL VELOCITY INDICATOR	AAU-18A
STANDBY COMPASS	AQO-5/A, AQU-5/A

Figure 1-2. Electronic Nomenclature

PART 2 — SYSTEMS

AIR INLET CONTROL SYSTEM

The air inlet control system (AICS) consists of two primary air inlets, one on each side of the fuselage at the intersection of the wing glove and fuselage. The purpose of the AICS is to decelerate free-stream air in flight to provide even, subsonic airflow to the engine throughout the aircraft flight envelope. Figure 1-3 shows the basic elements of the AICS mechanization.

The rectangular cross-sectional inlet sidewalls are spaced away from the fuselage to minimize boundary layer air ingestion and are highly raked to optimize operation for high angle-of-attack conditions. Inlet ramps are positioned by electrohydraulic actuators, which respond to fixed schedules in the AICS programmers. Separate probes, sensors, programmers, actuators, and hydraulic power systems provide completely independent operation of the left and right air inlet control systems. No pilot control is required during normal modes of operation. Electronic monitoring in the AICS detects failures that would degrade system operation and performance. AICS caution lights (L and R INLET, L and R RAMPS), and INLET RAMPS stow switches are shown in figure 1-4. Electrical power supply for AICS operation is from ac and dc essential no. 2 buses. Hydraulic power is supplied individually to the left AICS from the combined hydraulic system and to the right AICS from the flight hydraulic system.

Normal AICS Operation

Sectional side views of representative variable geometry inlet configurations with descriptive nomenclature are shown in figure 1-5. Inlet geometry is varied by three automatically controlled hinged ramps on the upper side of the inlets that are independently positioned to decelerate the air, thereby controlling the formation of shock waves in the external compression field, outside the inlet and regulate the amount and quality of air going to the engine. During ground static and low-speed (< Mach 0.5) operation, the inlet ramps are mechanically restrained in the stowed (retracted) position. The predominant airflow is concentrated about the area defined by the lower lip of the inlet duct and is supplemented by reverse air flow through the bleed door around the forward lip of the third ramp. As flight speed is increased to 0.35 Mach, hydraulic power is ported to the ramp actuators, but the ramps are not scheduled out of the stowed position until 0.5 Mach. (See figure 1-6.) The throat slot bleeds low-energy, boundary-layer air from the movable ramps.

At airspeeds greater than 0.5 Mach, the ramps program as a function of Mach for optimum AICS performance. At transonic speeds, a normal shock wave attaches to the second movable ramp. The third ramp deflects with the first two movable ramps to maintain proper throat-slot height for transonic and low supersonic flight.

At supersonic speeds, four shock waves compress and decelerate the engine air. The design results in substantially higher performance above Mach 2 than simpler inlet designs. The throat slot removes boundary-layer air and stabilizes the shock waves.

AICS Test

Two types of AICS tests are provided to check the general condition of the AICS and to pinpoint the appropriate system component causing detected failures.

AICS BUILT-IN TEST (BIT)

Built in test (BIT) in the AICS computer programmer is automatically and continually initiated within the programmer to check components of the AICS when the programmer is energized.

The operational status of the AICS depends on the detected failures of the systems major WRA's, for example, static and total pressure sensors, positioning errors in the No. 1, 2, and 3 ramps, programmer continuous end-to-end BIT, and loss of hydraulic pressure to any of the ramp servo cylinders. A detected failure of these major WRA's will cause the AICS to automatically transfer to a seriously degraded fail-safe-mode of operation indicated by the illumination of an INLET caution light.

Detected failures in the angle of attack, total temperature, engine fan speed, bleed door and/or mid-compression by-pass discrete, and out-of calibration detection of the Δ Pl, Ps, Pt sensors, will cause the AICS to revert to less serious fail-operational mode of operation. Nominal values of angle of attack, total temperature, and engine fan speed will be substituted for the failed values in the AICS programmer without illumination of an INLET caution light.

In all cases detected failures are continuously registered by inflight performance monitoring system and displayed on the horizontal situation (HSD) and the tactical information display (TID).

Note

In aircraft BUNO 159588 and subsequent, the left total temperature probe is not incorporated; however, the electrical circuits in the programmer remain. Total temperature associated acronyms should be ignored.

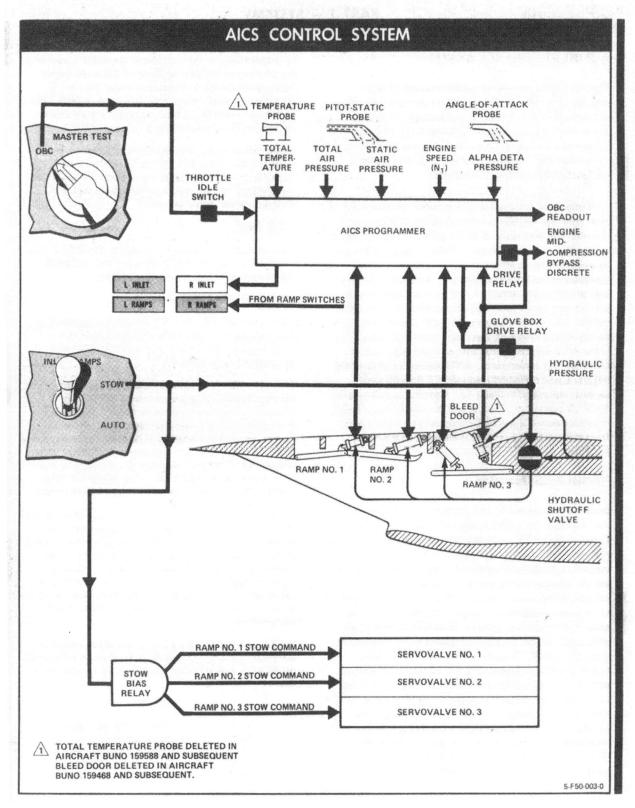

Figure 1-3. AICS Control System

NAVAIR 01-F14AAA-1

AICS CONTROLS AND INDICATORS

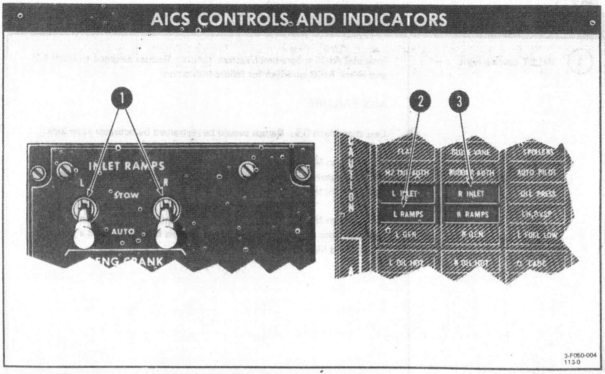

NOMENCLATURE	FUNCTION
① INLET RAMPS switches	AUTO - Inlet ramp position is determined by the AICS programmer. STOW - Electrically commands the respective inlet ramp actuator to the stow position; opens the appropriate hydraulic shutoff valve. **WARNING** • DO NOT takeoff with the INLET RAMPS switches in the STOW position. Hydraulic power is on and may drive the ramps out of the stow locks during certain servocylinder failure modes causing an engine stall. • If wing sweep advisory light is illuminated, cycling R AICS circuit breaker (LG2) may cause unintentional wing sweep unless WING SWEEP DRIVE No. 1 (LE1) and WING SWEEP DRIVE No. 2/MANUV FLAP (LE2) circuit breakers are pulled.
② RAMPS caution light	Indicates ramps not positioned in either stow or trail locks during critical flight conditions (see figure 1-7).

Figure 1-4. AICS Controls and Indicators (Sheet 1 of 2)

NOMENCLATURE	FUNCTION
(3) INLET caution light	Indicates AICS programmer/system failure: Reduce airspeed to Mach 1.2 and check AICS acronym for failure indication. AICS FAILURE Less than Mach 0.5: Ramps should be restrained by actuator stow locks. Greater than Mach 0.5: Ramp movement is restrained by trapped hydraulic pressure and mechanical locks, depending on Mach when INLET light illuminates. Greater than Mach 0.9: Ramp movement is minimized by actuator spool valves and the aerodynamic load profile in this Mach range and a RAMP light should illuminate.

Figure 1-4. AICS Controls and Indicators (Sheet 2 of 2)

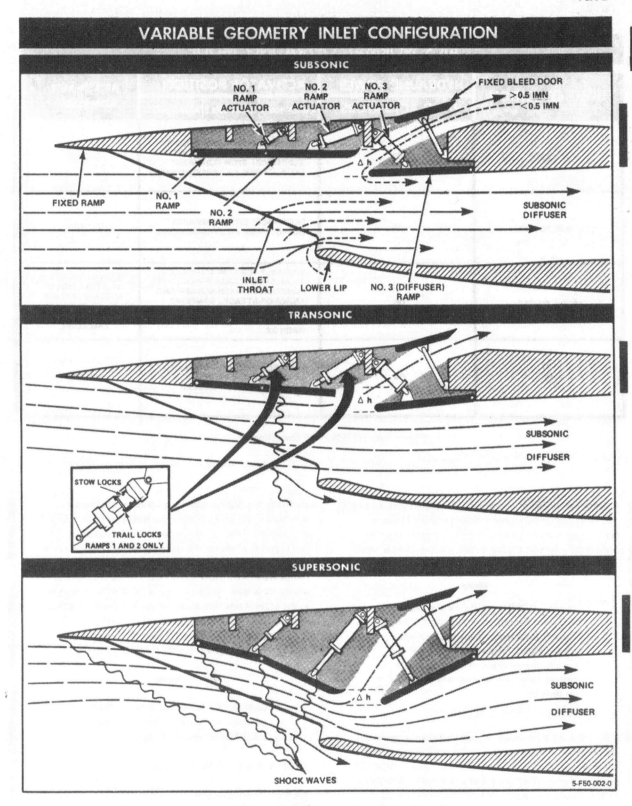

Figure 1-5. Variable Geometry Inlet Configuration

AICS NORMAL OPERATING MODE

FLIGHT CONDITION	HYDRAULIC POWER RAMP ACTUATORS	ACTIVATOR POSITION			AICS → MCB COMMAND
		RAMP NO. 1	RAMP NO. 2	RAMP NO. 3	
M < 0.35	OFF	MECHANICALLY RESTRAINED BY STOW LOCKS IN STOWED POSITION; ELECTRICAL STOW COMMANDS OUTPUT FROM AICS PROGRAMMER.			CLOSED
M > 0.35 TO < 0.5	ON	ELECTRICAL STOW COMMANDS OUTPUT FROM AICS PROGRAMMER.			CLOSED
M > 0.5 TO < 2.2	ON	VARIABLE POSITION SCHEDULED BY AICS PROGRAMMER AS A FUNCTION OF MACH NUMBER AND ANGLE-OF-ATTACK. RAMPS NO. 1 AND NO. 2 BEGIN POSITIONING AT MACH 0.5, RAMP NO. 3 BEGINS AT MACH 0.9.			CLOSED OR OPENED AS A FUNCTION OF ANGLE-OF-ATTACK BOUNDARIES SCHEDULE AND/OR AICS FAIL LOGIC.
M > 2.2	ON	VARIABLE POSITION SCHEDULED BY AICS PROGRAMMER AS A FUNCTION OF MACH NUMBER			OPENED

Figure 1-6. AICS Normal Operating Mode

In both fail modes of operation, detected failures are continuously registered by the inflight performance monitoring system and displayed with AIC acronyms on the horizontal situation display (HSD) and the tactical information display (TID) (figure 1-8).

Note

In aircraft BUNO 159588 and subsequent, and prior aircraft incorporating AFC 441, the left total temperature probe is not incorporated; however the electrical circuits in the programmer remain. Total temperature associated acronyms should be ignored.

AICS ONBOARD CHECK (OBC)

OBC, initiated by the pilot during post-start or ground maintenance checks, performs a dynamic check of the left and right AICS. In addition to the regular AICS BIT program, sensor calibration checks are made. The status of the programmer electronics and the ramp actuators are checked throughout an altitude and airspeed schedule as psuedo pneumatic inputs to the programmer are varied to simulate a flight sequence of maximum airspeed conditions and back to static sea level conditions within 65 seconds. This cycles the ramp actuators through their full range, illuminates the ramp lights, exercises the complete AICS for preflight failure detection and ensures the ramps are in their stow locks. OBC is the only way to ensure stow lock integrity since it verifies the ramps are in the stowed position and then removes ramp hydraulic power. Detected AICS failures are indicated by AIC acronyms or AIC acronym(s) with associated INLET caution light(s) displayed after completion of OBC.

Note

With INLET RAMP switches in STOW position, AICS OBC will fail test and INLET lights will illiminate.

AICS Failure Modes of Operation

AICS mode of operation following a BIT detected failure may be either fail-operational or fail-safe as shown in figure 1-8.

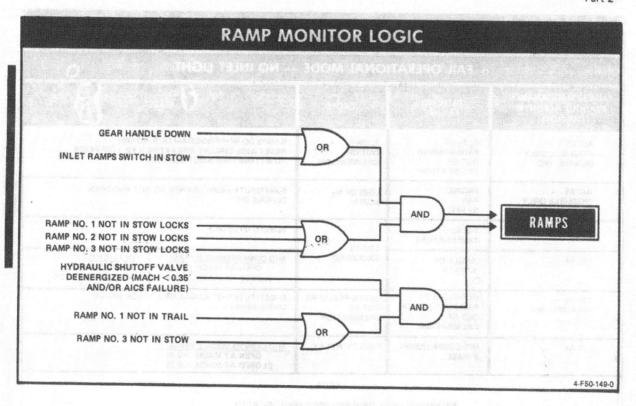

Figure 1-7. Ramp Monitor Logic

FAIL-OPERATIONAL

Failures in the air inlet control system are detected by the AICS programmer, which automatically initiates appropriate corrective action. Mode entry is indicated by the display of a fail-operational AIC acronym. The fail-operational mode results in no significant degradation in AICS operation and the mission can be continued without any flight restrictions or corrective action by the pilot.

FAIL-SAFE

The fail-safe mode results in significantly degraded AICS operation. Mode entry is indicated by the display of a fail-safe AIC acronym and illumination of the appropriate INLET caution light. Under these conditions, the AICS programmer provides an MCB discrete that opens the MCB valve when Mach is above 0.35 and closes the MCB valve when Mach is below 0.35. The programmer also provides a shutoff signal to close the ramps' hydraulic shutoff valve. If the hydraulic shutoff valve closes below Mach 0.9, the ramps are normally in a safe configuration (no. 1 ramp within trail locks, no. 3 ramp in stow locks and no. 2 ramp is restrained by trapped hydraulic pressure.) Engine operations may be successful below 1.2 IMN in this configuration; however corrective procedures shall be performed.

Note

Fail-safe operations result in a slight degradation of cruise and excess thrust performance due to the off-optimum configuration and MCB open at speeds greater than 0.35 IMN.

If the hydraulic shutoff valve closes above Mach 0.9, the ramps are normally in an unsafe configuration, and the appropriate RAMPS caution light will accompany the INLET caution light (figure 1.7). Above Mach 0.9, the no. 3 ramp normally begins programming below the actuator stow lock. When the fail-safe mode is entered above Mach 0.9, the unpowered no. 3 ramp will eventually move and may cause compressor stalls above 80% engine RPM. The aircraft shall be decelerated below 1.2 IMN and the appropriate INLET RAMPS switch shall be selected to STOW.

CAUTION

Do not select STOW at speeds greater than 1.2 IMN. Compressor stalls may occur due to ramp mispositioning.

AICS DETECTED FAILURE MODES

FAIL OPERATIONAL MODE — NO INLET LIGHT

FAILURE MAINTENANCE READOUT ACRONYM (SEE NOTE)	DETECTED FAILURE	CAUSE	RESULT
AIC S1 (POSSIBLE ONLY DURING OBC)	P_s, P_t, OR PROGRAMMER OUT OF CALIBRATION	P_s OR P_t OUT OF CALIBRATION	RAMPS DO NOT PROGRAM DURING OBC. RESET AICS CIRCUIT BREAKERS (LF2, LG2) PRIOR TO ATTEMPTING ANOTHER OBC.
AIC S1 (POSSIBLE ONLY DURING OBC)	ENGINE FAN N_1 RPM	LOSS OF N_1 SIGNAL	SUBSTITUTES 6300. RAMPS DO NOT PROGRAM DURING OBC.
AIC S3	TOTAL TEMPERATURE	LIMITS EXCEEDED	SUBSTITUTES 60°F
AIC S4	ANGLE OF ATTACK	LIMITS EXCEEDED	<u>MID-COMPRESSION BYPASS:</u> OPEN AT MACH > 0.35, CLOSED AT MACH < 0.35 / <u>IN FLIGHT:</u> SUBSTITUTES +2° ANGLE OF ATTACK
AIC S4 (DURING OBC)	ALPHA DELTA PRESSURE SENSOR OUT OF CALIBRATION	DELTA PRESSURE OUT OF CALIBRATION	SUBSTITUTES +2° ANGLE-OF-ATTACK VALUE UNTIL RESET
AIC A4	MID-COMPRESSION BYPASS	FAULTY RELAY	<u>MID-COMPRESSION BYPASS:</u> OPEN AT MACH > 0.35, CLOSED AT MACH < 0.35

NOTE

AIC SYMBOL HAS L OR R APPENDED (AICL, AICR) TO IDENTIFY ON WHICH SIDE FAILURE WAS DETECTED.

FAIL-SAFE MODE — INLET LIGHT ILLUMINATED

FAILURE MAINTENANCE READOUT ACRONYM (SEE NOTE)	DETECTED FAILURE	CAUSE	RESULT MACH < 0.5	RESULT MACH > 0.5
AIC P, P	AICS PROGRAMMER	FAILED END-TO-END BIT	HYDRAULIC SHUTOFF VALVE REMAINS CLOSED. RAMP ACTUATORS REMAIN MECHANICALLY RESTRAINED WITHIN STOW LOCKS, PROVIDED THEY FAILED WITHIN STOW LOCKS.	RAMP MOVEMENT IS RESTRAINED BY ACTUATOR MECHANICAL LOCKS IF FAILURE OCCURRED WITH RAMPS WITHIN LOCKS. OTHERWISE RAMP(S) MOVE SLOWLY WITH AERODYNAMIC LOADS.
AIC S1, P_s	STATIC PRESSURE	MINIMUM OR MAXIMUM LIMITS EXCEEDED		
AIC S2, P_t	TOTAL PRESSURE	MINIMUM OR MAXIMUM LIMITS EXCEEDED		
AIC A1	RAMP NO. 1	SUSTAINED COMMAND AND FEEDBACK ERROR		
AIC A2	RAMP NO. 2	SUSTAINED COMMAND AND FEEDBACK ERROR		
AIC A3	RAMP NO. 3	SUSTAINED COMMAND AND FEEDBACK ERROR		
AIC A1, A2, OR A3 (INLET CAUTION LIGHT EVENTUALLY ILLUMINATES > 0.5 MACH)	HYDRAULIC PRESSURE LOSS OF RAMP NO. 1, NO. 2, OR NO. 3	SUSTAINED ERROR DUE TO LOSS OF HYDRAULIC PRESSURE		RAMP(S) MAY MOVE IF FAILURE OCCURRED WITH RAMP(S) OUT OF MECHANICAL LOCKS. RAMP LIGHT WILL ILLUMINATE.
NONE (NO INLET CAUTION LIGHT < 0.5 MACH)		LOSS OF HYDRAULIC PRESSURE		

Figure 1-8. AICS Detected Failure Modes

STOW MODE OF OPERATION

STOW commands the appropriate hydraulic shutoff valve to open and provides a direct electrical signal to the ramp actuators, porting hydraulic pressure directly to the retract side of the actuator. When the ramps are retracted to the stow position, the RAMPS light will extinguish and the stow locks should remain engaged even if hydraulic power is subsequently lost. Once in the stow position, AICS programmer-detected electronic failures may be reset below Mach 0.5.

DUMP INHIBIT

Whenever the hydraulic shutoff valve closes (i.e., fail-safe mode entry), hydraulic spool valves in the ramp actuators sense the absence of pressure and block the actuator pressure and return ports, causing a hydraulic lock (dump inhibit). This feature reduces ramp movement when an AICS failure occurs and the ramps are not being restrained by mechanical actuator locks. Although dump inhibit prevents the ramp from rapidly extending and causing an engine stall, the ramps will still slowly move. Under normal circumstances, the pilot will have sufficient time to select the STOW position and prevent an engine stall. Flight test results show that with dump inhibit, the time interval between illumination of a RAMPS caution light and engine stall following an AICS failure was 15 to 40 seconds on the ground at MIL power, and approximately 50 seconds at 10,000 feet at MIL power.

RAMP ACTUATOR MECHANICAL LOCKS/POSITIONING

In addition to the actuator stow locks, the first and second ramp actuators have another set of latches (trail locks), that prevent further ramp actuator extension after a failure within these trail locks. The actuator stow and trail locks restrain actuator movement in tension only. Hydraulic pressure (500 psi) is required to disengage the lock finger latches.

Safe positioning of the ramp actuator (no. 1 ramp in trail locks, no. 2 ramp in or out of trail locks, no. 3 ramp in stow locks), is monitored by the ramp monitor logic shown in figure 1-7. A RAMPS caution light will illuminate anytime an INLET RAMPS switch is in stow, or the landing gear handle is down, and all RAMPS are not in stow. A RAMPS caution light also illuminates when the hydraulic shutoff valve is closed and the RAMPS are not in a safe configuration (figure 1-7). RAMPS lights will extinguish when a safe configuration is attained.

AICS FAILURE IN-FLIGHT OPERATION

Most AICS failures occurring inflight do not require rapid pilot response due to system design features for fail-safe operation. Inflight, the no.1 and 2 ramps tend to blow back to the stow position or are restrained within the trail locks due to aerodynamic loads. The hydraulic restriction of all ramps during loss of hydraulic power and after fail-safe mode entry, should prevent rapid ramp movement. Internal failure of an actuator, especially the no. 3 ramp actuator, may allow rapid ramp extension and cause engine stall. Additionally, failure to stow the ramps in a reasonable amount of time after INLET light illumination or inability to stow following a hydraulic system failure, may result in compressor stalls at high power settings (engine RPM above 80%.) Engine start attempts may not be successful unless the RAMPS are stowed (RAMPS caution light extinguished).

ENGINES

The aircraft is powered by two Pratt and Whitney TF30-P-414 dual axial flow compressor turbofan engines (figure 1-9) equipped with an afterburner for thrust augmentation. The installed engine thrust at military and maximum afterburner power is 10,875 and 17,077 pounds, respectively.

A low specific fuel consumption is characteristic of basic engine turbofan operation. However, afterburner operation is less efficient than contemporary turbojet engines. Each engine is slung in a nacelle with the thrust axis laterally offset approximately 4-1/2 feet from the aircraft centerline.

The buildup of engine components is identical regardless of which side of the aircraft the engine is installed. Engine suspension in the nacelles is provided by three mounts. The thrust mount is forward and the two aft engine mounts carry only vertical and horizontal loads. The ring clamp mating the engine to the inlet duct can take engine thrust loads in an emergency.

ENGINE CONSERVATION

Engine rpm and temperature changes, the magnitude of these changes, and the amount of time the engine is operated at high rpm and temperature (even when well within the specified operating range) have a direct effect on the service life of a turbine engine. High operating rpm and temperature reduce life by causing gas path component erosion, creep and thermal stress. Rpm and temperature changes reduce engine life by increasing gas path component thermal stress, accentuating component creep and using available low cycle fatigue life of discs and fan blades. The larger the change and the more frequently the change is made, the more engine life is shortened. Life is further shortened by the rpm excursions and higher steady rpm associated with afterburner operation.

CAUTION

Throttle control by the pilot is a fundamental factor in determining engine life. Every reasonable effort should by made to minimize throttle excursions which produce high operating temperature (TIT) and/or high rpm, and to limit the magnitude, rate, and frequency of throttle changes in all portions of the operating envelope.

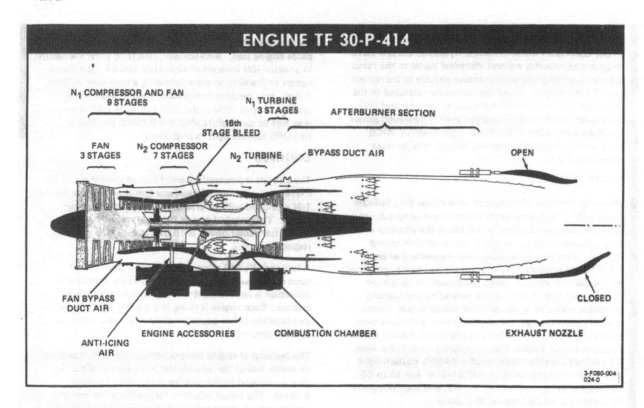

Figure 1-9. Engine TF30-P-414

Fan Section

The engine fan section is composed of three fan stages. The fan and low-pressure compressor rotor are driven as a single rotating unit by the last three turbine stages. Fan discharge air divides into two separate streams with approximately 46% of the airflow directed through the full-annular fan duct, bypassing the basic engine, to mix with basic engine airflow in the afterburner duct downstream of the turbine discharge. The core of fan discharge air (remaining 54%) is further compressed by six additional stages of the low-pressure compressor. Air discharge from the low-pressure compressor is directed aft through seven stages of high-pressure compression, driven through concentric shafting by the first-stage turbine. The low- and high-pressure compressors are completely free to rotate with no mechanical connection with each other. The high-pressure compressor rotor is speed-governed by the engine main fuel control and the low-pressure compressor is rotated by the three turbine stages at whatever speed will ensure optimum flow through the compressor. The power takeoff for engine accessory drive is geared to the high-pressure compressor rotor. Sixteenth-stage compressed air from the high-pressure compressor is used to support combustion and is diverted for air-conditioning and ancillary purposes.

Combustion Section

In the combustion section of the basic engine, compressed air and fuel are mixed and burned within eight annular-type combustion cans each of which contains four fuel ejector nozzles. Engine starting ignition is provided by a dual ignition system, with a spark igniter in each of the two bottom combustion chambers. Cross-ignition tubes provide flame propagation between combustion cans. The burning gases are directed rearward from the combustion section through the split, four-stage turbine where approximately 60% of the energy available is converted into torque to drive the engine compressors and accessories. The fan discharge air and basic engine turbine exhaust gases are mixed in the forward section of the afterburner duct. Afterburner fuel is injected through five circular manifolds with zones no. 1 and no. 4 located in the turbine exhaust core and zones no. 2, no. 3, and no. 5 positioned in the fan discharge stream. Afterburner ignition is initiated by a dual hot streak of overrich fuel mixture torching through the turbine stages to the zone no. 1 manifold and flame holder. With the annular fan duct serving as an insulation shroud for the basic engine, fan discharge airflow cools the liner in the afterburner duct and exhaust nozzle flaps.

Variable-Area Exhaust Nozzle

Turbine expansion pressure ratio and exhaust gas flow for both basic engine and afterburner operation are controlled by a variable-area, convergent-divergent exhaust nozzle. The nozzle is operated by four fuel pressure-operated actuators that drive a unison ring to position the nozzle flaps. For basic engine operation, the nozzle is closed to the minimum area (except for ground idle operation to reduce residual thrust), and in afterburning, the nozzle area is infinitely variable to a full open position, which represents a 110% increase in exit area.

Throttle Control

Two throttle levers for regulating engine thrust are on the left console of the forward cockpit. Unrestricted engine operation under independent control is afforded, however, normal symmetric thrust control is provided by collective movement of the throttle levers. Numerous engine control and subsidiary functions are performed by movement of the throttle levers within the full range of travel as shown in figure 1-10. The forward and aft throw of each throttle lever in the quadrant is restricted by hard detents at the OFF, IDLE, MIL and MAX (afterburner) positions. At the OFF and IDLE detents, the throttles are spring-loaded to the inboard position. At the MIL detent, the throttles can be shifted outboard to the afterburner sector or inboard to the basic engine sector of operation by merely overcoming a lateral breakout force. Lateral shifting of the throttles at the MIL detent does not affect engine control. Thus, placement of the throttle outboard at the MIL position provides a natural catapult detent to prevent unintentional retarding of the throttles during the launch. This, however, does not inhibit the selection of afterburner. The friction control lever on the outboard side of the quadrant permits adjustment of throttle friction to suit individual requirements. With the friction lever in the full aft position, no throttle friction is applied at the quadrant: increased throttle friction is obtained by forward movement of the lever.

A locking pin device prevents the left throttle from moving into the cut-off position when the right throttle is either traversing or at rest on the face of the right hand idle stop block.

THROTTLE CONTROL MODES

Manual, boost, and auto are the threee modes of throttle control over engine operation selectable by the THROTTLE MODE switch located outboard of the quadrant on the pilot's left console. The toggle switch must be lifted out of a detent to select MAN from BOOST or BOOST from MAN. The switch is solenoid held in the AUTO position upon successful engagement of AUTO mode. A functional schematic of throttle control modes, including system major components, is shown in figure 1-11. Except for the auto throttle computer and mode control switch, the throttle control system for each engine is completely redundant. Independent engine operation is possible in the manual or boost mode of throttle control; however, full system operation is necessary in the auto mode since operation under single engine control is impracticable due to asymmetric thrust considerations.

MANUAL THROTTLE MODE. The manual throttle is a degraded mode of operation and was designed as a backup system. Due to hysteresis and friction in the manual system, engine rpm may vary from the boost mode at a given throttle position. Positive selection of the OFF position will always be available.

In the manual mode of operation, movement of each throttle is mechanically transmitted to the respective engine cross shaft by a push-pull cable and a rack and sector mechanism. The cross shaft is mechanically linked to the engine main and afterburner fuel controls and the exhaust nozzle control unit. An electric clutch in the throttle servo actuator, which is splined to the cross shaft, is disengaged in the manual mode to reduce operating forces.

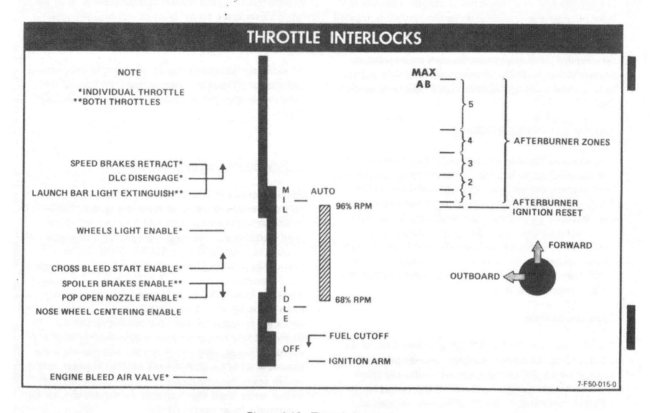

Figure 1-10. Throttle Interlocks

With the throttle friction lever in the OFF position, approximately 8 pounds of force per throttle must be applied at the grip to operate the throttles in the OFF to MAX range. Frictional forces increase to 10 pounds in the IDLE to OFF range.

BOOST THROTTLE MODE. The BOOST mode of throttle is used for normal operations. A force of 2 to 3 pounds at the grip is required to move each throttle throughout its range with the throttle friction lever OFF. Essentially, the boost mode provides electric throttle operation, with the push-pull cables serving as a backup control path. Throttle movement is detected by the throttle position sensor. The signal is resolved in the amplifier to provide positional followup commands to the actuator. Movement of the actuator rotates the engine cross shaft, which drives the push-pull cable.

If a boost system malfunctions, applying approximately 17 pounds at the throttle grip for 1.25 seconds automatically reverts the throttle control to the manual mode by disengaging the actuator electric clutch. In aircraft BUNO 161133 and subsequent and aircraft incorporating AFC 595, AYC 633, and AVC 2235, the throttle control reverts to manual mode in 0.25 second. In the event of a boost system malfunction, the throttle mode switch will remain in the BOOST detent. By manually placing the throttle mode switch in MAN and then back to BOOST, transient failures in the BOOST mode can be reset. Additionally, if an actuator seizes, a mechanical clutch in the actuator will slip when a force of approximately 50 pounds is applied at the throttle grip. This permits the pilot to override an actuator seizure. There is no visible warning of these anomalies; only the noticeable increase in the forces required to manipulate the affected throttle.

APPROACH POWER COMPENSATOR (AUTO THROTTLE MODE). The AUTO mode of throttle control is a closed loop system that automatically regulates basic engine thrust to maintain the aircraft at an optimum approach angle of attack for landing. All components of the throttle control system except the throttle position sensor are used in the AUTO mode of control. The angle-of-attack signal from the angle-of-attack probe on the left side of the forward fuselage is the controlling parameter within the auto throttle computer. Additional parameters are integrated within the computer to improve

NAVAIR 01-F14AAA-1

THROTTLE CONTROL

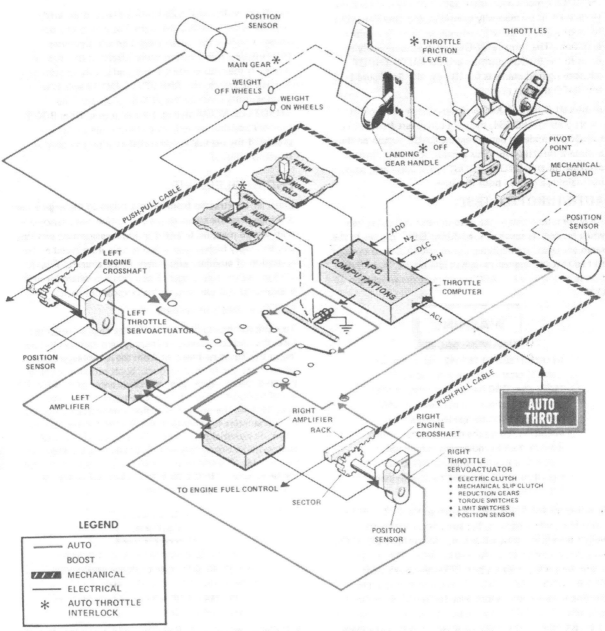

Figure 1-11. Throttle Control

response. The air temperature switch on the pilot's left console effects a computer gain change to compensate for temperature effects on engine thrust response. The switch can also be used to adjust the system for pilot preferred reaction rate. In order to engage the auto throttle, throttles must be between 78% to 96% RPM with weight off wheels, gear handle down and throttle friction off. With all conditions met, the throttle mode switch will be held by an electrical solenoid when placed in the AUTO position. The throttle control mode automatically reverts to the BOOST mode upon interruption of any interlock in the system or by manually overriding the throttles with a force of approximately 11 pounds per throttle in either direction. The throttle MODE switch automatically returns to the BOOST position and the AUTO THROT advisory light illuminates for 10 seconds. See figure 1-12 for AUTO throttle controls.

In aircraft BUNO 160909 and subsequent and aircraft incorporating AFC 554, the pilot can revert from AUTO to BOOST mode by depressing the CAGE button on the outboard throttle grip. This provides a smooth throttle override for an AUTO-to-BOOST mode approach, while maintaining a grip on both throttles.

AUTO-THROTTLE TEST

An automatic check of the auto throttle control system while on deck is accomplished during OBC. Signals to the servo actuators are inhibited during the OBC auto throttle test so that the engines remain at idle thrust. A malfunction is indicated by an APC acronym at the conclusion of OBC.

> **WARNING**
>
> In aircraft BUNO 159442 and earlier and those aircraft not incorporating AVC 1588, after OBC has been selected during ground operations, do not deselect until the program has completed the entire operation. When the disable signal, which inhibits throttle movement is removed, the APC will run through its BIT check and advance the throttles to greater than 80% RPM.

Rotating the MASTER TEST switch to the FLT GR DN position and depressing it, bypasses the auto throttle weight-on-wheels interlock and an end-to-end check of the auto throttles may be performed on deck. The throttles should be placed at about 85% RPM and the throttle MODE switch placed in AUTO. The throttles must be positioned above idle before selecting AUTO to ensure a valid test. Once AUTO is engaged, the control stick should be programmed fore and aft to check for the appropriate power response.

> **CAUTION**
>
> High power settings may result during aft stick deflection.

If the THROTTLE MODE switch does not remain engaged or the APC does not respond properly to indicated AOA and longitudinal stick movements, a malfunction exists in the auto throttle system.

In aircraft with IAFC 623/IAVC 2342 incorporated, depressing and holding the autopilot emergency disengage paddle switch with weight on wheels causes the throttle control system to be placed in the manual mode. If the auto mode was selected before depressing the paddle switch, the THROTTLE MODE switch will automatically move to the BOOST position. The THROTTLE MODE switch must be moved from BOOST to MAN position while holding the paddle switch depressed if the manual mode is desired after the paddle switch is released.

ENGINE BLEED AIR

Bleed air is extracted from various stages of the engine low- and high-pressure compressors to increase compressor operating stall margins, to perform engine-associated services, and to supply high-pressure and temperature bleed air for operation of auxiliary equipment in the aircraft. Figure 1-13 shows the four stages of compression where bleed air is extracted and summarizes the purposes of each.

Mid Compression Bypass

To minimize pressure distortion at the engine compressor face, the mid compression bypass system controls the porting of the 7th stage bleed air from the low-pressure compressor into the fan bypass duct. Such action is primarily directed towards decreasing the blade angle of attack in the initial stages of the low-pressure compressor; however, it does improve the quality of airflow downstream. Controlling parameters for operating the electric (dc essential No. 1 bus; dc essential No. 2 bus in aircraft BUNO 161168 and earlier not incorporating AFC 610) solenoid are shown in figure 1-14. The solenoid uses 16th stage bleed air to close the bleed valves. In the absence of electrical power, the device fails to the closed position.

> **CAUTION**
>
> Engine compressor stall margin is reduced by high aircraft angle of attack and/or yaw conditions, by firing of the internal gun, by operation in severe air turbulence, or by jet wash conditions.

A MID COMPRESSION BYPASS TEST panel (figure 1-15) is installed on the RIO's right console in aircraft BUNO

NAVAIR 01-F14AAA-1

AUTO THROTTLE CONTROLS AND INDICATOR

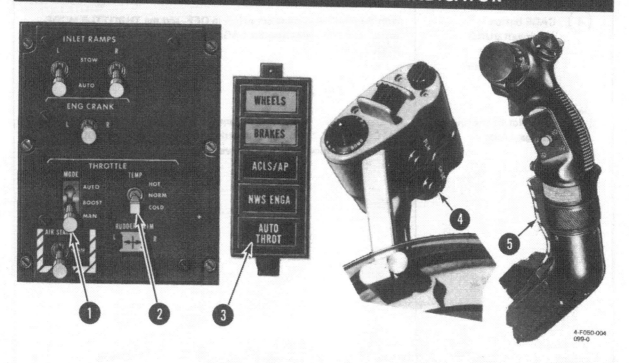

NOMENCLATURE	FUNCTION
(1) THROTTLE MODE switch	AUTO - Engine thrust is automatically regulated to maintain optimum angle of attack for landing by the throttle control computer. BOOST - Normal operating mode. Reduces effort required to move throttles manually with friction control aft. MAN - Movement of each throttle is mechanically transmitted to the respective engine cross-shaft by a push-pull cable.
(2) THROTTLE TEMP switch	Used with the AUTO throttle mode to effect throttle computer gain changes to compensate for air temperature. HOT - Above 80°F NORM - Between 40° and 80°F COLD - Below 40°F
(3) AUTO THROT caution light	Auto throttle mode is disengaged. During preflight check, remains illuminated for 10 seconds, then goes off and throttle mode switch automatically returns to BOOST position.

Figure 1-12. Auto Throttle Controls and Indicator (Sheet 1 of 2)

Section I
Part 2

NAVAIR 01-F14AAA-1

NOMENCLATURE	FUNCTION
(4) CAGE button (In aircraft BUNO 160909 and subsequent and aircraft incorporating AFC 554.)	With the pilot's weapon select switch in OFF, and the THROTTLE MODE switch in AUTO, depressing the CAGE button reverts the throttles to BOOST mode.
(5) Autopilot emergency disengage paddle	In aircraft BUNO 161157 and subsequent and aircraft incorporating IAFC 623/IAVC 2342, reverts throttle system from AUTO or BOOST mode to MAN mode only while depressed and with weight on wheels.

Figure 1-12. Auto Throttle Controls and Indicator (Sheet 2 of 2)

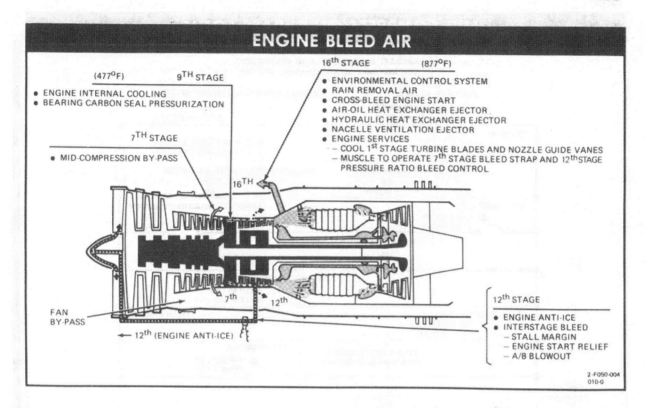

Figure 1-13. Engine Bleed Air

160921 and subsequent and aircraft incorporating AFC 592 and AVC 2212. The MID COMPRESSION BYPASS TEST panel is used during the Functional Checkflight (profiles A and B) test of the mid compression bypass. This checks electrical circuits to MCB, but does not verify that MCB is open. This panel includes left and right indicator test lights and a TEST/OFF switch. With TEST selected, the test light for the appropriate engine illuminates if power is available to the mid compression bypass valve.

Interstage Bleed

Engine speed transients impose an additional stall margin problem over that of steady-state operating conditions. The high-pressure compressor with its small mass and accessory drive horsepower extraction, characteristically accelerates and decelerates faster than the low-pressure compressor, with its larger mass and diameter. As a result of this transient operating imbalance, airflow and rotor speed in some parts of the compressor are no longer compatible and without corrective measures, a stall could occur. To alleviate stalls, interstage bleed air is extracted from the high-pressure compressor at the 12th stage to reduce the stall possibility during rapid engine decelerations and at low engine operating speeds. Additionally, an afterburner blowout signal will momentarily open the 12th stage bleed valves much the same as in the snap deceleration situation. The pressure ratio bleed control uses 16th stage bleed air as a medium for operating the 12th stage bleed valves.

Engine Anti-ice

The compressor inlet guide vanes and nose dome are susceptible to icing under a wider range of conditions, particularly at static or low speed with high engine rpm, than that which causes ice to form on external surfaces of the airframe. Ice formation at the compressor face can restrict engine maximum airflow, which results in a thrust loss, decreased stall margin, and dislodgement of ice, which can damage the compressor. The engine anti-icing system is designed to prevent the formulation of ice rather than de-ice the inlet guide vanes and nose dome. Hot bleed air (12th stage) is passed through the hollow compressor inlet guide vanes to the nose dome and discharged into the engine at the rotor hub. No inlet duct anti-icing is provided. Cockpit control of the engine anti-icing system is effected through the ENG/PROBE ANTI-ICE switch (figure 1-16) with three positions (OFF, AUTO, and ORIDE).

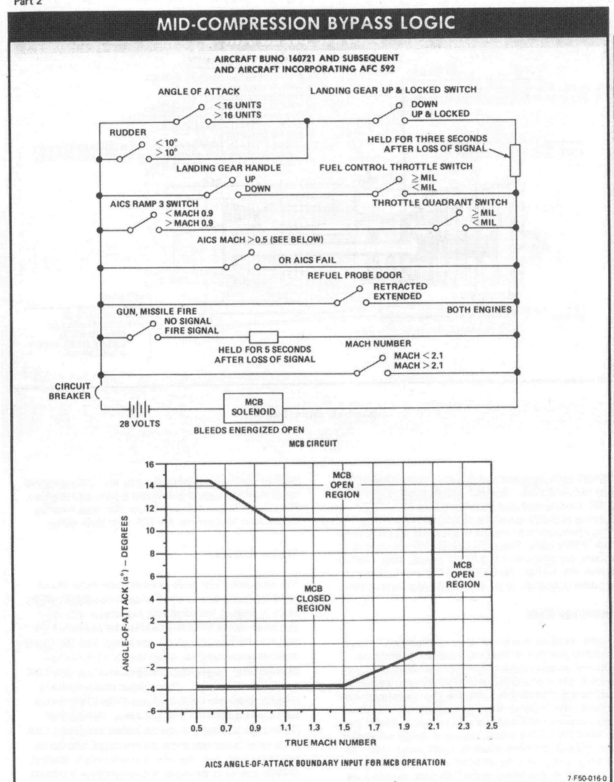

Figure 1-14. Mid Compression Bypass Logic

NAVAIR 01-F14AAA-1

Section I
Part 2

MID COMPRESSION BYPASS TEST PANEL

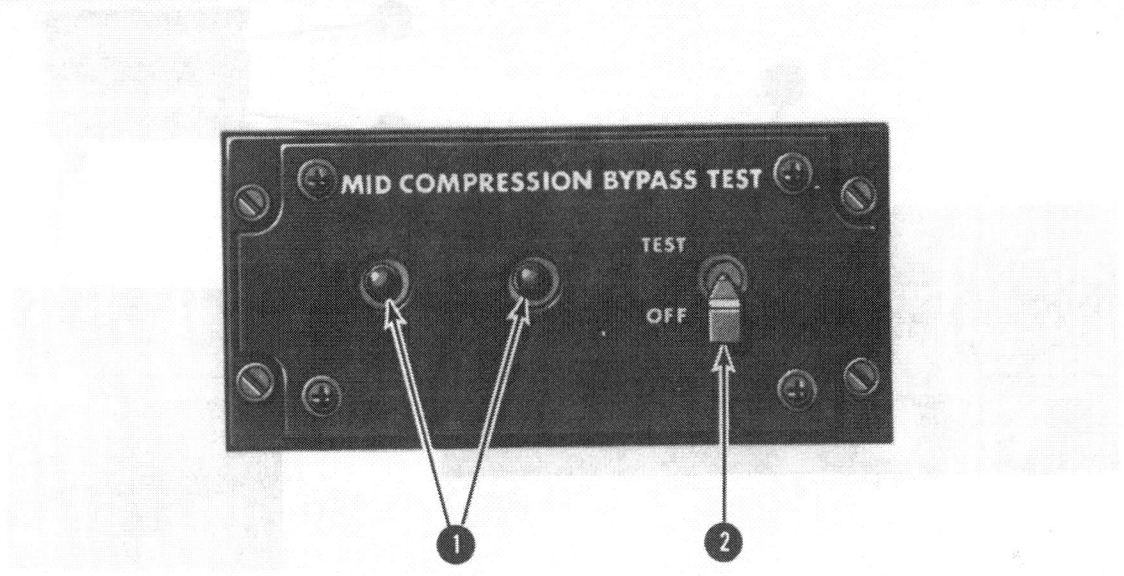

AIRCRAFT BUNO 160921 AND SUBSEQUENT AND AIRCRAFT INCORPORATING AFC 592 AND AVC 2212

NOMENCLATURE	FUNCTION
1. Left and right MID COMPRESSION BYPASS test lights	Illuminate when test switch is in the TEST position indicating mid compression bypass on. Also, when the IND LT/DDI BIT switch on the RIO's TEST panel is in the IND LT position.
2. MID COMPRESSION BYPASS TEST switch	TEST - Test light for appropriate engine(s) illuminates if power is available to mid compression bypass valve. OFF - Normal position, test circuit open.

Figure 1-15. Mid Compression Bypass Test Panel

1-23

Section I
Part 2

NAVAIR 01-F14AAA-1

ENGINE ANTI-ICE CONTROL

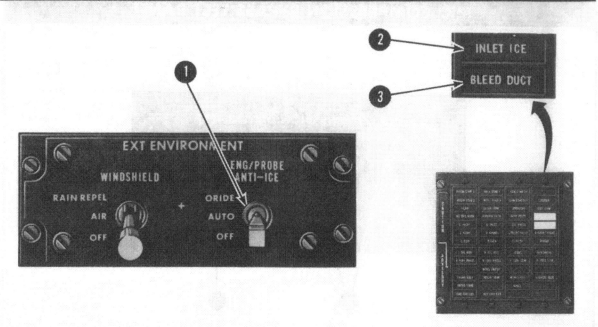

NOMENCLATURE	FUNCTION
(1) ENG/PROBE ANTI-ICE switch	ORIDE - Activates engine anti-ice system and turns on INLET ICE caution light independent of ice detector operation. Activates external probe heaters. AUTO - When icing is sensed, ice detector activates engine anti-ice system and turns on INLET ICE caution light. External probe heaters activated with weight off wheels. OFF - Engine anti-ice system and probe heaters shut off. INLET ICE caution light disabled.
(2) INLET ICE caution light	Illuminates when ice accumulates on ice detector with ENG/PROBE ANTI-ICE switch in the AUTO position or if ORIDE is selected. Does not illuminate with switch in OFF position.
(3) BLEED DUCT caution light	Illuminates when bleed air leak sensing elements detect temperatures greater than 575°F between engine and primary heat exchanger.

Figure 1-16. Engine Anti-Ice Control

1-24

During engine start, both engine anti-ice systems are disabled until the ENGINE CRANK switch returns to the OFF (center) position.

WARNING

Because of its adverse effects on engine stall margin and thrust, the engine anti-icing system should be used only during in flight and ground operations whenever icing conditions exist or are anticipated. The thrust loss associated with the use of the engine anti-icing system should be especially accounted for, as it degrades takeoff and waveoff performance.

The engine anti-icing control valve on the engine is electrically energized from the essential dc no. 2 bus through the ENG/PROBE/ANTI-ICE circuit breaker (RG2).

Miscellaneous Bleed Air Functions

The main supply of 16th stage bleed air is ducted out of the engine compartment to supply hot bleed air to the environmental control system. Additionally, 16th stage air is used for cross bleed engine starts, operation of the 7th and 12th stage bleed valves and, with weight on wheels, operation of the air-oil and IDG heat exchanger ejector.

Bleed Air Leak Detection

Bleed air leak sensing elements located along the outside of the bleed air ducts between each engine fire wall and the primary heat exchanger provide a cockpit indication of fire or overheating. When these sensing elements detect temperatures greater than 575°F, the BLEED DUCT caution light is illuminated. When the fire detection circuits in the engine compartments detect a leak in the high temperature bleed air duct, the appropriate L or R FIRE warning light is illuminated.

ENGINE IGNITION SYSTEM

The engine ignition system provides a redundant means of electrically igniting the atomized fuel-air mixture in the combustion cans during the engine start cycle, and automatic and manual airstart ignition in the event of an engine stall or flameout. Each engine incorporates a dual main ignition system consisting of an ignition alternator, ignition termination transformer automatic restart switch, ignition exciters, and spark igniter plugs, as illustrated in figure 1-17.

Electrical power for the engine ignition system is provided by an engine-driven, low-voltage (2 to 5 volts ac) alternator that has two independent current-generating circuits. In the deenergized mode, the ignition system is grounded in the ignition terminal transformer. When energized, the ground is removed to permit alternator current from redundant generating circuits to flow independently to two ignition exciter boxes which step up the low voltage output to 25,000 volts. Shielded ignition leads route the high-tension voltage output separately from the two exciter boxes to independent spark igniters in the lower two combustion cans. The atomized fuel-air mixture in the remaining six combustion cans is ignited by propagation of the flame through crossover tubes.

Selection of either engine for start with the ENG CRANK switch arms the redundant ignition circuitry of the selected engine so that high-energy ignition is automatically provided when the throttle is advanced from OFF to IDLE with the engine turning over. Above 8% RPM, ignition alternator voltage output is sufficient to effect a start. During the start cycle, electrical ignition is deenergized when the ENG CRANK switch spring returns to OFF (at about 45% RPM). Ignition is also terminated during an aborted start if the throttle is retarded to OFF.

The automatic restart switch on each engine automatically energizes the individual engine ignition system for approximately 30 seconds upon detection of a rapid decay in engine burner pressure. Additionally, the AIR START switch on the pilot's left console outboard of the throttle manually energizes the ignition system. Selecting the ON position provides continuous ignition for both engines if engine rotor speed is in excess of 8% RPM. The AIR START switch must be returned to NORM to stop ignition. Data on minimum airspeed to sustain sufficient windmill rpm for an airstart are provided in section V, Emergency Procedures.

ENGINE STARTING SYSTEM

Each engine is provided with an air turbine starter that may be pressurized from an external ground starting cart or by crossbleeding high-pressure bleed air from the other engine. Figure 1-18 shows the components associated with the engine starting system.

External Airstart

A low-pressure (50 psi) air source and 115 volt, 400-Hz ac power are required for engine start on the deck.

IGNITION SYSTEM

TF30-P-414 ENGINE

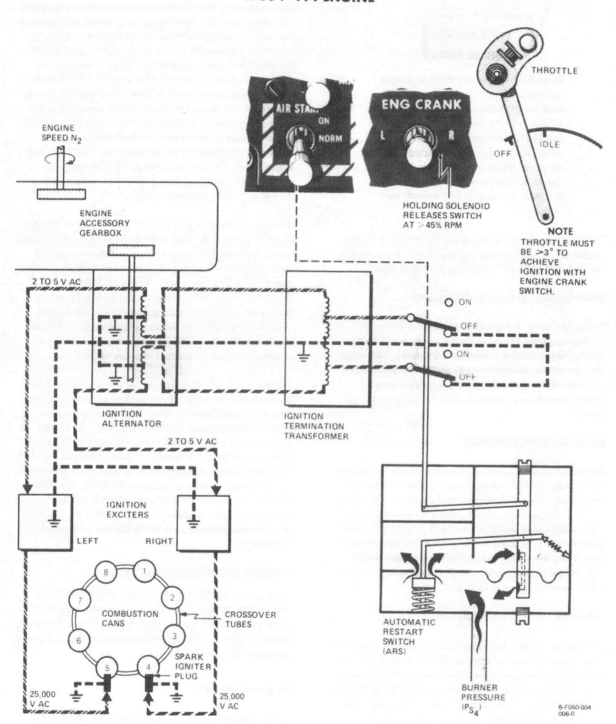

Figure 1-17. Ignition System

NAVAIR 01-F14AAA-1

Section I
Part 2

ENGINE CRANK SYSTEM

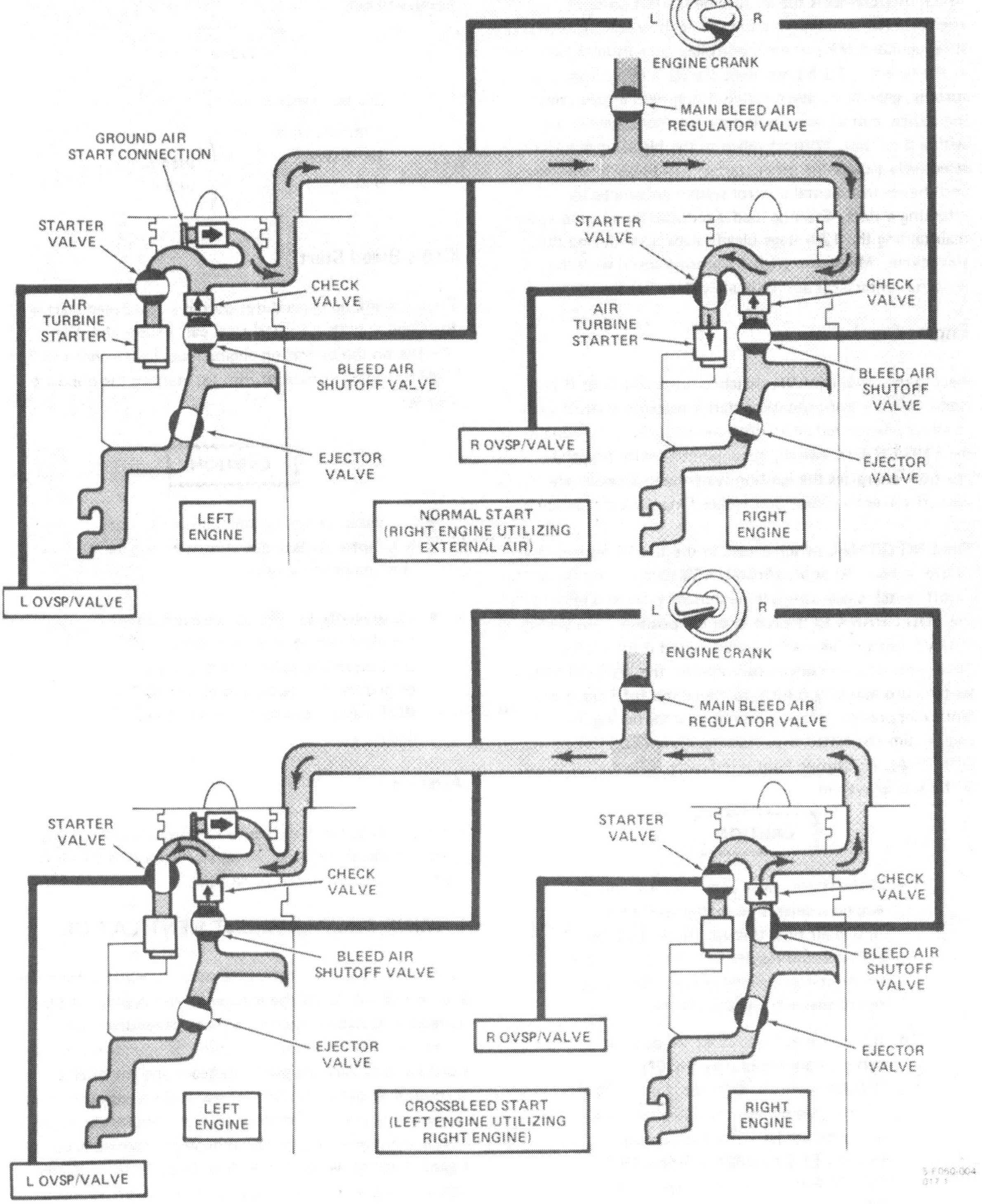

Figure 1-18. Engine Crank System

1-27

The air hose is connected to the aircraft fitting in the left sponson area, behind the main gear strut. Ground start air is ducted into a central bleed air (16th stage) manifold, which interconnects the air turbine starters on both engines. The air supply to each air turbine starter is pressure regulated (45 psi) and controlled by a shutoff valve at the turbine. Each pneumatic starter is composed of a turbine, gear train, overrunning clutch with a speed sensing device, and an overspeed disengagement mechanism with a shear pin. Shutoff valves in the bleed air manifold, selectively isolate the other starter, subsidiary bleed lines, and the environmental control system air supply for effecting a start. Starting load is reduced by automatically maintaining the 12th stage bleed valves open during the start cycle. Maximum engine motoring speed with the pneumatic starter is approximately 24% RPM.

Engine Crank

Placing the ENG CRANK switch in either the L or R position opens the corresponding starter pressure shutoff valve to allow pressurized air to drive the turbine. Additionally, the ENG CRANK switch, in conjunction with throttle position, energizes the ignition system and appropriate shutoff valves to condition the bleed manifold for starting.

The ENG CRANK switch is held in the L or R position by a holding coil. At approximately 45% rpm, a centrifugal cutoff switch closes the turbine shutoff valve and returns the ENG CRANK to the center or off position. An OVSP/VALVE caution light will be illuminated if the starter valve remains in the open position after the ENG CRANK switch automatically returns to the center (off) position. Since compressor overspeed is not a factor during the low engine rpm associated with starting, illumination of an OVSP/VALVE caution light is indicative of a malfunction in the starter system.

CAUTION

- If the starter valve does not close during engine acceleration to idle rpm, continued air flow through the air turbine starter could result in catastrophic failure of the starter turbine and possible damage to the airframe and engine.

- If the L or R OVSP/VALVE caution light is illuminated after the ENG CRANK switch is OFF, or if the ENG CRANK switch does not automatically return to the OFF position, ensure that the ENG CRANK switch is OFF by 55% RPM and select airsource to OFF to preclude starter overspeed.

This action, in turn, reconditions the bleed air manifold valves to permit 16th stage bleed air to flow to the environmental cooling system and ejectors in the engine compartment.

Note

Starter cranking limits:

1 minute or less on for 5 cycles

2 minutes on for 1 cycle

then 5 minutes off

Cross Bleed Start

Engine cranking procedures during a crossbleed start are the same as with a ground start cart except that the throttle on the operating engine must be advanced to 85% RPM to supply sufficient rpm for starting the opposite engine.

CAUTION

- To avoid starter turbine failure, do not attempt to crossbleed start with engine rpm greater than 5%.

- To prevent possible engine overtemperature during all crossbleed and air start attempts, select the operating engine for air source and return to BOTH after rpm stabilizes at idle or above.

Airstart

Setting AIR START switch to ON provides continuous ignition to both engines. After start, the AIR START switch must be returned to NORM.

ENGINE COMPARTMENT VENTILATION

Each engine compartment is completely isolated from the primary air inlet, and the efficiency and cooling of the variable-area exhaust nozzle are not dependent upon nacelle airflow. Therefore, within the bounds of the forward firewall (landing gear bulkhead) and the nozzle shroud, the cooling system for each engine compartment is a separate entity. Cooling requirements for the turbofan engine are minimized by the annular fan bypass duct. Figure 1-19 shows cooling airflow patterns through the engine compartment during ground and flight operations. Two air-cooled heat exchangers are also shown; however,

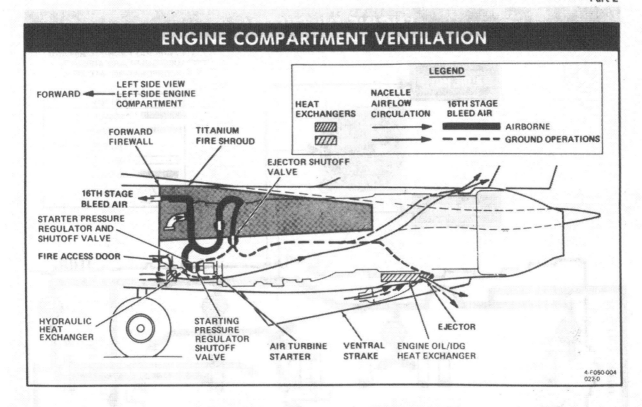

Figure 1-19. Engine Compartment Ventilation

only the hydraulic heat exchanger cooling airflow is associated with engine nacelle cooling. Fire access doors are on the outboard side of the nacelles at the forward end to permit insertion of fire suppressing agents by ground personnel in event of an engine compartment fire.

Engine In-Flight Ventilation

In-flight cooling of the engine compartment is accomplished by nacelle ram air scoops, circulating boundary-laryer air through the length of the compartment and expelling the air overboard through louvered exits, just forward of the engine nozzle shroud.

Engine Ground Ventilation

With weight on wheels, cooling airflow through the engine compartment is induced by the hydraulic heat exchanger ejector in the forward end of the compartment. Air enters through the nacelle ram air scoop on the left side, passes through the hydraulic heat exchanger, and is discharged into the engine compartment. The air flows through the full length of the nacelle to discharge overboard through a louvered port atop the nacelle on the outboard side of the vertical tail.

ENGINE OIL SYSTEM

Each engine has a self contained oil system which is pressure regulated to 45±5 psi. This provides pressure lubrication to the engine bearings and accessory drive (figure 1-20). Capacity of the system is 5 gallons, with 4 gallons usable. Normal oil consumption is 0.3 gallon per hour. The oil system permits engine operation under all flight conditions. During zero or negative g flight, the oil pressure may drop but will return to normal when positive-g flight is resumed.

The forward section of the engine accessory gear box serves as the oil stowage tank. Filters and dipsticks are

ENGINE OIL SYSTEM

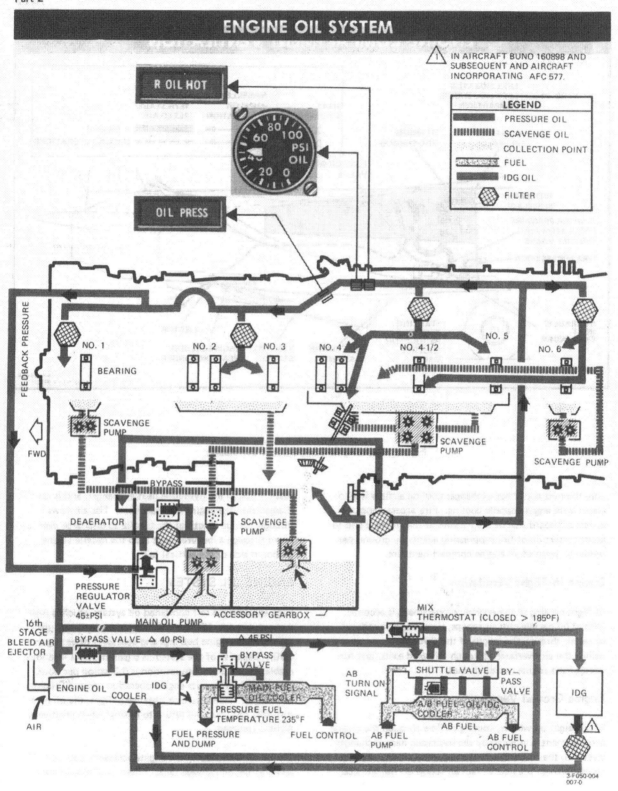

Figure 1-20. Engine Oil System

provided on both sides of the forward gearbox for checking the oil quantity and gravity servicing; however, only the outboard dipsticks should be used for checking the oil service level. Engine oil quantity should be checked within 15 minutes after engine shutdown to ensure proper servicing.

Oil Cooling

Oil from the main pump is pumped in series through the aircooler, main fuel-oil cooler, and afterburner fuel-oil cooler (if afterburner fuel pump is operating). If oil flow through any of the coolers becomes restricted, the pressure buildup opens the applicable bypass valve to permit unrestricted oil flow. Additionally, oil flow bypasses the main fuel-oil cooler if fuel temperature is too high. A temperature control valve opens to bypass the coolers after engine starting until the oil temperature rises above the preset value. After passing the coolers, the oil is then directed to the engine bearings and to the accessory gear box.

Scavenge pumps and drains in the bearing compartments and in the aft accessory gear box return the oil to the forward accessory gear box for reuse.

The efficiency of the air-oil heat exchangers can be compromised during engine ground operations by tailwind which can cause recirculation of hot ejector air into the ram air inlet scoop on the ventral fin.

OIL PRESSURE INDICATORS

The oil pressure indicating system operates on 26-volt ac power from the essential ac no. 2 bus through a transformer. The OIL PRESS caution light illuminates when either engine oil pressure drops below 25±3 psi. There are no cockpit indications or maintenance indicators of oil bypass or clogged filter except for visible oil pressure fluctuations and routine maintenance inspections.

OIL HOT CAUTION LIGHTS

In aircraft BUNO 160696 and earlier without AFC 579, the L and R OIL HOT caution lights illuminate when oil temperature exceeds 245°F. Illumination of an OIL HOT caution light is normal during hot weather ground operation. The light may also illuminate briefly if the throttles are reduced to IDLE following a MAX AB run long enough to produce high oil temperatures. If feasible, the best corrective action is to increase fuel flow through the fuel-oil cooler of the affected engine(s) by advancing the throttle(s).

In aircraft BUNO 160887 and subsequent and aircraft incorporating AFC 579, the preceding anomalies are eliminated. The L and R OIL HOT caution lights illuminate when oil temperature exceeds 316°F during increasing temperature and go off at 265°F minimum during decreasing temperature.

ENGINE INSTRUMENTS

Instruments for monitoring engine operation are on the pilot's left knee panel. (See figure 1-21.) Vertical-scale instruments are used for the integral dual (left and right) display of high-pressure rotor speed (RPM), turbine inlet temperature (TIT), and basic engine fuel flow. Adjacent to these instruments are left and right circular instruments for engine oil pressure and exhaust nozzle position. Takeoff checks at MIL thrust should display evenly matched tapes on corresponding vertical scale instruments, and all pointers on the circular instruments should be at the 9 o'clock position. Electrical power for all engine instruments is obtained from the essential no. 2 electrical buses. Data on engine operating limits are provided in part 4 of this section.

Engine RPM Indicator

A tachometer generator on the engine accessory gearbox provides a signal to the pilot's RPM indicator for displaying high-pressure-compressor rotor speed (RPM). The readout is in percent, with 100% equivalent to a high-pressure-compressor rotor speed of 14,583 RPM. No display is provided for the low-pressure-compressor rotor speed (N_1) since the two compressors are aerodynamically balanced; however, failure of the exhaust nozzle to close upon afterburner shutdown or during high-power, non-afterburning operation can induce an N_1 rotor over-speed condition due to the low back-pressure on the fan. This will illuminate the corresponding L or R OVSP/VALVE caution light.

Engine Oil Pressure Indicator

The engine oil pressure sensor is in the pressure supply line to the forward bearings on the engine. The system is pressure-regulated to 45±5 psi and the oil pressure at IDLE thrust settings should not be less than 30 psi. During engine start, the oil pressure rises correspondingly with rpm. In extremely cold weather, the rise of oil pressure during initial engine start will be sluggish. The OIL PRESS caution light (figure 1-21) illuminates whenever engine oil pressure is less than 25±3 psi. Oil pressure fluctuations of ±2 psi are allowable.

Note

During engine start or shutdown, and during emergency generator test, electrical transients may cause the L and R OVSP/VALVE lights to flicker.

Section I
Part 2

NAVAIR 01-F14AAA-1

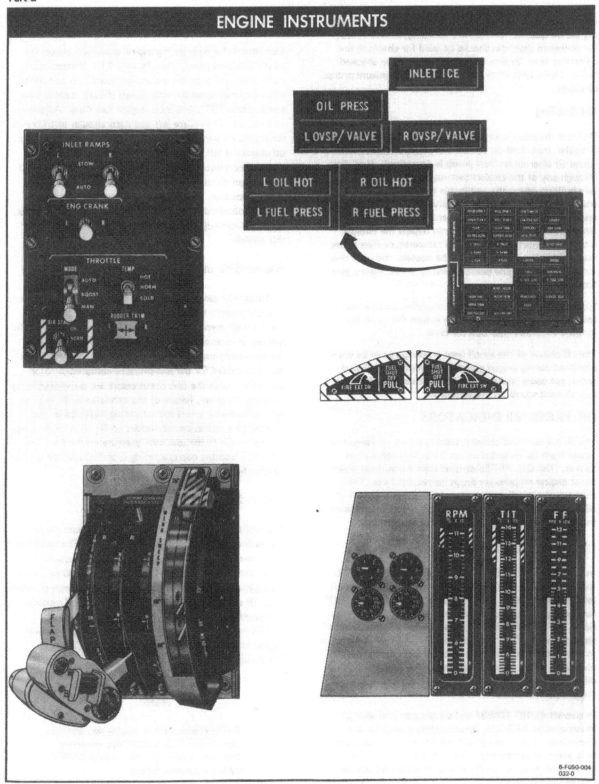

Figure 1-21. Engine Instruments

1-32

Engine TIT Indicator

The turbine inlet temperature (TIT) signal is not the result of a direct measurement on the engine but is indirectly computed from measurements of compressor-inlet temperature, compressor-discharge temperature, and turbine-exhaust temperature. The TIT limiter compares the TIT signal to the cockpit with a preset maximum allowable value (1,175°C) to limit TIT at high steady-state thrust. If either L or R engine TIT exceeds 1,215°±15°C, a switch in the indicator activates an overtemperature alarm tone in the pilot's headset. This tone continues for 10 seconds or until overtemperature condition is removed, whichever occurs first.

Engine Fuel Flow Indicator

The transmitter supplying the signal for the display of fuel flow in the cockpit is downstream of the main fuel control, so that it accurately reflects the fuel being supplied to the basic engine for combustion; no indication of fuel flow to the afterburner is provided.

Exhaust Nozzle Position Indicator

The nozzle position indicators display equal graduations between 0 to 5 to indicate relative position from full closed to full open, respectively. The following approximate readings indicate afterburner operating level:

- 0 - Basic engine
- 1.5 - Zone 1
- 2.4 - Zone 2
- 3.2 - Zone 3
- 4.0 - Zone 4
- 4.3 - 5.0 - Zone 5
- 5.0 - Pop open (throttle idle, weight on wheels)

Engine Instrument Test

Selecting the INST position and depressing the master test switch on the pilot's MASTER TEST panel drives the tapes on the vertical scale instruments. The RPM indicator should read 80%±2%; the TIT indicator, 1300°±20°; and the FF indicator 4300±100 lb hr. (figure 1-21). The engine overtemperature alarm tone activates during the test only if both TIT overtemperature warning switches activate properly.

FIRE DETECTION SYSTEM

The fire detection system provides a cockpit indication of fire or overheating in either engine compartment. There is a separate system for each engine compartment, each consisting of a thermistor-type sensing loop monitored by a transistorized control unit. The system is powered by 28 volts from the essential dc no. 2 bus. Figure 1-22 is a functional schematic of the system.

The sensing loop for each engine compartment consists of a 45-foot continuous tubular element routed throughout the entire length of the engine compartment on both sides above the nacelle door hinge line. The tube sheath, which is clamped in grommets to the engine compartment structure, contains a ceramic-like thermistor material in which are embedded two electrical conductors; one of the conductors is grounded at both ends of the loop. Electrical resistance between the two conductors varies inversely as a function of temperature and length, so that heating of less than the full length will require a higher temperature for the resistance to decrease to the alarm point. The L or R FIRE warning lights in the cockpit illuminate when the respective entire sensing loop is heated to approximately 600°F or when any 6-inch section is heated to approximately 1,000°F.

The fire alarm output relay to the light is a latching type which remains in the last energized position independent of power interruptions until the fault clears.

False alarms triggered by moisture in the sensing element and connectors, or by damage resulting in short circuits or grounds in the sensing element are unlikely due to the system design. Additionally, there is no loss or impairment of fire detector capability from a single break in the sensing element as long as there is no electrical short. With two breaks in the sensing element, the section between the breaks becomes inactive although the remaining end of segments remain active.

Fire detection circuits in the engine compartments detect a leak in the high-temperature duct and light the appropriate L or R FIRE warning light. Between each engine firewall and the primary heat exchanger, bleed-air leak-sensing elements are placed along the outside of the duct. When these sensing elements detect temperatures over 575°F, the BLEED DUCT caution light is illuminated.

Fire Detection Test

An integrity test of the fire detection system can be performed by selection of the FIRE DET position on the pilot's master test switch. The integrity test simultaneously checks the sensing element loops of both engine compartments for continuity and freedom from short circuits, and the fire alarm circuits and FIRE warning lights for proper functioning. Presence of a short circuit or control unit malfunction causes the warning light to remain out. Fire detection test is not available on the emergency generator.

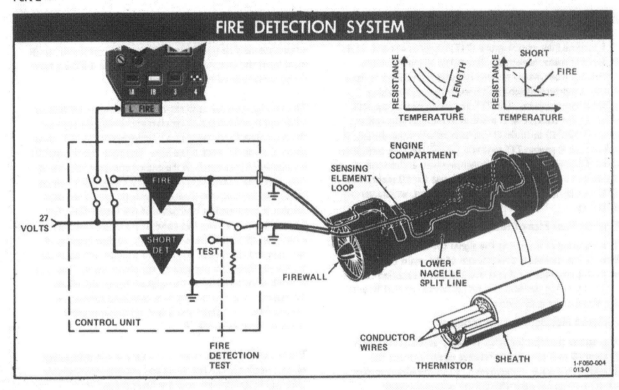

Figure 1-22. Fire Detection System

FIRE EXTINGUISHING SYSTEM

The fire extinguishing system is capable of discharging an extinguishing agent into either engine nacelle and its accessory section. The system consists of two containers of extinguishing agent, piping and nozzles to route and discharge the agent, cockpit switches to activate the system and advisory lights which alert the flight crew to a drop in system pressure beyond a predetermined level.

The fire extinguishing agent is a clean, colorless, odorless, and electrically nonconductive gas. It is a low toxicity vapor that chemically stops the combustion process. It will not damage equipment because it leaves no water, foam, powder, or other residue.

The retention time of an adequate concentration of the extinguishing agent in the engine compartment will determine probability of re-ignition, and therefore the probability of aircraft survival. At high airspeeds, where airflow through the engine compartment is increased, agent retention time is reduced.

The slower the airspeed at the time the extinguisher is fired, the higher the probability of fire extinction and the lower the probability of re-ignition.

Circuit breaker protection is provided on the RIO DC ESSENTIAL NO. 1 CIRCUIT BREAKER PANEL by the R FIRE EXT (64C) circuit breaker and the F FIRE EXT (65C) circuit breaker.

Fire Extinguisher Switches

The discharge switches for the fire extinguishing system are located behind the FUEL SHUT OFF handles. The FUEL SHUT OFF handle for the affected engine must be pulled to make the switch for that engine accessible. (See figure 1-23.) If the left or right fire extinguishing switch is activated, the contents of both extinguishing containers are discharged into the selected engine and its accessory section. Since it is a one shot system, both system advisory lights, ENG FIRE EXT and AUX FIRE EXT will illuminate and remain illuminated after container pressures drop below a preset level.

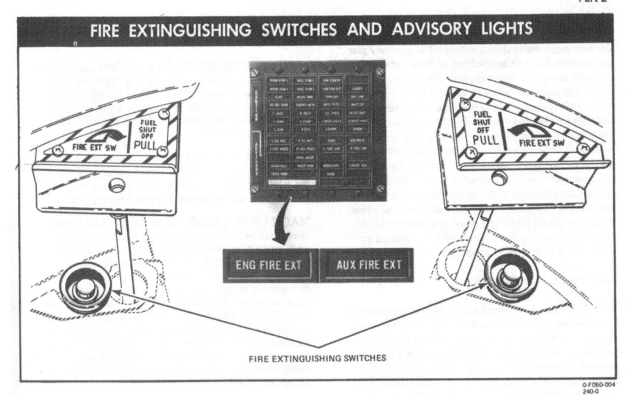

Figure 1-23. Fire Extinguishing Switches and Advisory Lights

Fire Extinguisher Advisory Lights

Two advisory lights are provided to indicate low pressure in the fire extinguishing agent containers. The lights, ENG FIRE EXT and AUX FIRE EXT, illuminate when container pressure drops 90 psi below a nominal pressure of 600 psi at 70°F. (See figure 1-23.)

Fire Extinguisher Test

The fire extinguishing system is tested by raising and rotating the MASTER TEST switch to the FIRE DET/EXT position and depressing the knob; the L FIRE AND R FIRE warning lights will illuminate. If there is a detection circuit problem the corresponding FIRE warning light will not illuminate. Simultaneously, the fire extinguishing system initiates a self-test indicated by either a GO or NO GO light. If all tests pass, the GO light illuminates; if the NO GO illuminates, or if both or neither GO and NO GO lights illuminate the system has not tested properly and a failure exists somewhere in the system.

ENGINE FUEL SYSTEM

The engine fuel system, which is identical for each engine, provides motive flow fuel to effect fuel transfer and metered fuel for combustion as a function of throttle commands and numerous engine operating parameters. A schematic of the engine control system is provided on page FO-7.

Motive Flow Fuel Pump

The gear-driven centrifugal pump on the engine accessory gear box returns high-pressure fuel to the fuselage and wing tanks to effect normal fuel transfer. The motive flow fuel is the medium used to power the turbine section of the engine boost pump in the respective sump tank. This fuel continues through control valves to ejector pumps in the respective fuselage and wing fuel tanks. There is no cockpit control for the motive flow fuel pump. Failure of the pump lights either R or L FUEL PRESS caution light and also reduces the rate of fuel transfer. Failure of a motive flow fuel pump will preclude afterburner operation above 15,000 feet MSL altitude but will not inhibit the transfer of fuel from any tank.

Main Fuel Pump

The two-stage, engine-driven pump on the accessory gear box provides 1st stage boosted fuel to the afterburner hydraulic pump and 2nd stage, high-pressure fuel to the engine fuel control. Failure of the centrifugal impeller (pump's first stage) will stop fuel flow to the AB hydraulic pump which may cause failure of AB hydraulic pump. Such a failure would cause the nozzle to float or oscillate: no cockpit indication is provided for failure of the main fuel pump first stage. Conversely, failure of the 2nd stage gear pump will result in a flameout due to insufficient pressure for operation of the hydromechanical main engine fuel control. Although no cockpit indication is provided for such a failure, it will be evident when attempting an air-start that no fuel flow registers with the throttle at IDLE and an engine speed greater than 15% RPM.

Main Engine Fuel Control

The main engine fuel control is a hydromechanical device that operates in tandem with the main fuel pump. Based on pilot commands through the throttles, the main fuel control on each engine automatically and independently establishes engine thrust for the numerous operating variables to which the engine is sensitive. Absolute cutoff to fuel to the engine is provided with the throttle less than one-half the travel between OFF and IDLE. At throttle settings of MIL or greater, engine output is limited by turbine inlet temperature, corrected airflow, burner-case pressure, and creep of rotating components for which the main engine fuel control provides automatic compensation. The control unit incorporates ground adjustments (specific gravity and minimum fuel flow) for use of JP-4 as an alternate for JP-5 type fuel.

MACH LEVER

The Mach lever provides a mechanical input to the main fuel control as a function of aircraft Mach number to control engine airflow at high airspeeds by providing a maximum and minimum rpm limitation. The Mach lever on each engine receives signals from the central air data computer (CADC), which varies its position linearly between 0.25 IMN and 2.5 IMN. Airflow is kept within a topping limit, which prevents excessive distortion at maximum thrust and a bottoming limit, which prevents inlet buzz at idle thrust settings

If the Mach lever system is receiving erroneous airspeed or angle of attack inputs, it may increase idle RPM to approximately 90% on one or both engines. Pulling the MACH LVR/L or R FUEL CONTR cb (LF1 or LG1) will deenergize the appropriate Mach lever, and the retract spring will drive the Mach lever actuator to the neutral position. Also, lowering the main landing gear will command both the left and right actuators to neutral.

MACH LEVER SHIFT. Mach lever shift increases engine idle speed in the subsonic and high AOA regime to provide increased stall margins. Below 0.9 IMN if AOA is increased to 16±1 units or more, the CADC input to the Mach lever trim system is replaced by a fixed value equivalent to 1.3 IMN.

MACH LEVER FAILURE. To minimize the consequences of a failure, the Mach lever is designed to fail to the minimum Mach condition. A landing gear down signal also provides such action as an overriding function.

MACH LEVER TEST. An automatic check of the Mach lever control unit and actuators can be initiated while on deck by selecting the MACH LEV position and depressing the MASTER TEST switch on the pilot's MASTER TEST panel. The test is normally conducted during the prestart checklist. The test simulates an input signal from the CADC to the Mach lever control unit and compares mach lever position with preset schedule. If the comparison indicates proper operation, the GO light on the Master test panel illuminates. NO GO light illumination may indicate a malfunction. Landing gear handle and weight-on-wheels switch interlocks prevent initiation of the test while airborne.

A NO GO light may be associated with BIT circuitry, system noise, or torque loads on the fuel control rather than an actual mach lever malfunction. In the event of a NO GO light, wait at least 30 seconds before initiating another test. If the test fails the engines may be started, however be alert for a possible RPM increase to 85% signalling an actual mach lever malfunction. The idle RPM increase function of the mach levers may be checked by selecting mach lever test with engines running. Engine idle RPM will increase in accordance with the type of mach lever installed. The CADC mach and AOA inputs are not checked by mach lever test.

Two types of mach lever controllers are presently used. With the Edison controller the engines will run up to 81 to 86% RPM and remain there until the master test switch is deselected. With the Bendix controller, the engines will run up to 81 to 86% RPM and then decrease approximately 15 to 20% RPM. With either controller it is normal for RPM to fluctuate. Correct engine response to this test is sufficient to accept the aircraft with or without a GO light.

CAUTION

Mach lever test with engines running will significantly increase intake suction and exhaust blast and should be performed in a clear, FOD-free area.

Mach lever shift can be checked while airborne by retarding both throttles to IDLE and exceeding 17 units AOA. An increase in RPM above 80% indicates a satisfactory check.

TIT LIMITER

The TIT limiter is a control device that limits TIT at high steady-state thrust to the preset maximum allowable value of 1,175°C. Turbine inlet temperature signals to the limiter are the same as displayed on the pilot's TIT indicator. Limiter authority approximates 2% RPM, with the upper limit set at the engine ground trimmed MIL power governor schedule.

FUEL PRESSURIZATION DUMP VALVE

Upon engine shutdown, the stoppage of metered fuel flow allows the fuel pressurization dump valve to open and drain fuel in the primary manifold overboard through the lower nacelle door. Forward and aft drains in the lower portion of the combustion chamber also route residual fuel overboard.

AFTERBURNER FUEL SYSTEM

The afterburner augments main engine thrust by injecting fuel through one to five manifold zones into the turbine and fan discharge airstream in the forward section of the afterburner duct. The afterburner increases the exhaust velocity at the nozzle. Initial afterburner ignition is provided by a hot-streak ignition system. Thrust augmentation can be modulated between 11,700 pounds and 17,000 pounds installed net thrust at static sea level conditions; the minimum thrust augmentation represents a 1,200 pound thrust increase over that developed by the main engine without augmentation. Rapid throttle movement through the full afterburner range results in automatic staging through the five zones in consecutive order at the maximum rate consistent with maintaining afterburner stability. Normal minimum acceleration time through the full afterburner range is approximately 8 seconds.

The efficiency of afterburning in the lower zones is on a par with that of turbojet engines of comparable thrust ratings. The efficiency of operation is significantly less with maximum thrust augmentation; thrust specific fuel consumption (pounds of fuel per pound of thrust) between MIL and MAX ratings differs by a factor of five.

WARNING

- In aircraft BUNO 158637 and earlier, terminate afterburner operations during high Q (greater than 650 KIAS) when either fuselage tank reading is less than 2500 pounds and feed tank on the same side is less than 1400 pounds. At high fuel flow demands, the outflow from the sump tank can exceed the gravity inflow rate from the wing box tank on the respective side. Under these conditions, adequate fuel level in the sump tanks depends on supplemental gravity fuel transfer from cells 2 and 5. This does not prevent the use of afterburner for waveoff or other limited operations at a low fuel state.

- With fuel in feed tank below 1,000 pounds, afterburner operation could result in afterburner blowout.

Note

Fuel dump operations with either engine in afterburner are prohibited. The fuel dump mast can be torched.

Afterburner Hydraulic Pump

The afterburner (AB) hydraulic pump is a two-stage pump that receives fuel from the main engine fuel pump interstage and boosts the pressure to provide a hydraulic medium for actuating the exhaust nozzle and supplying initial flow of metered fuel for afterburner operation. The pump supplies the first 3,500 pounds per hour of afterburner fuel flow metered in the afterburner fuel control.

Afterburner Fuel Pump

During nonafterburning operations or less than 3,500 pounds per hour afterburner fuel flow, the afterburner fuel pump impeller runs dry with the bearings lubricated by the engine oil system. Upon development of an afterburner fuel flow demand exceeding 3,500 pounds per hour, the afterburner fuel control directs engine boosted fuel feed to the centrifugal impeller of the afterburner fuel pump. Initial flow from the pump sends a signal to the afterburner fuel control transfer valve to provide a cutoff for afterburner hydraulic pump contribution of fuel to the afterburner metering system, and sends a signal to the afterburner fuel-oil heat exchanger to permit engine and CSD oil to flow through the cooler.

When afterburner metering flow demands are reduced to less than 3,500 pounds per hour, the afterburner fuel pump discharge ceases, engine and CSD oil ceases to flow through the afterburner fuel-oil heat exchanger, and the pump overboard and drain and vent valve unseats. Failure of the afterburner fuel pump will result in an afterburner blowout with greater than 3,500 pounds per hour afterburner flow demand; however, minimum zone afterburner operation with demands less than 3,500 pounds per hour will not be impaired.

Afterburner Fuel Control

The afterburner fuel control is a hydromechanical unit that contains five afterburner fuel metering units, one for each zone of operation, providing a near linear net thrust to throttle relationship at steady-state conditions between the minimum and maximum ranges of afterburning.

Initiation of afterburner operation requires that the high pressure compressor rotor speed be greater than 85% RPM, control ratio be near maximum, and the throttle sufficiently advanced in the afterburner range. An exhaust nozzle position feedback is used to schedule the rate of change of afterburner fuel flow during transient operation.

A sudden decrease in turbine discharge pressure initiates an afterburner blowout signal, which shuts off all afterburner fuel, momentarily deriches the main fuel control and opens 12th stage bleed until the exhaust nozzles close. A ground adjustment on the afterburner fuel control accounts for specific gravity differences between JP-4 and JP-5 fuels.

AFTERBURNER IGNITER VALVE

The afterburner ignitor valve controls the phasing of the hot-streak ignition and afterburner zone and manifold fuel flow to effect an afterburner light in response to afterburner fuel control signals and turbine discharge pressure. Initial afterburner ignition is provided by a hot-streak ignition system whereby a main squirt of raw fuel is injected into the lower combustion can that imparts a burst of flame into the forward portion of the afterburner. An auxiliary squirt of raw fuel is injected just aft of the fourth-stage turbine disk. Both the main and auxiliary squirts last only momentarily, so that subsequent to an unsuccessful afterburner light or an afterburner blowout, the afterburner igniter valve must be reset before a successful relight may be attempted. Standard procedure is to retard the throttle to the MIL detent and hesitate momentarily before advancing for a relight attempt. Failure of the afterburner igniter valve will most probably inhibit afterburner lights.

Variable-Area Exhaust Nozzle

The convergent-divergent geometry of the variable-area exhaust nozzle (figure 1-24) provides the optimum exhaust throat area. The iris action of the variable area nozzle is achieved by close-tolerance overlapping seals, which bridge adjacent leaf segments to provide a relatively smooth surface contour. Engine exhaust gases at the higher thrust settings are discharged through the nozzle throat at sonic velocity and are accelerated to supersonic velocity by the controlled expansion of the gases.

Nozzle area is infinitely variable between the closed (3.53 square feet) and full open (7.5 square feet) positions. Positioning of the nozzle is accomplished automatically by four synchronized hydraulic actuators that translate a circumferential unison ring interconnecting the translating leaf assemblies. High-pressure fuel discharged from the afterburner hydraulic pump is used as the hydraulic medium for positioning the actuators. The engine exhaust nozzle control unit is a hydromechanical computing device that regulates fuel to the actuator ports for setting the exhaust nozzle position in response to controlling parameters. A nozzle position feedback system, mechanically transmits unison ring position to the exhaust nozzle control unit. The nozzle position transmitter for the cockpit indication is mounted at the exhaust nozzle control feedback.

For on-deck operations, the nozzle is scheduled full open with the throttle within 1/4 inch of the IDLE position to reduce engine residual thrust to 450 pounds per engine. Otherwise, during ground or in flight engine operation in the nonafterburning region, the nozzle is maintained in the full closed position. Afterburner ignition and lightoff is accomplished with the nozzle opening slightly as afterburner is selected on the throttle in anticipation of a light and closes if light is not accomplished. Although nozzle position is a linear function of throttle position in the afterburner range, variations in nozzle position will occur in an attempt to maintain an optimum turbine pressure ratio under airflow and fuel flow conditions that differ from the nominal.

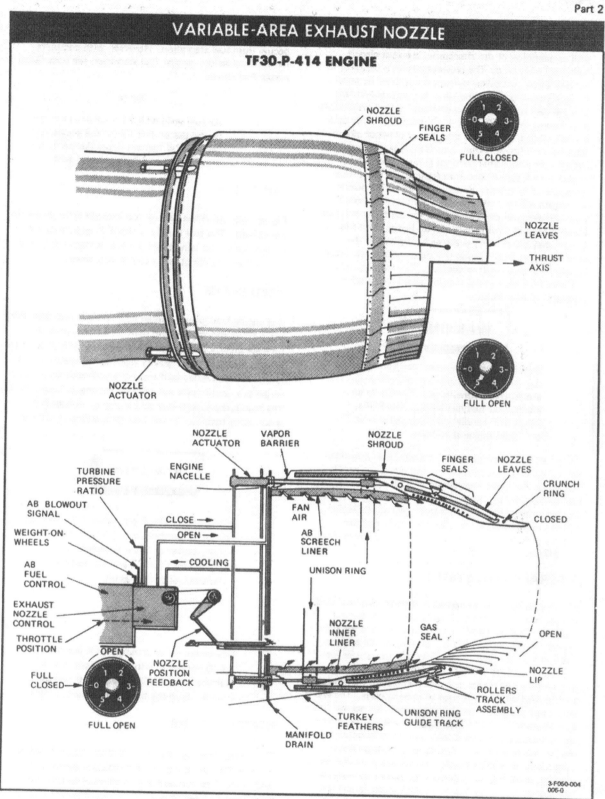

Figure 1-24. Variable-Area Exhaust Nozzle

An afterburner blowout on one engine results in thrust asymmetry that is dependent upon initial thrust setting as well as operation of the afterburner blowout circuit following a blowout. The blowout circuit is designed to prevent thrust loss below military thrust level by automatically closing the nozzle when fan suppression falls 25% below designed demand schedule. Afterburner blowouts from zone 1 or 2 result in less than 25% suppression and would not activate the afterburner blowout circuit. Afterburner blowouts from zone 3 may or may not activate the afterburner blowout circuit depending on engine zone 3 suppression trim settings. In these situations, selecting military power would close the nozzle and regain military rated thrust. The blowout circuit also provides overspeed protection for the N_1 compressor. Zones 1 through 3 blowouts that do not activate this system may result in N_1 overspeeds. Failure of the afterburner hydraulic pump or a pressure line causes the exhaust nozzle to float or oscillate. The exhaust nozzle will also float after inflight engine shutdowns that exceed 1-minute duration.

WARNING

Nozzle failure in the open position at military power or less may indicate either an afterburner hydraulic pump failure or an afterburner fuel line failure. The FUEL SHUTOFF handle should be pulled and the failed engine shut down.

Failure of the nozzle position feedback cable automatically shuts down the afterburner and closes the nozzle. Binding in the nozzle mechanism or a failed actuator can best be detected by reduced slew rates during full authority step commands such as the pop-open nozzle action during on-deck operations. Normal nozzle position slew rates through full authority are 1.5 seconds to close and 2.0 seconds to open.

AIRCRAFT FUEL SYSTEM

The aircraft fuel system normally operates as a split feed system, with the left and aft tanks feeding to the left engine and the right and forward tanks feeding the right engine. (Refer to pages FO-8.) Except for the external tanks, the system uses motive flow fuel to transfer fuel. The supply of high-pressure fuel from engine-driven motive flow fuel pumps operates fuel ejector pumps to transfer fuel without the need of moving parts. The system is not dependent on electrical power for normal fuel transfer and feed. Total internal and external fuel quantity indication is provided, with a selective quantity readout for individual tanks. Fuel system management requirements are minimal under normal operation for feed, transfer, dumping, and refueling. Sufficient cockpit control is provided to manage the system under failure conditions. The aircraft fuel system is designed so that all usable fuel will normally be depleted under two-or single-engine operating conditions before an engine flame-out occurs from fuel starvation. However, with complete motive flow failure, engine fuel starvation can occur with usable fuel aboard.

Note

All fuel weights in this manual are based on the use of JP-5 fuel at 6.8 pounds per gallon, JP-4 fuel at 6.5 pounds per gallon or, JP-8 fuel at 6.7 pounds per gallon.

Fuel Tankage

Figure 1-25 shows the general fuel tankage arrangement in the aircraft. The fuel supply is stored in eight separate fuselage cells, two wing box cells, two integral wing cells, and (optional loading) two external fuel tanks.

FEED TANKS

The engine feed group, consisting of the left and right box beam tanks and the left and right feed tanks, span the fuselage slightly forward of the mid-center of gravity. Fuel in each box beam tank gravity flows to its respective feed tank. The feed tanks (self-sealing) are directly connected to the box beam tanks and contain the engine boost pumps. The feed tanks supply fuel to the engine. A negative-g check valve traps fuel in the feed tank during negative-g flight.

WARNING

Zero- or negative-g flight longer than 10 seconds in afterburner or 20 seconds in military power or less will deplete the fuel feed tanks (cells 3 and 4), resulting in flameout of both engines.

Note

Deselection of afterburner with less than 1,000 pounds in either feed group may illuminate the FUEL PRESS light due to uncovering of the boost pump inlet.

FORWARD TANK

The forward fuselage fuel tank is in the center fuselage, between the inlet ducts and immediately ahead of the feed tanks. The forward tank is partitioned into two bladder cells (1 and 2) that are interconnected by open

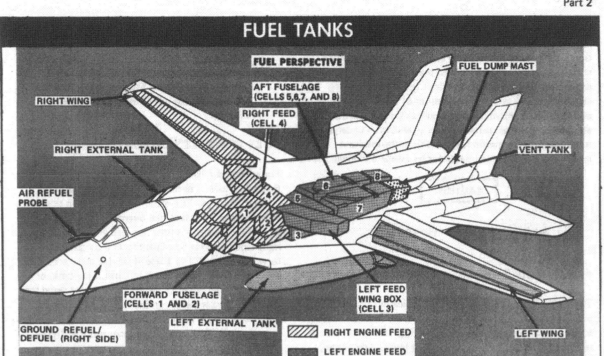

Figure 1-25. Fuel Tanks

ports at the top for vent and overflow purposes. Flapper valves at the base provide for forward-to-aft fuel gravity transfer.

AFT TANK

The aft fuselage fuel tank group is partitioned into four bladder cells (5, 6, 7, and 8) and a vent tank. The forward-most cell in the aft tank group (cell 5) lays laterally across the center fuselage. Extending aft are two coffin-shaped tanks that contain two cells (6 and 8) on the right side and one cell (7) plus an integral fuel vent tank on the left side. The coffin tanks straddle the center trough area, which contains the control runs, Sparrow missile launchers, and electrical and fluid power lines. All fuel cells in the aft tank group are interconnected by one-way flapper valves at the base for aft-to-forward fuel gravity transfer.

WING TANKS

There are integral fuel cells in the movable wing panels between the front and aft wing spars. Because of the wing sweep pivot location and the extensive span (20 feet) of the wing tanks, wing fuel loading provides a variable aft center-of-gravity contribution to the aircraft longitudinal balance as a function of wing sweep angle. Each wing panel consists of the integral fuel cell, which is designed to withstand loads due to fuel sloshing during catapulting and extreme rolling maneuvers with partial or full wing fuel. Fuel system plumbing (transfer and refuel, motive flow, and vent lines) to the wing tanks incorporate telescoping sealed joints at the pivot area to provide normal operation independent of wing sweep position.

EXTERNAL TANKS

Fuel, air, electrical, and fuel precheck line connectors are under the engine nacelles for the external carriage of two fuel tanks. Check valves in the connectors provide an automatic seal with the tank removed. Although the location is designated as armament stations 2 and 7, no other store is designed to be suspended there so that the carriage of external fuel tanks does not limit the weapon-loading capability of the aircraft. Suspension of the drop tanks and their fuel content has an insignificant effect on the aircraft longitudinal center of gravity, and, even under the most adverse asymmetric fuel condition, the resultant movement can be compensated for by lateral trimming.

CAUTION

See section I, part 4 for external tank limitations.

Fuel Quantity System

The fuel-quantity measurement and indication system provides the flightcrew with a continuous indication of total internal and external fuel remaining, a selective readout for all fuel tanks, independent low fuel detection, and automatic fuel system control features.

WARNING

To prevent fuel spills from overfilled vent tank caused by failed level control system, set WING/EXT TRANS switch to OFF if left tape reading reaches 6200 pounds or right tape reading reaches 6600 pounds. If either fuel tape reading is exceeded, aircraft shall be downed for maintenance inspection.

Note

Fuel in the vent tank is not gaged.

The quantity measurement system uses dual-element, capacitance-type fuel probes to provide the flightcrew with a continuous display of fuel quantity remaining. Fuel thermistor devices and caution light displays provide a backup FUEL LOW level indicating system, independent of the capacitance-type gaging system. Additionally, the pilot is provided with a BINGO set capability on the fuel quantity indicator to preset the total quantity level for activation of a BINGO caution light.

Note

Fuel quantity system malfunctions that result in erroneous totalizer readings will invalidate the use of the BINGO caution light.

FUEL QUANTITY INDICATORS

The pilot and RIO fuel quantity indicators are shown in figure 1-26 with a definitive breakdown of tape and counter readings. The white vertical tapes on the pilot's indicator show fuselage fuel quantity. The left tape indicates fuel quantity in the left feed and aft fuselage; the right tape indicates fuel quantity in the right feed and forward fuselage. The L and R labeled counters display either feed (box beam and feed tank), wing tank, or external tank fuel quantity on the side selected using the QTY SEL rocker switch on the fuel management panel. The rocker switch is spring loaded to the FEED position. The pilot's TOTAL quantity display and the RIO's display indicate total internal and external fuel.

Note

The RIO's fuel quantity indicator is a repeater of the pilot's total fuel indicator. The difference between the two should not exceed 300 pounds.

FUEL LOW CAUTION LIGHTS

A L FUEL LOW or R FUEL LOW caution light illuminates with 1000±200 pounds of fuel remaining in the respective feed tank. The RIO is provided with a single FUEL LOW caution light that illuminates with one or both of the pilot's FUEL LOW caution lights.

Each FUEL LOW caution light is illuminated by two thermistors operating in series. One set of thermistors is in the right feed tank and cell no. 2. The other set of thermistors is in the left feed tank and cell no. 5.

WARNING

- If the thermistors in either cell no. 2 or no. 5 remain covered during a fuel transfer failure, it is possible to partially deplete the feed tank without illuminating the respective FUEL LOW caution light.

- When both FUEL LOW caution lights illuminate, less than 1 minute of fuel is available if both engines are operating in zone five afterburner.

NAVAIR 01-F14AAA-1

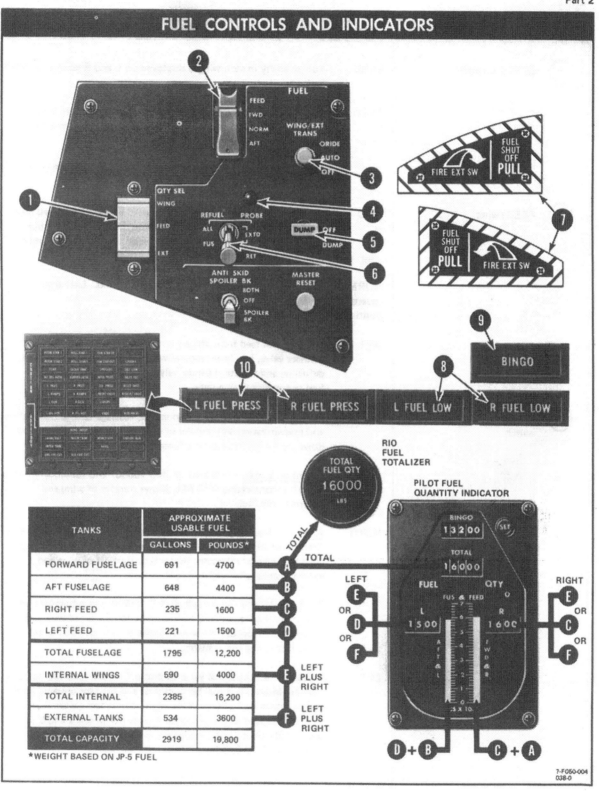

Figure 1-26. Fuel Controls and Indicators (Sheet 1 of 3)

NOMENCLATURE	FUNCTION
① QTY SEL switch	WING - Fuel quantity in each wing is displayed on L and R counter of pilot's fuel quantity indicator. FEED - Rocker switch spring returns to FEED when not held in WING or EXT. FEED tank fuel quantity displayed on L and R counter of pilot's fuel quantity indicator. EXT - Fuel quantity in each external tank displayed on L and R counter of pilot's fuel quantity indicator.
② FEED switch	FWD - Both engines feed from right and forward tanks. Opens feed tank interconnect valve, box beam vent valves, motive flow isolation valve, defueling and transfer selector valve, and shuts off motive flow fuel to all aft tank pumps. NORM (guarded position) - Right engine feeds from forward and right tanks. Left engine feeds from aft and left tanks. AFT - Both engines feed from aft and left tanks. Opens feed tank interconnect valve, box beam vent valves, motive flow isolation valve, defueling and transfer selector valves, and shuts off motive flow fuel to forward tank pumps.
③ WING/EXT TRANS switch	ORIDE - <u>Airborne</u> - Allows transfer of wing fuel, fuselage tank pressurization, and pressurization and transfer of external tanks with landing gear down, and with electrical malfunction in transfer system. <u>Weight on Wheels</u> - In aircraft BUNO 160887 and subsequent and aircraft incorporating AFC 541, allows transfer of wing and external tank fuel. AUTO - <u>Airborne</u> - Normal position. Wing fuel is automatically transferred. Transfer of external fuel and fuselage pressurization is automatic with landing gear retracted. Automatic shut off of wing and external tanks when empty. <u>Weight on Wheels</u> - In aircraft BUNO 160887 and subsequent and aircraft incorporating AFC 541, automatic transfer of wing and external tank fuel cannot be accomplished; switch must be set to ORIDE for wing fuel transfer. OFF - Closes solenoid operated valve to shut off motive flow fuel to wing and also inhibits external tank transfer and fuselage pressurization. Spring return to AUTO position when master test switch is actuated in INST position, and when either thermistor in cell 2 and 5 is uncovered, when DUMP is selected, and when REFUEL PROBE switch is in ALL EXTD position.

Figure 1-26. Fuel Controls and Indicators (Sheet 2 of 3)

NOMENCLATURE	FUNCTION
④ In-flight refueling probe transition light	Illuminates whenever probe cavity forward door is open during retraction or extension of probe.
⑤ DUMP switch	OFF - Dump valve closed. DUMP - Opens a solenoid operated pilot valve, which ports motive flow fuel pressure to open the dump valve and allows gravity fuel dump overboard from cells 2 and 5. Wing and external tank transfer automatically initiated. Dump electrically inhibited with weight on wheels or speed brakes not fully retracted.
⑥ REFUEL PROBE switch	ALL EXTD - Extends refueling probe. Shuts off wing and external tank fuel transfer to permit refueling of all tanks. FUS EXTD - Extends refueling probe. Normal transfer and feed. Used for practice plugins, fuselage-only refueling, or flight with damaged wing tank. RET - Retracts refueling probe.
⑦ Left and right FUEL SHUT OFF PULL handles	Pulling respective handle manually shuts off fuel to that engine. Push forward resets shutoff valve open.
⑧ L and R FUEL LOW caution lights (Also single light on RIO CAUTION panel.)	Fuel thermistors uncovered in aft and left or forward and right feed group. Illuminates with approximately 1000 pounds remaining in individual feed group.
⑨ BINGO caution light	Illuminates when total fuel quantity indicator reads lower than BINGO counter value.
⑩ L and R FUEL PRESS caution lights	Indicates insufficient discharge pressure (less than 9 psi) from respective engine boost pump.

Figure 1-26. Fuel Controls and Indicators (Sheet 3 of 3)

FUEL QUANTITY INDICATION TEST

Actuation of master test switch in the INST position causes the fuel quantity indicator to drive to 2000±200 pounds and illuminates the FUEL LOW caution lights. The test can be performed on the ground or in flight. The test does not check the fuel probes or the thermistors. A test of the BINGO set device can be obtained concurrently with the INST test by setting the BINGO level at greater than 2000 pounds. In this case, the BINGO caution light will illuminate when the totalizer reading decreases to a value less than the BINGO setting.

Fuel Feed

Figure 1-27 shows aircraft fuel system components associated with fuel feed operations.

An engine boost pump is submerged in each feed tank to provide boosted fuel pressure (10 to 50 psi) to each engine. The boost pump discharge capacity is sufficient to supply boosted fuel pressure to the engine main fuel pump, afterburner fuel pump, and the motive flow fuel pump under all operating conditions. The motive flow fuel pumps draw fuel from each engine feed line and produce a supply of high-pressure fuel that is used to operate the boost pumps and transfer system. This fuel drives the boost pump turbine, which is shafted to the boost pump impeller. The motive flow continues through the turbine to the fuselage and wing transfer ejector pumps. The boost pump is provided with a flexible pendulum inlet that seeks the fuel content of the feed tank under various aircraft g conditions.

Under normal (split feed) operation, the engine boost pump in each of these tanks supplies fuel to its respective engine and the engine feed cross flow valve is closed. If boost output pressure drops to less than 9 psi, the L or R FUEL PRESS caution light will illuminate and the engine cross flow valve opens so that the boost pump in the other feed tank supplies boosted fuel to both engines. With a FUEL PRESS light illuminated, do not use afterburner above 15,000 feet. Operation of the crossfeed is automatically initiated by either boost pump discharge pressure switch, which also illuminates the respective L or R FUEL PRESS caution light.

With both L and R FUEL PRESS lights illuminated, the engines will operate on suction feed and fuel transfer to engine feed tanks may be limited to gravity flow only. This restricts engine operation to below 25,000 feet at cruise power setting and to below 15,000 feet at MIL power. Wing fuel transfer may not be available. Aircraft attitude may affect fuel transfer from forward or aft into the feed tank. Nose high attitudes will encourage transfer from the forward tanks; nose low attitudes will encourage fuel transfer from the aft tanks. Fuselage fuel imbalance cannot be corrected by using the FUEL FEED switch if no motive flow is available. Operation with the dual failure should be limited to crusie conditions.

WARNING

Complete loss of motive flow will result in the feed tank interconnect and the engine crossfeed valve remaining in the closed position, thus isolating the forward and aft systems. Consequently, single engine operation will cause fuel on the opposite side to be unavailable.

If two-engine fuel flow demands exceed flow capacity of the single boost pump, the pump bypass valves in either feed tank will allow suction feed. Thus, under failure conditions, engine feed continues uninterrupted, but the total engine feed will now come from either the forward or aft tank, leading to a fuselage fuel imbalance.

To correct fuel imbalance, cockpit control is provided by the FUEL FEED switch on the fuel management panel. This switch controls the sequence of fuel consumption between the forward and aft tanks and, thereby, the distribution of fuel in the aircraft. It does not alter the conditions for each engine feeding from its respective feed tank. In the guarded NORM position, the sump tank interconnect valve and motive flow isolation valve, normally closed, are subject to automatic opening by the uncovering of either thermistor device in the forward (cell no. 2) or aft (cell no. 5) fuel tanks. Selection of either FWD or AFT positions on the FUEL FEED switch opens the feed tank interconnect, motive flow isolation valves, box beam vent valves and shuts off motive flow fuel to and stops transfer from the ejector pumps on the side opposite switch selection. Opening the motive flow isolation valve permits normal operation of all fuselage motive flow actuated components with a single motive flow system failure.

NAVAIR 01-F14AAA-1

Section I
Part 2

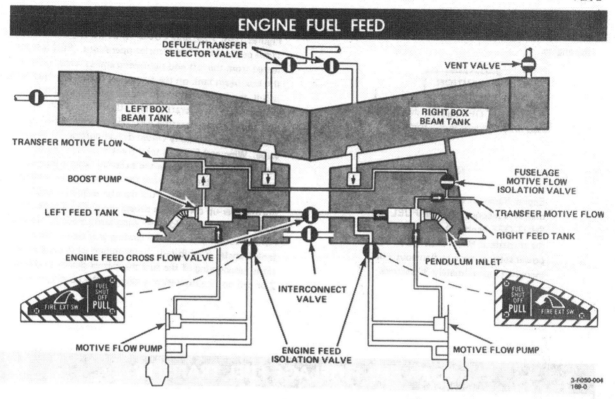

Figure 1-27. Engine Fuel Feed

WARNING

- Do not select FWD or AFT position with the fuel FEED switch unless at least one of the fuel fuselage quantity tapes indicates less than 4,500 pounds. The FEED switch opens the feed tank interconnect valve and could cause overfilling or imbalance of fuel. Overfilling fuselage tanks will result in venting fuel overboard.

- During afterburner operations, the NORM position shall be selected. FWD or AFT positions could deplete fuel in feed tanks.

- Aircraft attitude will have a significant influence on the direction of fuel movement if FWD or AFT is selected. Nose down will transfer fuel forward and nose up will transfer fuel aft.

Note

The preceding selections are not required normally but provide control over the fuel distribution between the forward and aft tanks for abnormal conditions which dictate corrective actions.

Fuel Shutoff Handles

Individual engine fuel feed shutoff valves in the left and right feed lines at the point of nacelle penetration are connected by control cables to the FUEL SHUTOFF handles on the pilot's instrument panel. During normal operation, the handles should remain pushed in so that fuel flow to the engine fuel feed system is unrestricted. If a fire is detected in the engine nacelle, the pilot should

1-47

pull (approximately 3 or 4 inches) the FUEL SHUT OFF handle on the affected side to stop the supply of fuel to the engine.

| CAUTION |

FUEL SHUT OFF handles should be used only in emergency situations.

Note

Engine flameout will occur approximately 4 seconds after the FUEL SHUT OFF handle(s) is pulled with the throttle at MIL. With lower power settings time to flameout will increase (approximately 30 seconds at IDLE).

Fuel Transfer

Figures 1-28, 1-29, and 1-30 show aircraft system components pertinent to fuel transfer operations. Fuel is transferred from the left and right wing and external tanks to the box beam tank on the respective side; if the feed tank is full, the fuel overflows to the aft and forward tanks, respectively. In aircraft BUNO 159588 and subsequent and aircraft incorporating AFC 451 cross transfer is prevented by the normally closed defueling/transfer selector valve. Wing and fuselage tank transfer is effected by motive flow fuel, whereas the external tanks use pressure regulated bleed air from the environmental control system to transfer fuel. Fuel transfer sequence under normal gear-up conditions causes the external tanks to complete transfer before the wing tanks transfer. Under normal conditions with the landing gear down, the drop tank transfer is inhibited. Independent of manual selections, uncovering of the fuel thermistor device in cell no. 2 or cell no. 5 automatically releases the holding solenoid

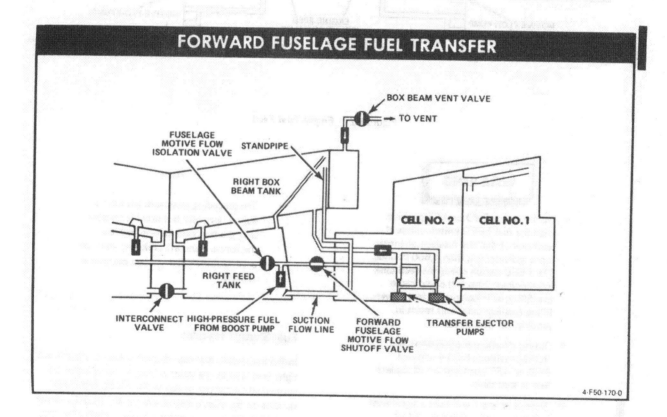

Figure 1-28. Forward Fuselage Fuel Transfer

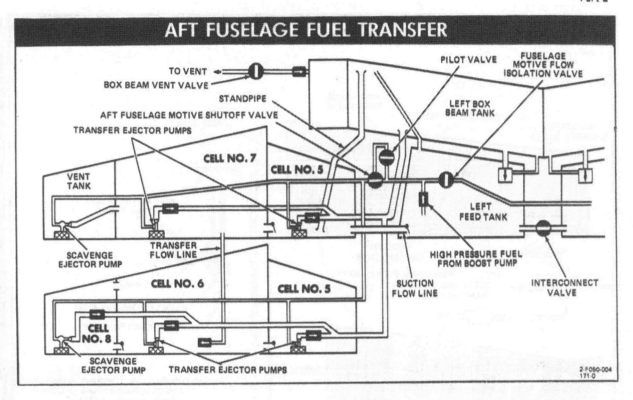

Figure 1-29. Aft Fuselage Fuel Transfer

for maintaining the WING/EXT TRANS switch in the OFF position. The switch returns to the AUTO position. After 5 seconds the WING/EXT transfer switch can be reselected to OFF if desired. This feature minimizes the probability of causing engine fuel starvation with usable fuel aboard due to mismanagement of the fuel switches. The 5-second delay enables reselection of the OFF transfer condition in the event of a wing motive flow line failure or battle damage to the wings.

FORWARD FUSELAGE FUEL TRANSFER

Fuel transfer from the forward tank (cell no. 2) into the right wing box tank is accomplished by two transfer ejector pumps that together have a maximum pumping capacity of 17,700 pounds per hour.

An electrically actuated motive flow shutoff valve in the forward tank controls operation of the ejector pumps. Selecting the AFT position on the fuel FEED switch supplies electrical power to shut off motive flow fuel to the forward tank ejector pumps; however, when NORM or FWD is selected, power is deactivated and and the valve is open.

AFT FUSELAGE FUEL TRANSFER

Fuel from the aft fuselage tanks is transferred collectively by four transfer ejector pumps; two pumps in cell no. 6 and 7 and two pumps in cell no. 5. Together these ejector pumps provide a maximum pumping capacity of 35,400 pounds per hour. As in the case of the forward and right fuselage tanks, the gravity transfer system in the aft tanks operates by means of a large suction transfer line. Gravity flow paths from cells no. 6, 7, and 8 to cell no. 5 and from cell no. 7 to 6 are also provided. An electrically actuated pilot valve for the motive flow shutoff valve in the left feed tank controls operation of ejector pumps in the aft tanks. Selection of the FWD position on the fuel FEED switch supplies electrical power to shut off motive flow

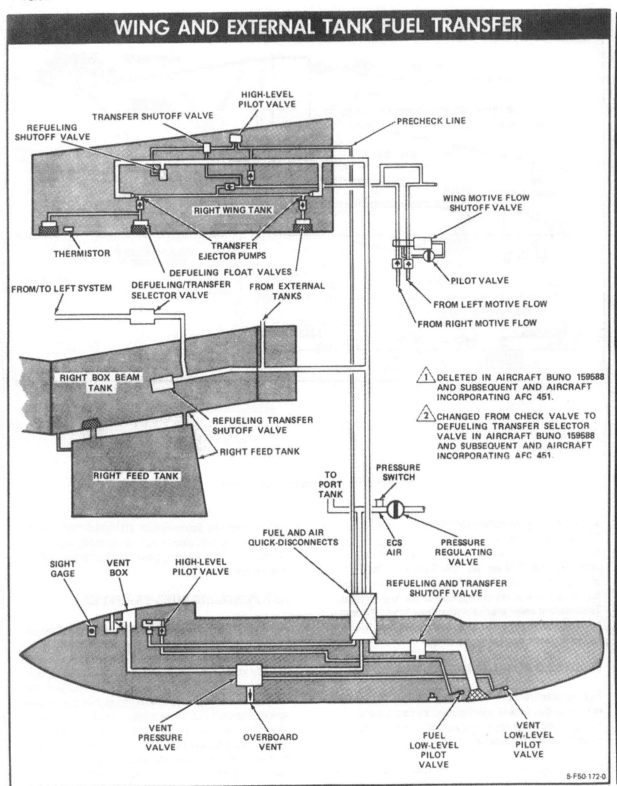

Figure 1-30. Wing and External Tank Fuel Transfer

fuel to the aft and vent tank ejector pumps; however, when NORM or AFT is selected, power is deactivated and the valve is opened.

WING FUEL TRANSFER

Fuel transfer from the left and right wing tanks to the box beam tank on the respective side is accomplished by two transfer ejector pumps in each wing, which provide a nominal transfer capacity of 14,000 pounds per hour per side. An electrically actuated pilot valve for the wing motive flow shutoff valve in the forward tank controls operation of all wing ejector pumps. With the WING/EXT TRANS switch in the AUTO position, the wing motive flow shutoff valve is open (power deactivated) until it automatically energizes the close upon depletion of fuel in both wing tanks. This automatic valve shutoff feature is energized by the uncovering of fuel thermistors at the wing tips and can be overridden by pilot selection of ORIDE on the WING/EXT TRANS switch. Selection of OFF on the WING/EXT TRANS switch directs electrical power to close the valve and shutoff motive flow fuel to the wings independent of wing fuel quantity.

In aircraft BUNO 160887 and subsequent and aircraft incorporating AFC 541, a weight-on-wheels inhibit function prevents opening of the wing motive flow shutoff valve. To transfer wing fuel during ground operations, the WING/EXT TRANS switch must be set to ORIDE to bypass the weight-on-wheels function.

Note

In aircraft BUNO 157981 thru 159468, turning WING/EXT TRANS switch to OFF when fuselage imbalance is greater than 1,000 pounds may prevent a further imbalance.

All wing fuel is transferred normally to its respective feed tank. In aircraft BUNO 159468 and earlier aircraft without AFC 451, either wing tank can transfer to either feed tank through the restricter check valve and connecting plumbing. In aircraft BUNO 159588 and subsequent and aircraft incorporating AFC 451, if the fuselage level control system fails, selecting FWD or AFT with the FEED switch also opens the defueling and transfer selector valves which allows crossfeed transfer. Fuel from the wing enters the fuselage tanks through the fuselage level control system, which will terminate transfer if the fuselage is full.

In aircraft BUNO 159588 and subsequent and aircraft incorporating AFC 451, the loss of motive flow fuel on one side will cause wing fuel transfer to cease on that side.

Selection of the FEED switch to the affected side will open the motive flow isolation valve, supplying motive flow fuel to both wings.

In aircraft BUNO 159468 and prior not incorporating AFC 451, the system incorporates a wing motive flow automatic crossfeed valve which permits transfer from both wings with a single motive flow system failure.

Activation of fuselage fuel dump automatically initiates wing fuel transfer in sequence after external tank transfer by automatically moving the WING/EXT TRANS switch to the AUTO position if selected OFF. Positioning the REFUEL PROBE switch to the ALL/EXTD position also releases the solenoid holding the WING/EXT TRANS switch in the OFF position.

EXTERNAL FUEL TRANSFER

Fuel transfer from the external tanks is accomplished by engine bleed air extracted downstream of the primary heat exchanger, which is pressure-regulated to approximately 25 psi. The maximum fuel transfer rate from each external tank is approximately 750 pounds per minute. Fuel transfer is automatically sequenced with the landing gear retracted. The external tanks will transfer before the wing tanks. Because of variations in air pressure reaching the tanks and/or positioning of the external tank refueling and transfer shutoff valve, the rate and sequence of transfer may vary.

Activation of fuel dump automatically initiates external tank fuel transfer with the landing gear retracted.

Pressurized bleed air (25 psig) enters each external tank through the fuel and air disconnect, through the vent valve, and into the tank. Fuel flows out through the fueling and transfer shutoff valve to each box beam tank through the fuselage level control system. Activation of the fuel and vent low level pilot valves at the bottom of the tank results in closure of the fueling and transfer shutoff valve and a recycling of the vent valve to relieve tank pressure when the tank is empty.

External tank transfer can be checked on the deck by placing the WING/EXT TRANS switch to the ORIDE position, or selecting the FLT GR UP position with the MASTER TEST switch and noting depletion of external tank fuel quantity. In addition, when FLT GR UP is selected, the GO/NO GO light on the MASTER TEST panel is illuminated by a pressure switch in the aircraft's pressure line leading to the external tanks and indicates status of line pressure. Since the FLT GR UP position serves to bypass the landing gear down interlock in the external tank transfer circuit, the WING/EXT TRANS switch may remain in the AUTO (normal) position for this check.

Note

- Verifying tank operation by observing fuel transfer is both time consuming with a full fuselage fuel load and aggravates fuel slosh loads in the external tanks during catapult launch.
- Engine RPM above idle (approximately 75%) may be required to provide sufficient bleed air pressure for a satisfactory check.

Fuel Dump

Figure 1-31 shows aircraft fuel system components associated with fuel dump operation. Fuel dump standpipes in the forward (cell no. 2) and aft (cell no. 5) fuselage tanks are connected to the fuel dump manifold at the dump shutoff valve. The manifold extends aft to the fuselage boattail. Actuation of the fuel DUMP switch to the DUMP position supplies power (dc essential no. 2) to open the solenoid-operated pilot valve, which ports motive flow fuel pressure to open the dump shutoff valve with weight off the main landing gear and the speed brakes retracted.

The fuel DUMP switch circuit is deactivated on deck or with speed brakes extended. Fuel dump with the speed brakes extended is inhibited due to the resulting flow field disturbance, which would result in fuel impingement on the fuselage boattail and exhaust nozzles. The speed brake switch is electrically bypassed during a combined hydraulic system failure, enabling the pilot to dump fuel when the speed brakes are floating. The electrical bypass is accomplished whenever the combined pressure falls below 500 psi.

Note

- The FUEL FEED/DUMP circuit breaker is on the pilot's right knee circuit breaker panel.
- Dump operations with either engine in afterburner should be avoided since the fuel dump mast discharge will be torched.
- After terminating fuel dump, wait approximately 1 minute to allow residual fuel in the fuel dump line to drain before extending speed brakes or lighting afterburners.

FUEL VENT AND DUMP

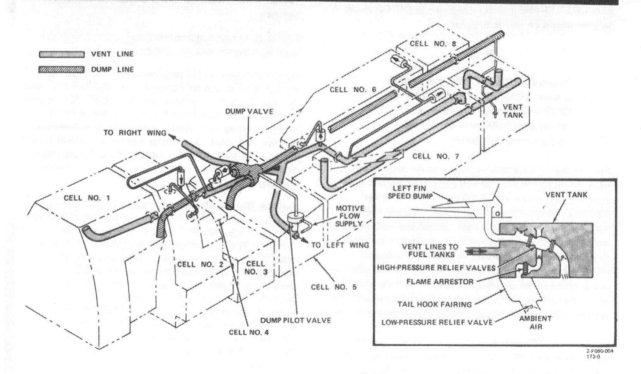

Figure 1-31. Fuel Vent and Dump

Fuel in the wings and external tanks is dumped by transferring to the fuselage. When the fuselage fuel dump circuit is activated, wing and external tank transfer to the box beam tanks is automatically initiated. Fuel dump is by gravity flow with a nominal discharge rate of 1,500 pounds per minute. The dump rate is affected by aircraft pitch attitude and total fuselage fuel quantity with discharge flow inhibited at nose down conditions. The standpipes in the fuel cells control the minimum fuel dump level in the tanks, which under normal operations (feed tanks full) is approximately 4,000 pounds.

Internal Tank Pressurization and Vent

The internal fuel vent system is shown in figure 1-31. In aircraft BUNO 159468 and earlier aircraft without AFC 451, it is nonpressurized (except by ram air) open vent type of system.

In aircraft BUNO 159588 and subsequent and aircraft incorporating AFC 451, engine bleed air from the 25 psi external tank pressure system is reduced to 1.75 psi by a fuselage pressure regulator and distributed to all tanks through the fuselage vent system. This air is automatically supplied when the landing gear handle is UP or the WING/EXT TRANS switch is in ORIDE. When the WING/EXT TRANS switch is in OFF the low pressure bleed air is cut off.

In flight, the vent tank is maintained at a positive pressure up to 2.50 psi maximum. This pressure is fed by connecting lines to all internal tanks. These connecting lines are routed to provide venting to both the forward and aft end of each fuselage tank so it can function as both a climb and dive vent. Venting of the box beam tanks is controlled by solenoid-operating valves, which when closed, provide suction transfer through the gravity flow paths in cells no. 2 and 5 to the feed tanks.

Fueling and Defueling

Figure 1-32 shows the refueling system. The aircraft is equipped with a single-point refueling system, which enables pressure filling of all aircraft fuel tanks from a single receptacle. The receptacle is at the recessed ground refuel and defuel station, behind a quick-access door on the lower right side of the forward fuselage. The maximum refueling rate is 450 gallons per minute at a pressure of 50 psi. Since ground and air refueling connections use a common manifold, the refueling sequence is the same.

Standpipes refuel the aft and forward fuselage tanks by overflow from the left and right box beam tanks. A high-level pilot valve at the high point of the forward tank shuts off the fuselage refueling valve in the right box beam tank when the forward tank group is full. Fuel flows from the left box beam tank to cell no. 5, after which it overflows to the right side, then the left side. A high-level pilot valve at the high point of the left box beam tank and aft tank (cell no. 7) shuts off the fuselage refueling valve in the left box beam tank when the aft tank group is full. Individual wing and external tank filling is accomplished by flow through a shutoff valve in each tank.

CAUTION

Gravity refueling of the aircraft fuel system should be accomplished only under emergency situations. While performing such an operation, avoid introducing contaminants into the fuel tanks or damaging the fuel quantity probes and wiring.

PRECHECK SYSTEM

Ground refueling control is by two precheck selector valves and a vent pressure gage adjacent to the refueling receptacle on the ground refuel and defuel panel. The precheck valves functionally test high-level pilot valve operation incident to ground pressure refueling; the valves separately check the pilot valves in the fuselage tanks and the wing and external tanks. In addition to this precheck function, the precheck valves can be used for ground selective refueling of only the fuselage or all tanks. Since the precheck valves, which are manually set by the ground crew, port pressurized servo fuel to the high-level pilot valves and subsequently to the shutoff valves, no electrical power is necessary on the aircraft to perform ground refueling operations. Additionally, ground refueling control without engines running is completely independent of switch positioning on the fuel management panel. The direct-reading vent pressure indicator monitors pressure in the vent lines. The gage consists of a pointer on a scale having two bands, one green and one red.

The green band indicates a safe pressure range (0 to 4 psi) and the red band indicates an unsafe range (4 to 8 psi).

CAUTION

During ground refueling operations, the direct-reading vent pressure indicator shall be observed and refueling stopped if pressure indicates in the red band (above 4 psi).

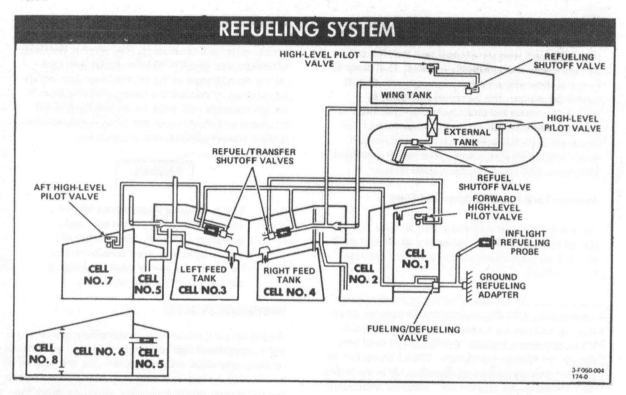

Figure 1-32. Refueling System

IN-FLIGHT REFUELING

Note

Refer to section III, part 4 for in-flight refueling procedures.

The in-flight refueling system permits partial or complete refueling of the aircraft fuel tanks while in flight. The retractable refueling probe has an MA-2 type nozzle, which is compatible with any drogue type refueling system. A split refueling system is provided with fuel routed into the left and right box beam tanks for initial replenishment of feed tank fuel. Selectable fuel management controls dictate the extent of further distribution to the wing tanks, external tanks and/or fuselage tanks. The maximum refueling rate is approximately 475 gallons per minute at a pressure of 57 psi.

To prevent fuel fumes from entering the cockpit through the ECS due to possible fuel spills during in-flight refueling, select L ENG air source.

- Maximum airspeed for extension or retraction in-flight of the refueling probe is 400 KIAS (0.8 IMN).

- Maximum airspeed for refueling from KC-97 or KC-135 aircraft is 255 KIAS.

Note

- With the in-flight refueling probe extended, the pilot's and RIO's altimeter and airspeed and Mach indicators will show erroneous indications because of changes in airflow around the pitot static probes.

- Flight operations with in-flight refueling probe door removed are not authorized due to effects of water intrusion, exposure to elements, and structural fatigue to electrical and hydraulic hardware assemblies.

IN-FLIGHT REFUELING PROBE. The retractable in-flight refueling probe is in a cavity on the right side of the forward fuselage section, immediately forward of the pilot's right vertical console panel.

Extension of the refuel probe is provided through redundant circuits by the REFUEL PROBE switch. A hydraulic actuator within the probe cavity extends and retracts the probe. The probe actuator is powered by the combined hydraulic system. It can be extended and retracted by means of the hydraulic hand pump in the event of combined system failure.

Note

- To extend or retract the refueling probe using the hydraulic hand pump requires essential dc no. 2 electrical power and requires approximately 25 cycles of the pump handle.

- Probe retraction is not available if the FUEL P/MOTIVE FLOW ISOL V (P-PUMP) circuit breaker (RG1) is pulled.

The red probe transistion light immediately above the REFUEL PROBE switch illuminates whenever the probe cavity forward door is not in the closed position. Since the closed door position is indicative of both the probe retracted and extended position, the light serves as a probe transition indicator as well as a terminal status indicator. The probe external light illuminates automatically upon probe extension with the EXT LTS master switch ON.

IN-FLIGHT REFUELING CONTROLS. Regardless of fuel management panel switch positioning, at low fuel states the initial resupply of fuel is discharged into the left and right wing box tanks. The split refueling system to the left and right engine feed group provides for a relatively balanced center of gravity condition during refueling. Selective refueling of the fuselage or all full tanks is provided on the REFUEL PROBE switch with the probe extended. In the FUS/EXTD position, normal fuel transfer and feed is unaltered. This position is used for practice plug-ins, fuselage only refueling or return flight with a damaged wing tank. The ALL/EXTD position shuts off wing and external drop tank transfer to permit the refueling of all tanks.

HOT REFUELING

Hot refueling can be accomplished with the refueling probe extended or retracted. If the probe is extended, control of the tanks to be refueled is accomplished in the same manner as during in-flight refueling. If the probe is not extended, select WING/EXT TRANS switch to OFF to refuel all tanks. Select ORIDE to refuel the fuselage only.

Automatic Fuel Electrical Controls

AUTOMATIC LOW LEVEL WING TRANSFER SHUTOFF

A thermistor is located at the low point in each wing cell. When both are uncovered, a discrete electrical signal is generated, and through a control, the wing motive flow shutoff valve is energized and closes, terminating all wing transfer. If either or both thermistors are again submerged, wing transfer resumes.

Failure of this override system could result in a wing transfer failure. Selection of WING/EXT TRANS switch to ORIDE removes all power from the wing motive flow shutoff valve, permitting it to open.

AUTOMATIC FUEL LOW LEVEL OVERRIDE

Under normal operating conditions, the forward and right fuselage tank complex is isolated from the aft and left tank. This is necessary for proper longitudinal center-of-gravity control and battle damage conditions. However, as fuel depletion progresses to the point of feed tank only remaining, it becomes mandatory that the tanks be connected to maintain an equal balance. To accomplish this, two thermistors are located at the low points in cells no. 2 and 5, and when either is uncovered (approximately 1,700 to 2,000 pounds per side) the following operations are electrically performed.

- Feed tank interconnect valve is opened.
- Motive flow isolation valve is opened.
- Box beam vent valves are opened.
- Engine crossfeed valve is opened.
- WING/EXT TRANS switch is energized to move from OFF to AUTO. This signal is maintained for 5 seconds.

- In aircraft BUNO 159588 and subsequent and aircraft incorporating AFC 451, defuel and transfer selector valves are opened.

WARNING

Uncovering either thermistor in cells no. 2 or no. 5 will only move the WING/EXT TRANS switch from OFF to AUTO, but under no circumstances will it override a wing transfer failure.

ELECTRICAL POWER SUPPLY SYSTEM

In normal operation, ac power is supplied by the engine-driven generators. This ac power is converted by two transformer-rectifiers (T/R) into dc power. (Refer to page FO-9.) One generator is capable of assuming the full ac power load and one T/R is capable of assuming the full dc power load. Additionally, a hydraulic-driven emergency generator provides an independent backup supply of both ac and dc power for electrical operation of essential buses. Ground operation of all electrically powered equipment is provided through the supply of external ac power to the aircraft. Switching between power supply systems is automatically accomplished without pilot action; however, sufficient control is provided for the flightcrew to selectively isolate power sources and distribution in emergency situations. See figure 1-33 for a functional description of the control switches. All electrical circuits are protected by circuit breakers accessible in flight to the pilot and RIO.

Normal Electrical Operation

MAIN GENERATORS

Two engine-driven, oil-cooled, integrated-drive generators (IDG) produce the normal 115 to 200-volt, 400-Hz, three phase ac electrical power. The normal rated output of each generator is 75 kVA, which is sufficient to individually assume the complete electrical load of the aircraft. Each main ac generator is controlled by a separate switch on the pilot's MASTER GEN control panel. Indication of a main power supply malfunction is provided by a L GEN and R GEN caution light. Cooling of the IDG oil is by a combined fuel-engine oil-IDG oil heat exchanger. Should an excessive amount of heat be developed in an IDG, a thermal (400°F) actuated device automatically decouples the drive clutch, which immobilizes the IDG. There are no provisions for recoupling the IDG unit in flight.

CAUTION

Failure of the weight-on-wheels circuit to the in-flight mode while on the deck will cause the loss of ECS engine compartment air ejector pumps, causing a subsequent CSD disconnect and illumination of the GEN light.

Generator output voltage and frequency are individually monitored by generator control units (GCU), which prevent application of internally generated power to the aircraft bus system until the generator output is within prescribed operating limits. With the main generator switch in the NORM position, the applicable generator is self-excited, so that during the engine start cycle, it automatically comes on the line at approximately 48% RPM under normal lead conditions. Likewise, during engine shutdown, the GCU automatically trips the generator off the line as the power output decreases below prescribed limits at approximately the same engine speed.

During normal operations, the generator control switches remain in the NORM position continuously. However, subsequent to an engine shutdown, stall, or flame-out in flight where the GCU has tripped the generator off the line, the reattainment of normal engine operation will not automatically reset the generator unless the engine speed decreases below about 30% N_2 RPM. If a transient malfunction or condition causes the generator to trip, the generator must be manually reset by cycling the applicable generator control switch to the OFF/RESET position then back to NORM.

When normal reset cannot be accomplished, the TEST position on the generator control switch, allows the generator to be excited but not connected to the aircraft buses. In test, a CSD, generator, or GCU failure causes the GEN light to remain illuminated. If the light goes out, the problem is in the distribution system.

TRANSFORMER-RECTIFIERS

Two transformer-rectifiers convert internal or external ac power to 28-volt dc power. A single TRANS/RECT advisory light on the pilot's annunciator panel provides failure indication for one or both transformer-rectifiers. No flightcrew control is exercised over transformer-rectifier operation aside from controlling the ac power supply or circuit breakers for the power converters. The transformer-rectifiers have a rated output of 100 amperes each. Each unit is capable of assuming the complete dc electrical load of the aircraft. Forced air cooling is provided with engines running to dissipate the heat generated by the power converters.

GENERATOR PANEL

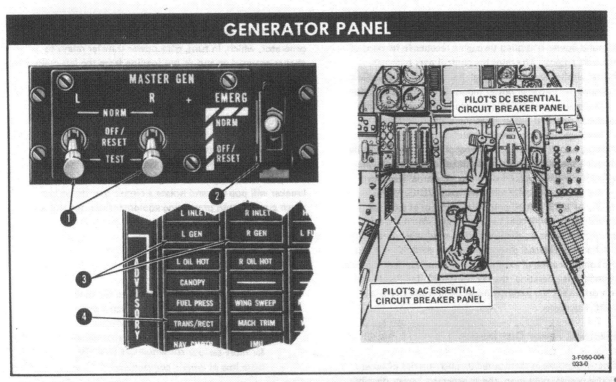

NOMENCLATURE		FUNCTION
① MASTER GEN switch (lock lever)	NORM -	Connects the generator to the main buses through the line contactor.
	OFF/RESET -	Disconnects generators from the buses. Reset the generator if tripped by an overvoltage, undervoltage, or fault condition.
	TEST -	The generators are energized but are not connected to the buses. Provides a means to analyze a system malfunction indicated by a generator caution light when an attempt to reset a generator is unsuccessful.
② EMERG generator switch	NORM -	Safety guard down. Emergency generator is automaticlly connected to the essential buses if both dc power sources fail.
	OFF/RESET -	Safety guard must be lifted. Disconnects the emergency generator from the essential buses. Resets the generator if tripped by an undervoltage or underfrequency condition.
③ L GEN and R GEN caution lights		GEN caution lights are on the pilot's caution/advisory light panel. Each light is tied to its respective main ac contactor and is powered by the essential bus no. 2. Illumination of the L GEN or R GEN caution light indicates that the corresponding generator is not operating due to a fault in the generator, generator control unit, or electrical distribution system.
④ TRANS/RECT advisory light		A TRANS/RECT advisory light is on the lower half of the pilot's caution/advisory indicator panel. Illumination of the TRANS/RECT advisory light indicates that the operable transformer-rectifier is powering the total dc load, or that a dual transformer-rectifier failure has occurred.

Figure 1-33. Generator Panel

EXTERNAL POWER

Ground power is applied through a receptacle forward of the port engine. The pilot has control over external power application only through hand signals to the plane captain. An external-power monitor prevents application of external power that is not within tolerances, and disconnects external power from the buses if undervoltage, overvoltage, underfrequency, overfrequency, or phase-reversal occurs. Power can be reapplied to the aircraft by pressing the reset button adjacent to the receptacle, provided it is within prescribed limits. External electrical power is automatically inhibited from VDIG, MDIG, AICS, APX-76, CADC and CSDC without external air conditioning connected to the aircraft. When the left generator comes on the line during start, it automatically disconnects external power. Although there is no direct cockpit indication of external power being applied after one generator is operating, the HYD TRANSFER PUMP will not operate if the external power plug is still in the aircraft receptacle.

Electrical Power Distribution

Electrical power is distributed through a series of buses. Under normal operation, the ac generator power distribution is split between the left and right main ac buses. Failure of either main ac generator trips a tie connector to connect both buses to the operative generator. If the bus tie fails to trip when the generator goes bad, the respective transformer-rectifier will not be powered and the indication of this double failure will be a L GEN or R GEN caution light and a TRANS/RECT advisory light. The left and right main ac buses in turn supply ac power directly to the respective transformer-rectifiers, and the left main ac bus also supplies power to both essential ac buses under normal operation.

External power is distributed through the aircraft electrical system in the same manner as main generator power. Like the main ac generators, dc power distribution from the two transformer-rectifiers under normal operations is split between the left and right main dc buses. Failure of either transformer-rectifier trips the respective tie contactor to connect both main dc buses to the operative transformer-rectifier. The TRANS/RECT advisory light provides a direct indication of dc bus tie status. An interruption-free dc bus interconnects the left and right main dc buses to provide a continuous source of dc power with failure of either main ac generator and/or transformer-rectifier. The left main dc bus additionally supplies power to both essential dc buses under normal operations. Power to the AFCS bus is normally supplied from the interruption-free dc bus; however, with an output failure from both transformer-rectifiers, the AFCS bus load is automatically transferred to the essential no. 2 bus. Loss of main dc power automatically activates the emergency generator, which, in turn, trips power transfer relays to change essential ac and dc bus loading from the left main ac and dc buses to the emergency generator, regardless of main generator output status.

CIRCUIT BREAKERS

Individual circuit protection from an overload condition is provided by circuit breakers, which are all located in the cockpits for accessibility in flight. The appropriate circuit breaker will pop out and isolate a circuit that draws too much current, thus preventing equipment damage and a possible fire.

> **CAUTION**
>
> Do not repeatedly reset or forcibly hold circuit breakers depressed. Such a condition is indicative of an overload condition or malfunction, which may produce far more serious consequences than the mere loss of certain components.

Cockpit circuit breaker panels are shown on pages FO-10 and FO-11. Circuit breakers in the pilot's cockpit comprise the majority of those required for essential aircraft systems. The circuit breakers are arranged in rows and are oriented so that the white banded shaft of popped breakers is readily visible for flightcrew surveillance. Panels, rows, and columns of breakers are identified to facilitate breaker location and designation. Placards adjacent to the breakers identify individual circuit breakers by affected components; amperage ratings are indicated on top of each circuit breaker.

CIRCUIT BREAKER LOCATION. The alphanumeric system for locating circuit breakers in the aircraft is as follows:

The panels in the RIO's cockpit are labeled 1 through 8 starting left-aft and proceeding clockwise. (See figure 1-34.) Thus, panels 1-4 are on the RIO's left and panels 5-8 are on his right. The pilot's left and right knee panels are designated L and R, respectively. The first digit in the three-part locator is the alphanumeric, which identifies the circuit breaker panel.

The second part is a letter and designates the row in which the circuit breaker will be found. The top row is designated A, the next row lower is B, etc.

NAVAIR 01-F14AAA-1

CIRCUIT BREAKER ALPHANUMERIC INDEX

3F6	AC ESS BUS NO. 2 FDR PH A	7E3	ANGLE OF ATTK IND DC
4F1	AC ESS BUS NO. 2 FDR PH B	4F5	ANLATTK/TOTAL TEMP HTR
4F4	AC ESS BUS NO. 2 FDR PH C	7C1	ANN PANEL PWR
214	ACM LT/SEAT ADJ/STEADY POS LT	7F3	ANN PILOT PANEL AUX PWR/TR-ADVSY
4B6	ACM PNL LT/INS SYNC	8C2	ANN PNL/DDI DIM CONT
6D5	ACM PNL PWR	8G2	ANT LOCK EXCIT
7B1	AFCS BUS FDR	1F2	ANT SVO HYD PH A
RB2	AFCS/NOSE WHEEL STEER	1F4	ANT SVO HYD PH B
3C6	AHRS PH A	1F6	ANT SVO HYD PH C
4C1	AHRS PH B	211	ANTICOLL/SUPP POS/POS LTS
4C5	AHRS PH C	7C2	ANTI-ICE CONTR/RAIN RPL/HK CONTR
LF2	AICS L	RG2	ANTI-ICE/ENG/PROBE
215	AICS L HTR	7E1	ANTI-SKID/R AICS LK UP PWR
7E2	AICS L LK UP PWR/EMER GEN TEST	7E7	APN-154
6A6	AICS L RAMP STOW	5C4	APX-72 AC
LG2	AICS R	6F7	APX-72 DC
218	AICS R HTR	816	APX-72 TEST SET
7E1	AICS R LK UP PWR/ANTI-SKID	7C7	ARA-63 ILS DC
6A5	AICS R RAMP STOW	3A6	ARA-63 ILS PH A
8A4	AIM-54 MSL	4A3	ARA-63 ILS PH B
8G4	AIM-7 BATT ARM	4A5	ARA-63 ILS PH C
8G3	AIM-7 R BATT ARM	6B3	ARC-159 NO. 1
RC2	AIR SOURCE CONTR	6D1	ARC-159 NO. 2
8B6	ALE-39 CHAFF/FLARE DISP	4E6	AS/ARA-48 ANT
8B5	ALE-39 SEQ 1 & 2 SQUIBS	1J2	ASW-27 (AC)
7C8	ALPHA COMP/PEDAL SHAKER	LA3	AUTO PITCH DRIVE TRIM
4B3	ALPHA HTR	1J1	AUTO THROT AC
2G2	ALQ-100 PH A	8B7	AUTO THROT DC
2G5	ALQ-100 PH B /3\	7G3	AUX FLAP/FLAP CONTR
2G8	ALQ-100 PH C	6B7	AWG-15 DC
6C1	ALQ-126 DECM /2\	5B5	AWG-15 PH A NO. 1
6B6	ALT LOW WARN	5B4	AWG-15 PH A NO. 2
8D1	AMC BIT/R DC TEST	5D5	AWG-15 PH B NO. 1
8G1	AMCS ENABLE	5D4	AWG-15 PH B NO. 2
114	AN/ALR-45	5F5	AWG-15 PH C NO. 1
8F5	AN/ALR-50	5F4	AWG-15 PH C NO. 2
3D7	AN/ARA-50		
7D6	AN/ARA-50	3D3	BARO ALTM AC
1D2	AN/AWG-9 CMPTR PH A	7C6	BARO ALTM DC
1D5	AN/AWG-9 CMPTR PH B	4F6	BDHI INST PWR/TACAN (AC)
1D6	AN/AWG-9 CMPTR PH C	7E8	BDHI/TACAN (DC)
2G3	AN/AWG-9 PUMP PH A	7F6	BINGO/OXY CAUTION
2G6	AN/AWG-9 PUMP PH B	7F2	BLEED AIR/L OIL HOT
2G7	AN/AWG-9 PUMP PH C	4A4	BLEED DUCT AC
8H6	AN/AWW-4	8D5	BRAKE ACCUM SOV /1\
1B3	AN/AWW-4 PH A		
1C4	AN/AWW-4 PH B		
1B6	AN/AWW-4 PH C	7A1	CABIN PRESS
3C3	ANGLE OF ATTK IND AC	7C5	CAN/LAD CAUTION/EJECT CMD IND

EFFECTIVITY

/1\ In aircraft BUNO 161168 and subsequent and aircraft incorporating AFC 622.

/2\ In aircraft BUNO 161168 and subsequent.

/3\ In aircraft BUNO 161168 and subsequent, ALQ-100 PH A, B, C circuit breakers are labeled MSL HTR PH A, B, C.

Figure 1-34. Circuit Breaker Alphanumeric Index (Sheet 1 of 5)

Section I
Part 2

NAVAIR 01-F14AAA-1

LA2	CHAN 1 CADC PH A	RF2	FLT CONTR AUTH DC
LB2	CHAN 1 CADC PH B	2A1	FLT HYD BACKUP PH A
LC2	CHAN 1 CADC PH C	2C1	FLT HYD BACKUP PH B
LH2	CHAN 2 CADC	2E1	FLT HYD BACKUP PH C
3D1	COMB HYD PRESS IND	3D2	FLT HYD PRESS IND
8A5	CONTR/DISPL SUBSYS	2H2	FORM/TAXI LT
1G2	CONTR/DISPLAY PH A	LF1	FUEL CONTR L/MACH LVR
1G3	CONTR/DISPLAY PH B	LG1	FUEL CONTR R/MACH LVR
1G6	CONTR/DISPLAY PH C	RD1	FUEL FEED/DUMP
8D2	COUNTING ACCEL or COUNTING ACCEL/TARPS	7F7	FUEL LOW CAUTION
7D5	CSDC	RC1	FUEL MGT PNL
3B7	CSDC PH A	RG1	FUEL P/MOTIVE FLOW ISOL V (P-PUMP)
4B2	CSDC PH B	7F1	FUEL PRESS ADVSY
4B5	CSDC PH C	3F2	FUEL QTY IND AC
		7D3	FUEL QTY IND DC
6B5	DC ESS NO. 1 FDR	7E4	FUEL TRANS ORIDE
7A2	DC ESS NO. 2 BUS FDR	7F9	FUEL VENT VALVE
7G7	DC L TEST/RUDDER TRIM		
8D1	DC R TEST/AMC BIT	2H1	GAS TEMP LIM/R PH A TEST
1I1	DDI AC	2H4	GAS TEMP LIM/R PH B TEST
8C2	DDI/ANN PNL DIM CONTR	2H8	GAS TEMP LIM/R PH C TEST
8G5	DDI DC/ASW-27	7F5	GEN L CAUTION
7B7	DISPLAY PWR	7F4	GEN R CAUTION
RD1	DUMP/FUEL FEED	LE3	GLOVE VAN CONTR
8E7	DYHR UNIT	8F4	GND PWR/COOLING INTERLK
		7G1	GND ROLL BRAKING/SPOILER POS IND
6C1	ECM DESTR	8D6	GND TEST/MACH LVR BIT
7D4	ECS TEMP CONT DC	8C3	GUN ARMED POWER
7C5	EJECT CMD IND/CAN-LAD CAUTION	8B3	GUN CLR/GUN CONTR PWR DC
8H4	ELEC COOLING	2H5	GUN CONTR PWR AC
6B2	EMER FLT HYD AUTO	8B3	GUN CONTR PWR DC/GUN CLR
6B1	EMER FLT HYD MAN	8I3	GYRO PWR
7E2	EMER GEN TEST/L AICS LK UP PWR		
8I2	EMERG GEN CONTR	7C2	HK CONTR/RAIN RPL/ANTI-ICE CONTR
RB1	ENG/ANTI-ICE/VALVES	1B1	HSD/ECMD PH A
3E2	ENG FUEL FLOW/L IND/OVSP	1B4	HSD/ECMD PH B
3E2	ENG L FUEL FLOW IND/OSVP	1B7	HSD/ECMD PH C
3C1	ENG L N2 TACH	8C4	HUD CAMERA DC
4C6	ENG L OIL PRESS/NOZ IND	1F1	HUD CAMERA PH A ⎫
7D1	ENG OIL COOL	1D4	HUD CAMERA PH B ⎬ △4
RG2	ENG/PROBE/ANTI-ICE	1E5	HUD CAMERA PH C ⎭
3E4	ENG R FUEL FLOW IND	3C7	HUD PH A
3C2	ENG R N2 TACH	4C2	HUD PH B
4D6	ENG R OIL PRESS/NOZ IND	4C4	HUD PH C
7F10	ENG STALL TONE	1A1	HV PWR SUP PH A
RE1	ENG START	1A2	HV PWR SUP PH B
7G10	EXT LT CONTR	1A3	HV PWR SUP PH C
7C4	FIRE L DET/LT	7E6	HYD PRESS IND
7C3	FIRE R DET/LT	7G11	HYD PUMP SPOILER CONT
7G3	FLAP CONT/AUX FLAP	7E5	HYD VALVE CONTR
3E3	FLAP IND/TAIL/RUDDER		
RE2	FLAP/SLAT CONTR SHUTOFF	LD3	ICE DET
LD1	FLT CONTR AUTH AC	6F3	ICS NFO

EFFECTIVITY

△4 In aircraft BUNO 161279 and subsequent and aircraft incorporating AFC 618.

Figure 1-34. Circuit Breaker Alphanumeric Index (Sheet 2 of 5)

NAVAIR 01-F14AAA-1

6F2	ICS PILOT	LG1	MACH LVR/R FUEL CONTR
1I7	IFF A/A AC	LD2	MACH TRIM AC
8F6	IFF A/A DC	RD2	MACH TRIM DC
7C7	ILS ARA-63 DC	LE2	MANV FLAP/WING SWP. DRIVE NO. 2
3A6	ILS ARA-63 PH A	6B4	MASTER ARM
4A3	ILS ARA-63 PH B	8H5	MASTER TEST
4A5	ILS ARA-63 PH C	8H3	MECH FUZING STA 1/8
7G9	INBD SPOILER CONTR	8H2	MECH FUZING STA 3/4
4B6	INS SYNC/ACM PNL LT	8H1	MECH FUZING STA 5/6
4A1	INST BUS FDR △5	7A4	MID CPRSN BYP L
5D3	INST BUS FDR △6	7A5	MID CPRSN BYP R
2I6	INST LT	8D7	MID CPRSN BYPASS PWR
1I2	INTEG TRIM AC	7G5	MLG HANDLE RLY NO. 1
8F3	INTEG TRIM DC	7G4	MLG HANDLE RLY NO. 2
4A2	INTRF BLANKER	6F5	MLG SAFETY RLY NO. 1
8I1	INTRPT FREE DC BUS FDR NO. 1	6F4	MLG SAFETY RLY NO. 2
8C6	INTRPT FREE DC BUS FDR NO. 2	8D3	MONITOR BUS CONTR
8A6	IR SUBSYS	8I5	MOTOR FIRE A
1H2	IR/TV PH A	8I4	MOTOR FIRE B
1H3	IR/TV PH B	1G1	MSL AUX PH A
1H6	IR/TV PH C	1G4	MSL AUX PH B
		1G7	MSL AUX PH C
6E3	JETT 1	8A1	MSL AUX SUBSYS
6E2	JETT 2	1C2	MSL HTR PH A
7D7	JULIET-28	1C5	MSL HTR PH B △7
		1C6	MSL HTR PH C
LF2	L AICS	2G2	MSL HTR PH A
2I5	L AICS HTR	2G5	MSL HTR PH B △2
6A6	L AICS RAMP STOW	2G8	MSL HTR PH C
8G4	L AIM-7 BATT ARM		
7G7	L DC TEST/RUDDER TRIM	1J3	NAV IMU PH A
3E2	L ENG FUEL FLOW IND/OVSP	1J5	NAV IMU PH B
3C1	L ENG N2 TACH	1J6	NAV IMU PH C
4C6	L ENG OIL PRESS/NOZ IND	1I3	NAV PWR SUP PH A
7C4	L FIRE DET/LT	1I5	NAV PWR SUP PH B
6C5	L FIRE EXT	1I6	NAV PWR SUP PH C
LF1	L FUEL CONTR/MACH LVR	2I2	NFO CONSOLE LT
7F5	L GEN CAUTION	7A3	NLG STRUT LCH BAR ADVSY
1A5	L MAIN XFMR/RECT	RB2	NOSE WHEEL STEER/AFCS
7A4	L MID CPRSN BYPASS		
7F2	L OIL HOT/BLEED AIR	7F2	OIL L HOT/BLEED AIR
3E6	L PH A TEST/P. ROLL TRIM	7D2	OIL R HOT/OVSP CAUTION
4E2	L PH B TEST/P. ROLL TRIM	8C5	OUTBD SPOILER CONTR
4E5	L PH C TEST/P. ROLL TRIM	2A2	OUTBD SPOILER PUMP PH A
4D1	L PITOT STATIC HTR	2B2	OUTBD SPOILER PUMP PH B △8
3B1	L TIT IND	2C2	OUTBD SPOILER PUMP PH C
7A3	LCH BAR NLG STRUT ADVSY	7D2	OVSP CAUTION/R OIL HOT
2H10	LIQUID COOLING CONTR AC	7F6	OXY/BINGO CAUTION
8B4	LIQUID COOLING CONTR DC	3F3	OXY QTY IND
7G8	MACH LEVER SHIFT	3E6	P. ROLL TRIM/L PH A TEST
8D6	MACH LVR BIT/GND TEST	4E2	P. ROLL TRIM/L PH B TEST
LF1	MACH LVR/L FUEL CONTR	4E5	P. ROLL TRIM/L PH C TEST

EFFECTIVITY

△2 In aircraft BUNO 161168 and subsequent.

△5 In aircraft not incorporating AFC 610.

△6 In aircraft BUNO 161270 and subsequent and aircraft incorporating AFC 610.

△7 In aircraft BUNO 161168 and subsequent, circuit breakers are labeled ALQ-126 PH A, B, C.

△8 In aircraft BUNO 160887 and subsequent and aircraft incorporating AFC 523, the three OUTBD SPOILER PUMP circuit breakers are replaced by one circuit breaker (2B2).

Figure 1-34. Circuit Breaker Alphanumeric Index (Sheet 3 of 5)

Section I
Part 2
NAVAIR 01 F14AAA-1

3E6	PH A L TEST/P. ROLL TRIM	7G7	RUDDER TRIM/L DC TEST
2H1	PH A R TEST/GAS TEMP LIM	3E7	RUDDER TRIM PH A
4E2	PH B L TEST/P. ROLL TRIM	4E1	RUDDER TRIM PH B
2H4	PH B R TEST/GAS TEMP LIM	4E4	RUDDER TRIM PH C
4E5	PH C L TEST/P. ROLL TRIM	2I4	SEAT ADJ/ACM LT/STEADY POS LT
2H8	PH C R TEST/GAS TEMP LIM	1E2	SEMIREG PWR SUP PH A
2I3	PILOT CONSOLE LT	1E4	SEMIREG PWR SUP PH B
7B3	PITCH COMPTR DC	1E6	SEMIREG PWR SUP PH C
LB1	PITCH COMPTR AC	1H1	SOL PWR SUP PH A
RA2	PITCH ROLL TRIM/SPD BK ENABLE	1H5	SOL PWR SUP PH B
4D1	PITOT L STATIC HTR	1H7	SOL PWR SUP PH C
4D2	PITOT R STATIC HTR	RA2	SPD BK/P-ROLL TRIM ENABLE
2I1	POS/ANTICOLL/SUPP POS LTS	7G11	SPOILER CONT HYD PUMP
RG2	PROBE/ENG/ANTI-ICE	7G9	SPOILER CONT INBD
4F2	PROBE LT	7G1	SPOILER POS IND/GND ROLL BRAKING
		8E4	STA 1 AIM-9 COOL PWR
LG2	R AICS	6D4	STA 1 AIM-9 REL PWR
2I8	R AICS HTR	6C7	STA 1 REL PWR A
7E1	R AICS LK UP PWR/ANTI-SKID	6C6	STA 1 REL PWR B
6A5	R AICS RAMP STOW	8E6	STA 1A AIM-9 PWR
8G3	R AIM-7 BATT ARM	2H7	STA 1A AIM-9 PWR PH B
8D1	R DC TEST/ACM BIT	8E5	STA 1B AIM-9 PWR
T 8G7	RECON ECS CONT DC	2H9	STA 1B AIM-9 PWR PH C
T 2G4	RECON GCS CONT AC	1E1	STA 1/8 AIM-7 PH A
T 2B1	RECON HTR PWR PH A	1E3	STA 1/8 AIM-7 PH B
T 2D1	RECON HTR PWR PH B	1E7	STA 1/8 AIM-7 PH C
T 2F1	RECON HTR PWR PH C	6D7	STA 2, 3, & 4 REL PWR A
T 1F4	RECON POD	6D6	STA 2, 3, & 4 REL PWR B
T 8F7	RECON POD CONT	8H2	STA 3/4 MECH FUZING
T 8F1	RECON POD DC PWR NO. 2	1D1	STA 3/6 AIM-7/AIM 54 PUMP PH A
T 8F2	RECON POD DC PWR NO. 1	1D3	STA 3/6 AIM-7/AIM-54 PUMP PH B
3E4	R ENG FUEL FLOW IND	1D7	STA 3/6 AIM-7/AIM-54 PUMP PH C
3C2	R ENG N2 TACH	8H2	STA 4/3 MECH FUZING
4D6	R ENG OIL PRESS/NOZ IND	1C1	STA 4/5 AIM-7 PH A
7C3	R FIRE DET/LT	1C3	STA 4/5 AIM-7 PH B
6C4	R FIRE EXT	1C7	STA 4/5 AIM-7 PH C
LG1	R FUEL CONTR/MACH LVR	6E7	STA 5, 6, & 7 REL PWR A
7F4	R GEN CAUTION	6E6	STA 5, 6, & 7 REL PWR B
2E3	R MAIN XFMR RECT	8E1	STA 8 AIM-9 COOL PWR
7A5	R MID CPRSN BYPASS	6D3	STA 8 AIM-9 REL PWR
7D2	R OIL HOT/OVSP CAUTION	6E5	STA 8 REL PWR A
2H1	R PH A TEST/GAS TEMP LIM	6E4	STA 8 REL PWR B
2H4	R PH B TEST/GAS TEMP LIM	8E3	STA 8A AIM-9 PWR
2H8	R PH C TEST/GAS TEMP LIM	2I7	STA 8A AIM-9 PWR PH B
4D2	R PITOT STATIC HTR	8E2	STA 8B AIM-9 PWR
3B2	R TIT IND	2I9	STA 8B AIM-9 PWR PH C
5D6	RADAR ALTM	1E1	STA 8/1 AIM-7 PH A
8A3	RADAR SUBSYS NO. 1	1E3	STA 8/1 AIM-7 PH B
8A2	RADAR SUBSYS NO. 2	1E7	STA 8/1 AIM-7 PH C
7C2	RAIN RPL/ANTI-ICE CONTR/HK CONTR	8H3	STA 8/1 MECH FUZING
5F6	RED FLOOD LT	7F8	STARTER CONTR VALVE
7B2	ROLL COMPTR DC	5A5	STBY ATTD IND PH A
LA1	ROLL COMPTR AC	5C5	STBY ATTD IND PH B
3E3	RUDDER/TAIL/FLAP IND		

The bracket with note applies to RECON items — TARPS aircraft only.

EFFECTIVITY

△9 TARPS aircraft only.

Figure 1-34. Circuit Breaker Alphanumeric Index (Sheet 4 of 5)

214	STEADY POS LT/ACM LT/SEAT ADJ	7G6	WHEELS POS IND
211	SUPP POS/ANTICOLL/POS LTS	2H6	WHITE FLOOD LT
		3B3	WING POS IND AC
3D6	TACAN ARN-84	7G2	WING POS IND DC
7E8	TACAN/BDHI	LE1	WING SWEEP DRIVE NO. 1
4F6	TACAN/BDHI INST PWR	LE2	WG SW DR NO. 2/MANUV FLAP
3E3	TAIL/RUDDER/FLAP IND	8B2	WSHLD DEFOG CONTR
2H2	TAXI/FORM LT		
4D4	TEMP CONT AC	1A5	XFMR RECT L MAIN
7D4	TEMP CONT ECS DC	2E3	XFMR RECT R MAIN
3B1	TIT L IND		
3B2	TIT R IND	7B6	YAW SAS A
7F3	TR-ADVSY/PLT ANN PANEL AUX PWR	LC1	YAW SAS A PWR SUP
1A5	TRANSFORMER REC L MAIN	7B5	YAW SAS B
2E3	TRANSFORMER REC R MAIN	LC3	YAW SAS B PWR SUP
		7B4	YAW SAS M
6F6	UHF CONTR	LB3	YAW SAS M PWR SUP
7E9	UHF NO. 2		
4D5	UTILITY LT	3F7	26 VAC BUS FDR
		1B2	28 VDC PWR SUP PH A
3B6	VDI PH A	1B5	28 VDC PWR SUP PH B
4B1	VDI PH B	1B8	28 VDC PWR SUP PH C
4B4	VDI PH C		

Figure 1-34. Circuit Breaker Alphanumeric Index (Sheet 5 of 5)

The third part is a number and designates the column in which the circuit breaker will be found. The innermost column of each panel 1, 2, 7, and 8 or aftmost column on each panel 3, 4, 5, 6 L and R is designated "1", the next outboard/forward column is "2", etc.

Note

- Panel no. 1 Row A, the column numbering is different from Rows B-J.
- Panel no. 2 Rows A - F, the column numbering is different from Rows G-I.

Degraded Electrical Operation

EMERGENCY GENERATOR

The emergency generator provides a limited but independent backup source of ac (5 kVA, 115/200 volts) and dc (50 amperes, 28 volts) power for flight-essential components. It is driven by combined system hydraulic pressure.

With normal combined hydraulic system operation, the emergency generator powers the essential ac and dc no. 1 and no. 2 buses and the AFCS dc bus. Operation of the generator is automatically initiated with the loss of dc electrical power (loss of both the main ac and/or dc power sources). Approximately 1 second elapses from the time of automatic initiation before the generator delivers rated power to flight essential ac and dc buses.

Pilot control of the emergency generator is through the guarded EMERG switch on the MASTER GEN control panel. With the switch in the NORM position, the emergency generator is automatically activated when all main ac or dc power is lost. The OFF/RESET position of the switch provides the pilot with the capability of isolating emergency electrical power from the aircraft buses (as in the case of an electrical fire) or resetting the generator.

EMERGENCY POWER DISTRIBUTION

An emergency generator control unit monitors the emergency generator output. If it senses that the emergency generator cannot supply power within the proper frequency and voltage tolerances, it disconnects the essential ac and dc essential no. 2 and the dc AFCS buses from the emergency generator. It is possible that this could happen if the combined hydraulic system is not operating normally. If combined hydraulic pressure subsequently recovers, the emergency generator switch must be cycled through OFF/RESET and back to NORM to regain the essential no. 2 and AFCS buses.

Note

The combined hydraulic pressure gage becomes inoperative when the emergency generator output is reduced to 1 kVA. Under these circumstances the pilot has no direct indication when the hydraulic pressure recovers.

The exact pressure at which the emergency generator is unable to power all three buses is dependent on the load placed on the generator, and can vary from 2,000 to 1,100 psi indicated. If the emergency generator is required, and there is a hydraulic emergency that could lower combined system operating pressure, the essential ac and dc no. 2 and AFCS buses can be powered with lower hydraulic pressure by reducing the electrical load, such as turning off the HUD and not jettisoning ordnance.

Note

When the emergency generator is operating with one main hydraulic system inoperative, large hydraulic flow requirements for flight controls may cause loss of the essential ac and dc no. 2 and AFCS buses. This will cause loss of engine instrumentation (RPM, TIT, and fuel flow). To regain these buses the emergency generator switch must be cycled through OFF/RESET to NORM after the hydraulic pressure recovers. In aircraft BUNO 161270 and subsequent and aircraft incorporating AFC 610, engine instruments are powered by essential ac bus no. 1. Engine instruments will be available or restored at lower engine RPM. In all aircraft, the airspeed at which engine instrumentation is restored (either automatically or by pilot cycling the emergency generator switch) could be higher than minimum airspeed.

EMERGENCY GENERATOR TEST

An operational check of the emergency generator can be accomplished any time the combined system is pressurized and at least one main generator is on the line by selecting the EMERG GEN position on the master test switch and depressing the switch. This provides 28 volts dc to activate the emergency motor-generator and checks the tie contactors by connecting electrical power to the essential ac and dc buses. The GO light on the MASTER TEST panel indicates a satisfactory check. A malfunction in the emergency generator operation is indicated by the NO GO light.

HYDRAULIC POWER SUPPLY SYSTEMS

The aircraft employs two main, independent, engine-powered hydraulic systems, supplemented by two electro-hydraulic power modules, a bidirectional transfer unit, and a cockpit handpump. The systems are pressurized to 3,000 psi and use MIL-H-83282 or MIL-H-5606 hydraulic fluid circulated through stainless steel and titanium lines. The electrohydraulic power modules enable on-deck, engines-off operation of certain hydraulically powered components at a reduced rate, using external ac electrical power in lieu of an external hydraulic power cart. Hydraulic power systems controls and indicators are shown in figure 1-35. The components serviced by each hydraulic power system are shown on page FO-12.

Flight and Combined Systems

ENGINE DRIVEN PUMPS

The flight and the combined systems are each pressurized by engine driven pumps. The flight hydraulic system pump is driven by the right engine and the combined hydraulic system pump by the left engine. Each of the main systems in normally pressurized to 3,000±100 psi at any time the respective engine is operating.

A HYD PRESS caution light illuminates when the discharge pressure from either engine-driven hydraulic pump falls below 2,100 psi; thereafter, the light goes out when pressure in both systems via the engine-driven pumps exceeds 2,400 psi. If the HYD PRESS caution light has been illuminated by low pressure in one main system, pressure failure in the other system will not cause the MASTER CAUTION light to illuminate again. The COMB and FLT gages on the hydraulic pressure indicator reflect system pressure provided by either the engine-driven pumps or the hydraulic transfer pump. With both systems normally pressurized to 3,000 psi, the gage needles form a horizontal line.

HYDRAULIC TRANSFER PUMP (BIDIRECTIONAL PUMP)

In order to assure the continuance of main system hydraulic pressure with an engine or engine-driven pump inoperative, a second source of pressure is provided by the hydraulic transfer pump. This unit consists of two hydraulic pumps, one in each of th main hydraulic systems, interconnected by a common mechanical shaft. Thus, a pressure deficiency in one system is automatically augmented using pressure in the other system as the motive power. The result is bidirectional transfer of energy without an interchange of system fluid. The efficiency of the pump is such that a 3,000 psi system on one side will pressurize the other system to approximately 2,400 to 2,600 psi.

In order to prevent damage to the hydraulic transfer pump with the loss of system fluid on one side and to conserve hydraulic power in the remaining good system, the pump is automatically secured when pressure less than 500 psi is detected on either side of the pump for 10 seconds. In addition, the pilot can manually shut off the hydraulic transfer pump by lifting the guarded HYD TRANSFER PUMP switch aft on the right outboard console, and selecting the SHUTOFF position.

CAUTION

If pressure in either system remains below 500 psi for 10 seconds, immediately lift the guard and select the SHUTOFF position with the HYD TRANSFER PUMP switch. Failure of the hydraulic transfer pump to automatically shut off after 10 seconds below 500 psi may cause the driving system to cavitate and overheat.

NAVAIR 01-F14AAA-1

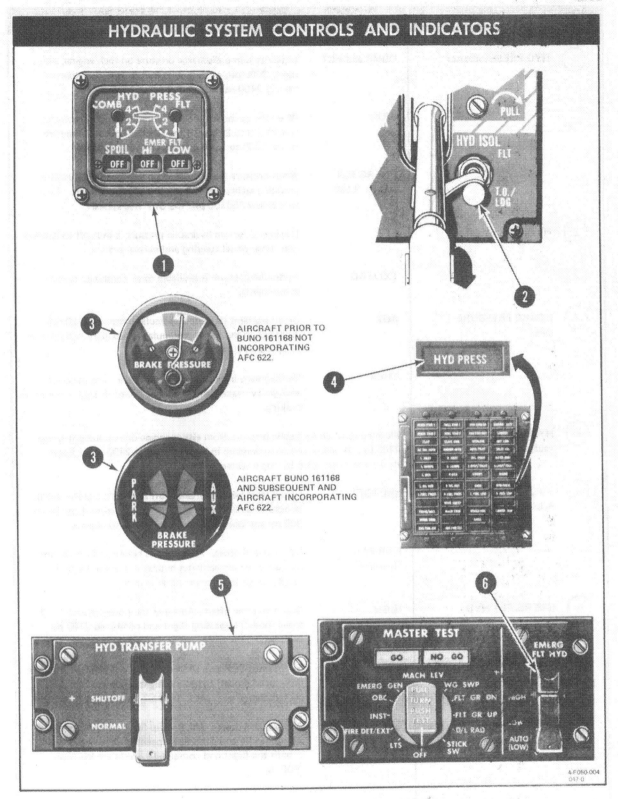

Figure 1-35. Hydraulic System Controls and Indicators (Sheet 1 of 2)

NOMENCLATURE	FUNCTION	
① HYD PRESS indicator	COMB and FLT	Indicates pump discharge pressure on each engine, normally 3000 psi, or hydraulic transfer pressure approximately 2400 psi.
	SPOIL	When the outboard spoiler hydraulic module is pressurized (1950 to 2050 psi) the ON flag appears. Pressure below 1900 to 1800 psi: the OFF flag appears.
	EMERG FLT HI LOW	When pressure from the backup flight control hydraulic module reaches 500±50 psi, the ON flag appears. Pressure below 350±50 psi: the OFF flag appears.
② HYD ISOL switch	FLT	Combined system hydraulic pressure is shut off to landing gear, nose wheel steering and normal braking.
	T.O./LDG	Hydraulic pressure is available to all combined system components.
③ BRAKE PRESSURE gage	AUX	Green segment indicates hydraulic pressure (1150±50 3000 psi) in the brake accumulator; auxiliary braking may be applied by ruddertoe pedals.
	EMER	Red segment indicates 950 to 1150 psi. The parking/emergency brake handle must be pulled to apply emergency braking.
④ HYD PRESS caution light	Illuminates when hydraulic pressure from either engine-driven pump is below 2100 psi. It will go out with pressure in both systems at 2400 psi or above, if pressure is provided by engine-driven pumps.	
⑤ HYD TRANSFER PUMP switch	SHUTOFF	Guard must be lifted. Shuts off hydraulic transfer pump. Should be activated when hydraulic pressure drops below 500 psi and does not rise again within 10 seconds.
	NORMAL (Guarded)	Safety guard down. Pressure loss below 2100 psi in one hydraulic system activates hydraulic transfer pump to supply pressure from the other system.
⑥ EMERG FLT HYD switch	HIGH	Guard must be lifted. Activates the power module (high speed model) bypassing flight and combined 2100 psi switches.
	LOW	Guard must be lifted. Activates the backup power module (low-speed model) bypassing flight and combined 2100 psi switches.
	AUTO (LOW)	Safety guard down. The backup flight control system is automatically activated (low speed mode) when pressure in both the flight and combined systems are less than 2100 psi.

Figure 1-35. Hydraulic System Controls and Indicators (Sheet 2 of 2)

With ground electrical power connected to the aircraft the hydraulic transfer pump is deactivated and can only be energized by a switch on the ground check panel. Normally with both engines running the hydraulic transfer pump is off. However with less than 2,100 psi hydraulic pump discharge pressure from either system, the pump will automatically come on and supply hydraulic power to the faulty system. The pilot has no direct control over the direction of pump flow, the system automatically shifts in the direction that supplemental power is required. Because of the location of the flight and combined system pressure switches, the pressurization contribution of the hydraulic transfer pump is reflected on the hydraulic pressure indicator and may produce slight pressure fluctuations. Failure of either hydraulic system (<2100 psi) will cause the HYD PRESS caution light to illuminate and remain illuminated with operation of the hydraulic transfer pump. If the failed system discharge pressure is restored to normal operating pressure (>2400 psi) by the engine-driven pump this HYD PRESS light will go out and the hydraulic transfer pump will shut off.

COCKPIT HANDPUMP

A manually operated pump handle is provided as a supplementary source of power for ground operations with engines shut down and as a backup for the loss of combined system fluid and/or pressure. It is an extendable handle in the pilot's cockpit between the left console and ejection seat. Forward and aft stroking of the handpump operates a double-acting wobble pump. The pump, which draws fluid from the combined system return line, recharges wheel brake accumulator pressure when the landing gear handle is down. It also serves as a backup means of extending the in-flight refueling probe.

The handpump is the only means of pressurizing the radome fold actuator, an operation that must be manually selected and the radome unlock on deck from the nose wheel well. The operation rate using the handpump power source is a function of the number of components selected. The recommended rate of operation is approximately 12 cycles per minute (a cycle is a complete forward and aft movement of the pump handle).

Hydraulic Power Distribution

The distribution of hydraulic power in the flight and combined systems is shown on page FO-12. Except for the left empennage control surfaces, the flight system services only those components on the right side of the aircraft and does not penetrate into the wings. The combined system distribution is more extensive throughout the aircraft, yet its services are predominently concentrated to the left side and extend to the inboard sections of the movable wing panels and to the landing gear. Although the flight and combined systems are completely independent of each other, in certain components both pressure sources are used without an interchange of fluid. Both systems operate in parallel to supply power for operation of the primary flight control surfaces, (except spoilers), and stability augmentation actuators: if one system fails, the other can continue to supply pressure for operation (with reduced power capability of such components). If either or both main hydraulic systems should fail, backup sources provide the capability for safe return flight and landing.

Major components in the combined and flight hydraulic power supply systems are shown on page FO-12. Each system has a piston-type reservoir and filter module in the sponson aft of the main landing gear strut on the respective side (combined-left; flight-right). Protrusion of mechanical pins on each filter module indicates a clogged filter.

HYDRAULIC PRIORITY VALVES

The combined and flight hydraulic systems each incorporate two priority valves (1,800 psi and 2,400 psi) shown on page FO-12. Hydraulic fluid will not pass through the one-way priority valves unless the input pressure exceeds the cracking threshold of the valve. Basically, the 2,400 psi priority valves give priority of the individual engine-driven pump discharge pressure to the primary flight controls (horizontal tails, rudders, inboard spoilers) and stability augmentation actuators. Conversely, the 1,800 psi priority valves give priority to the remaining systems on the other side (inlet ramps, wing sweep, glove vanes, etc.) with pressure supplied by the hydraulic transfer pump. Under such circumstances, the pilot should be aware of the hydraulic energy available and demands of the various system components. Large and abrupt control commands can rapidly consume total energy with the engine(s) at IDLE speed. For example, during a single-engine landing rollout, if excessive horizontal tail movements are commanded, the nosewheel steering and wheel brake operation could be temporarily lost.

NORMAL HYDRAULIC ISOLATION

The combined system incorporates isolation circuits to limit distribution of flight essential components. With the LDG GEAR handle in the UP position, normal isolation may be selected by the pilot to prevent loss of hydraulic fluid in the event of material failure or combat damage to the isolated systems. Normal isolation electrically shuts off hydraulic pressure to wheel brakes, antiskid, landing gear, and nosewheel steering. It is activated by placement of the HYD ISOL switch on the landing gear panel to the FLT position. Placement of the gear handle to the DN position mechanically cams the HYD ISOL switch to the T.O./LDG position or the pilot can manually select it before lowering the landing gear. Such action returns all combined-system components to normal operation.

WARNING

After landing gear retraction, leave HYD ISOL switch in T.O./LDG position. Selecting FLT position allows hydraulic pressure to decay in the main landing gear uplock actuators. A zero to negative g maneuver can cause the main landing gear to be released from the uplocks and subsequently rest on the landing gear doors. The application of positive g in this configuration could cause inadvertent extensions of one or both main landing gear.

Section I
Part 2

NAVAIR 01-F14AAA-1

OUTBOARD SPOILER SYSTEM

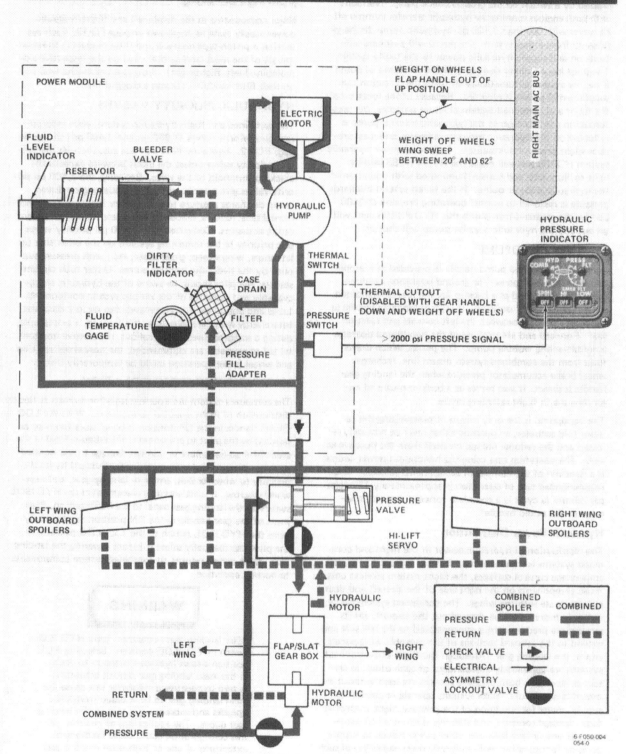

Figure 1-36. Outboard Spoiler System

1-68

Outboard Spoiler System

The outboard spoilers are powered by a separate closed-loop system, independent of the main hydraulic systems. (See figure 1-36.) An electrohydraulic power module supplies hydraulic pressure for outboard spoiler deflection, and provides a backup power source for the main flaps and slats. Outboard spoiler operation is electrically inhibited at wing sweep angles greater than 57°, and the power module is deactivated at wing sweep angles greater than 62°.

A thermal cutout circuit secures the system in the event of overheating. Normal operation is automatically restored when fluid temperature falls below the prescribed limit. The thermal cutout circuit is disabled with the gear handle down and weight-off-wheels to prevent overtemperature shutdowns during take-off or landing. To avoid overheating due to prolonged ground operations the outboard power module is deactivated with the flap handle up when on internal electrical power with weight-on-wheels.

Electrical power for the outboard spoiler system motor is supplied from the right main ac bus. The module can be activated using external ac electrical power. With the module pressurized, the ON flag appears in the SPOIL window at the bottom of the hydraulic pressure indicator; otherwise, an OFF indication is displayed in the window.

Reservoir servicing level is shown by an indicator rod protruding from the integral power package. A fluid temperature gage which registers current and retained peak system temperatures is on the power module. Protrusion of a red tipped pin on the integrated filter package is an indication of a clogged filter.

FLAP AND SLAT BACKUP OPERATION

Although normal operation of the main flap and slat segments is powered by a combined system motor on the flap and slat gearbox, an auxiliary motor powered by the outboard spoiler system is geared to the same shaft to provide for emergency operation (retraction and extension) of the mainflaps and slats at a reduced rate. Failure of combined system pressure activates the auxiliary motor to drive the flap and slat gearbox when selected by the normal flap handle or maneuvering flap thumbwheel.

Backup Flight Control System

The backup flight control system consists of a two-speed electrohydraulic power module that provides fluid energy to operate the horizontal tails and rudders at a reduced rate. (See figure 1-37.) Emergency power provides sufficient pitch, roll, and yaw control for return flight and landing with both main hydraulic power systems inoperative.

The module may be operated in two modes: emergency and ground test. In the emergency mode, the module is controlled by the EMERG FLT HYD switch, on the MASTER TEST panel. The switch has three positions: AUTO LOW, HIGH, and LOW mode. Electric power to the motor is supplied by the right main ac electrical bus through the FLT HYD BACKUP PH A, PH B, and PH C circuit breakers located on right main ac circuit breaker panel (number 7) in the rear cockpit. Loss of both engine-driven electrical generators eliminates in flight use of the module.

CAUTION

Never use the three phase circuit breakers (PH A, PH B, PH C) to start or shut off backup module as damage to the motor may result. These circuits must be engaged prior to any system test. Automatic control of the module is provided by the closing of both flight and combined hydraulic system pressure switches. Since the switches are set at 2,100 psi, both flight and combined hydraulic system pressures must drop below 2,100 psi before the module is turned on in the auto low mode. Once in this automatic mode of operation, the module cannot be turned off unless either or both flight and combined systems are pressurized above 2,100 psi. The EMERG FLT HYD switch is used to select the low or high mode. Either of these positions overrides the circuitry of the auto low mode and the module will remain on even if either or both system pressures become pressurized above 2,100 psi. When the module pump reaches 500 psi the on flag appears in the selected window at the bottom of the hydraulic pressure indicator.

In the low-speed mode, the system can operate indefinitely and should be used for maximum range and endurance. The high-speed mode should be used before transitioning to the landing configuration or during evasive maneuvering.

CAUTION

If either the flight or the combined hydraulic system pressure drops below 2,100 psi and the HYD PRESS caution light does not illuminate, immediately lift the guard and select LOW (center) position (clean) and HIGH priority to dirty up with the EMERG FLT HYD switch.

Section I
Part 2
NAVAIR 01-F14AAA-1

BACKUP FLIGHT CONTROL SYSTEM

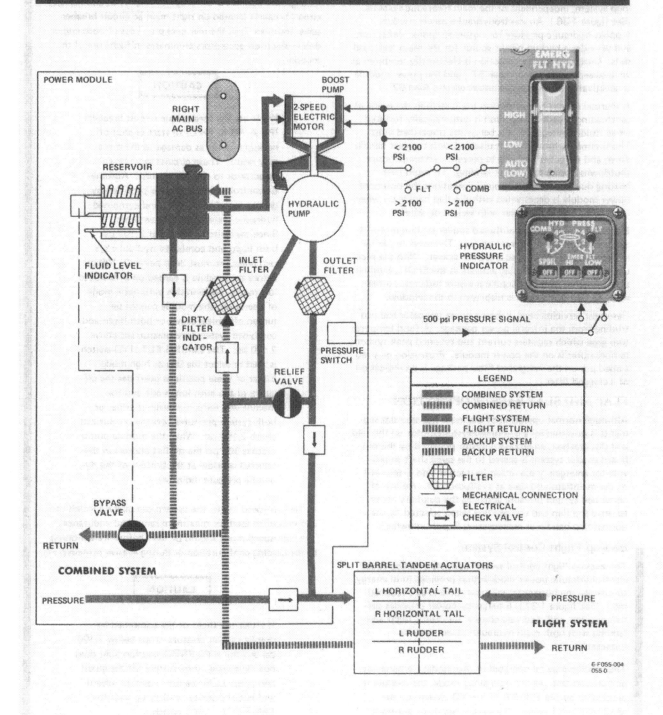

Figure 1-37. Backup Flight Control System

1-70

The ground test mode of operation is controlled by the AUX HYD CONT switch on the ground test panel in the rear cockpit. In this mode, the module operates in the high mode only. Ground test from the rear cockpit is electrically inhibited when the aircraft is on internal electrical power. For ground inspection purposes, protrusion of a red-tipped button on either the inlet or outlet filter cases is a positive indication of a dirty filter. Both such indications may be observed through an access door on the underside of the aft fuselage.

CAUTION

- Since the backup module cannot fulfill the system flow rate demands, all surface demands must be performed slowly and cautiously in order not to exceed the output rate of the system. Excessive system demands will cause the pump to cavitate and the motor to overheat. Checks should be made slow enough to ensure continuous on indication in the hydraulic pressure indicator. When the fluid temperature rises to 180°F the thermal cutoff switch shuts the module off. There are instances when the motor may overheat and fail prior to the hydraulic fluid temperature reaching 180°F. This condition can occur if the combined system is not serviced properly. The backup module reservoir has a small volume capacity, 1,000 CC (61 cubic inches) when full, but will decrease in volume to 500 CC (30.5 cubic inches) when the aircraft is not in use. In order to avoid depleting this reservoir below 500 CC (30.5 cubic inches) and cavitating the pump and overheating the motor, the combined and brake system accumulators must be hydraulically charged. If the accumulators are not charged prior to starting the module, depletion of the reservoir hydraulic fluid will occur. If this occurs too frequently, system damage and failure may result.

- Ensure both hydraulic system pressures indicate zero for a satisfactory backup module check.

PNEUMATIC POWER SUPPLY SYSTEMS

The pneumatic power supply systems consist of three independent, stored pneumatic pressure sources for normal and auxiliary operation of the canopy and for emergency extension of the landing gear. The high-pressure bottles for normal canopy operation and another for the emergency landing gear extension are ground-charged through a common filter connection in nosewheel well to 3,000 psi at 70°F ambient temperature. Individual bottle pressure is registered on separate gages on the right side of the nosewheel well. An auxiliary canopy open N_2 bottle, filter valve, and gage is on the turtleback behind the cockpit to allow opening the canopy from the cockpit or ground. Charges may be compressed air; however, pressurized dry nitrogen is preferred because of its low moisture content and inert properties.

Normal Canopy Control

The bottle that supplies a pressurized charge for normal operation of the canopy is on the right side of the forward fuselage, inboard of the air refuel probe cavity. Expenditure of bottle pressure for normal operation of the canopy is controlled by three (pilot, RIO, and ground) canopy control handles. A fully charged bottle provides approximately 10 complete cycles (open and close) of the canopy before reaching the minimum operating pressure of 225 psi.

Auxiliary Canopy Open Control

The auxiliary canopy air bottle supplies a pneumatic charge to translate the canopy aft so that the counter-poise action of the canopy actuator facilitates opening. It is on the turtleback behind the canopy hinge line.

Activation of the auxiliary mode can be effected from either of the three (pilot, RIO, or ground) canopy control handles. After activation of the auxiliary open mode, the control system will not return to the normal mode of operation (canopy will lower but will not translate forward) until the auxiliary selector valve on the aft canopy deck is manually reset (lever in vertical position). Servicing of the auxiliary canopy air bottle is through the small access panel immediately behind the canopy on the turtleback. The reservoir is normally serviced to 3,000 psi at 70°F ambient temperature. A fully charged bottle provides more than 20 operations in the auxiliary open mode. Minimum preflight pressure is 800 psi.

Emergency Gear Extension

The bottle that supplies the pneumatic force for a single emergency extension of the landing gear is on the right side of the nosewheel well. Expenditure of bottle pressure is controlled by a twist-pull operation of the landing gear handle. Minimum bottle pressure for accomplishing emergency extension of the gear to the down-and-locked condition is 1,800 psi. Normal preflight bottle pressure is 3,000 psi at 70°F.

Note

- Emergency extension of the landing gear shall be logged in the Yellow Sheet (OPNAV Form 3760-2).

- Once the landing gear is extended by emergency means, it cannot be retracted while airborne and must be reset by maintenance personnel.

- Use of emergency gear extension results in loss of nosewheel steering.

CENTRAL AIR DATA COMPUTER (CADC)

The central air data computer (CADC) is a single-processor digital computer with a separate independent analog back-up wing sweep channel. It is capable of making yes and no decisions, solving mathematical problems, and converting outputs to either digital or analog form as required by each aircraft system. The CADC gathers, stores, and processes pitot pressure, static pressure, total temperature and angle-of-attack data from the aircraft airstream sensors. (See figure 1-38.) It performs wing sweep, glove vane, and flap and slat schedule computations, limit control and electrical interlocks, failure detection, and systems test logic. Major systems that depend on all or part of these CADC functions are shown in figure 1-39.

Note

In aircraft BUNO 159637 and subsequent and aircraft incorporating AFC 417 and AVC 1992, the CADC is an improved version of the super single processor (CP-1166B/A).

CADC Tests

BUILT-IN TEST (BIT)

Built-in test (BIT) capabilities provide continuous monitoring of the CADC and its inputs and outputs. The failure indicator matrix (figure 1-40) tabulates the functions that are monitored and associated fail indications.

ONBOARD CHECKOUT (OBC)

The CADC performs a self test during onboard checkout (OBC) only with weight on wheels. When OBC is initiated, normal air data inputs are locked out and in their place constants from the computer memory are received. Self-test detected failures may be manually reset by pressing the MASTER RESET pushbutton.

Pressing the MASTER RESET pushbutton for 1 second resets transient failures in the CADC. Activating the master reset circuit recycles the failure detection process in the CADC. This recycling process puts off the caution, advisory, and warning light(s) and may take as long as 10 seconds to check out the status of the system. If a failure exists, the light(s) will illuminate again. If a transient failure existed, the light(s) will remain off.

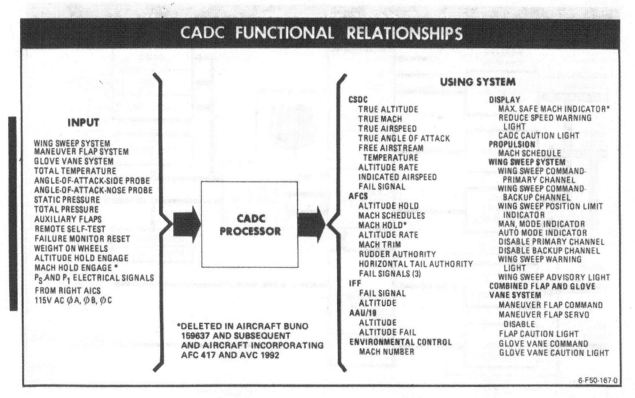

Figure 1-38. CADC Functional Relationships

The following caution, advisory, and warning lights are activated by the CADC:

CADC	REDUCE SPEED
FLAP	WING SWEEP (advisory)*
GLOVE VANE	WING SWEEP (warning)**

*If the WING SWEEP advisory light does not recycle when MASTER RESET pushbutton is depressed, the light is activated by the wing flap and glove vane controller.

**If the WING SWEEP warning light does not recycle when MASTER RESET pushbutton is depressed, the light is activated by the emergency WING SWEEP handle being out of detent.

Three independent CADC fail signals drive the automatic flight control system (AFCS) failure detection circuits. If these signals exist, the AFCS will illuminate the following lights:

- CADC fail signal to pitch computer: no light

- CADC fail signal to yaw computer: RUDDER AUTH, HZ TAIL AUTH

- CADC fail signal to roll computer: MACH TRIM

Pressing MASTER RESET pushbutton will also update the wing sweep, flap, and glove vane commands to their respective feedback signals. As a result, there may be movement in the wings, maneuver flaps, and glove vanes when MASTER RESET pushbutton is depressed.

1-73

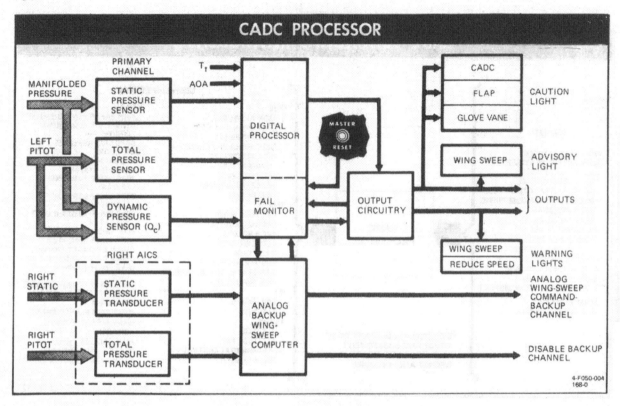

Figure 1-39. CADC Processor

Note

Gun firing effects on CADC computations may cause the RUDDER AUTH and HZ TAIL AUTH caution lights and the MACH TRIM advisory light to illuminate. If these lights illuminate immediately after gun firing, press the MASTER RESET pushbutton to extinguish them.

WING SWEEP SYSTEM

The variable geometry of the wing sweep system provides the pilot with considerable latitude for controlling wing lift and drag characteristics to optimize aircraft performance over a broad flight spectrum.

Under normal operating conditions, the wings are automatically positioned to the optimum sweep angle for maximum maneuvering performance. The pilot can selectively position the wings at sweep angles aft of optimum.

A mechanical backup control system is provided for emergency and oversweep operations. Details of the wing sweep system are shown on page FO-13.

The outboard location of the wing pivot reduces the change in longitudinal stability as a function of wing sweep angle. Two independently powered, hydromechanical screwjack actuators, mechanically interconnected for synchronization, position the wings in response to pilot or CADC commands. In flight, the wings can be positioned between 20° and 68° wing leading edge sweep angle. On the deck, the range is extended aft to 75° (oversweep position) to reduce the span for spotting. Such authority results in a variation of wing span from approximately 64 to 33 feet.

Cavities above the engine nacelles and the midfuselage accommodate the inboard portions of the wing panels as they sweep aft. Sealing of the underside is by a wiper seal and air bag. The bag is pressurized by engine bleed air. Air bag pressure is released during oversweep to avoid overloading of the flap mechanism. An overwing fairing encloses the wing cavity and provides a contoured seal along the upper surface of the wing for the normal range of

NAVAIR 01-F14AAA-1
Section I
Part 2

CADC PROCESSOR INDICATORS

FAILURE \ INDICATION	CADC	WING SWEEP ADVISORY	WING SWEEP WARNING	FLAP	GLOVE VANE	REDUCE SPEED	RUDDER AUTH	HZ TAIL AUTH	MACH TRIM	REMARKS
CADC - P_s/P_t SENSOR COMPARE DIGITAL PROCESSOR	●	●		●	●	●	●	●	●	WING-SWEEP LIMIT BUG MAY BE INACCURATE, MAXIMUM SAFE MACH INDICATOR MAY BE INACCURATE, MANEUVER FLAPS, GLOVE VANE INOPERATIVE, ALTIMETER TO STANDBY.
CADC WING-SWEEP COMMAND (SINGLE FAILURE)	●	●								
CADC WING-SWEEP COMMAND (DUAL FAILURE)	●	●	●							AUTO WING SWEEP INOPERATIVE
WING SWEEP (SINGLE FAILURE)		●								
WING SWEEP (DUAL FAILURE)		●	●							AUTO WING SWEEP INOPERATIVE
CADC MANEUVER FLAP COMMAND	●			●						MANEUVER FLAPS VIA THUMBWHEEL INOPERATIVE
MANEUVER FLAP - COMMAND AND SERVO MISCOMPARE				●		✶				MANEUVER FLAPS VIA THUMBWHEEL INOPERATIVE
MANEUVER FLAP - HYDRAULIC VALVE AND/OR ACTUATOR MISCOMPARE				●		✶				
MANEUVER FLAP - HANDLE AND/OR HYDRAULIC VALVE MISCOMPARE				●						
AUXILIARY FLAP AND MANUEVER FLAP MISCOMPARE.				●						
AUXILIARY FLAP ASSYMETRY				●						
CADC RUDDER OR STABILIZER COMMAND AUTHORITY	●						●	●		
ANGLE-OF-ATTACK SIGNAL	●									ANGLE-OF-ATTACK DISPLAY NOT PRESENT ON HUD DURING LANDING MODE
TOTAL TEMPERATURE SIGNAL	●									AUTO PILOT CAUTION LIGHT ILLUMINATES IF IN ALTITUDE HOLD, ALTITUDE HOLD WILL BE DISENGAGED, VERTICAL SPEED NOT PRESENT ON HUD DURING TAKEOFF AND LANDING MODE
CADC WING-SWEEP INDICATOR OUTPUT	●									WING SWEEP INDICATOR INACCURATE
GLOVE VANE COMMAND	●				●					GLOVE VANE INOPERATIVE, GLOVE VANE WILL RETRACT IF EXTENDED
GLOVE VANE - ACTUATOR AND COMMAND MISCOMPARE					●					GLOVE VANE INOPERATIVE; WILL RETRACT IF EXTENDED
ECS FAILURE AND MACH > 0.25.	●									CABIN TEMPERATURE MAY RISE AFTER LANDING. COOLING AIR ADVISORY LIGHT MAY ILLUMINATE
ECS FAILURE AND MACH > 0.4.	●									
CADC-DIGITAL DATA TO CSDC	●									ALTITUDE AND MACH NOT DISPLAYED ON HUD. ANGLE-OF-ATTACK DURING LANDING DISPLAY NOT ON HUD. DURING TAKE OFF AND LANDING VERTICAL SPEED NOT ON HUD.
MACH LEVER OUTPUT	●									ENGINE MAY STALL MACH > 1.5.
ALTITUDE HOLD OUTPUT ALTITUDE RATE OUTPUT	●									AUTO PILOT CAUTION LIGHT ILLUMINATES IF IN ALTITUDE HOLD. ALTITUDE HOLD WILL BE DISENGAGED
MACH TRIM OUTPUT	●								●	
ALTITUDE TO PRESSURE ALTIMETERS	●									ALTIMETERS SWITCH TO STANDBY

✶ REDUCE SPEED LIGHT WILL ILLUMINATE IF AIRSPEED > 225 KIAS AND FLAPS ARE GREATER THAN 4°.

Figure 1-40. CADC Processor Indicators

in-flight sweep angles. The left and right overwing fairing actuators are pressurized by the combined and flight hydraulic systems, respectively.

Wing Sweep Performance

Maximum wing sweep rate (approximately 15°/s) is adequate for most transient flight conditions; however, wing sweep rate can be significantly reduced or stalled by negative g or large positive g excursions. Sufficient capability has been provided in the system, consistent with the sustained performance capabilities of the aircraft. Failure of either the combined or flight hydraulic systems permits the wings to move at a reduced rate.

Note

- The overwing fairings and flaps are susceptible to a high frequency (60 cycles per second), low amplitude oscillation that can be felt in the cockpit. This overwing fairing and flap buzz is normal and is influenced by rigging of the fairings and air in the hydraulic systems.

- Overwing fairing and flap buzz is usually encountered between 0.9 and 1.4 IMN.

Wing Sweep Modes

Normal control of the wing sweep position in AUTO, AFT, FWD, and BOMB modes is by the four-way mode switch on the inboard side of the right throttle grip. (See figure 1-41.) As an emergency mode of control, changes in wing sweep position can be selected manually with the emergency WING SWEEP handle on the inboard side of the throttle quadrant. The handle is connected directly to the wing sweep hydraulic valves. The command source for positioning the wings depends upon the mode selected by the pilot or in certain cases, is automatically selected. Electrical and mechanical wing-sweep command paths are shown on page FO-13. Wing sweep modes are shown in figure 1-42.

WARNING

Any time hydraulic pressure is on, the wings can be moved inadvertently. When positioning the wings during ground operation other than pilot post start or post landing checklist procedures, use the emergency WING SWEEP handle to minimize the possibility of moving the wings inadvertently.

Note

- When positioning the wings, do not command opposite direction until wings have stopped in original commanded position (all sweep modes) to increase motor life.

- The optimum wing position (triangular index) and the AUTO/MAN flags may be unreliable when the CADC caution light is illuminated.

AUTO MODE

Selection of the AUTO mode is made by placing the four-way switch in the upper detented position AUTO, which permits the CADC wing sweep program to position the wings automatically. The program positions the wings primarily as a function of Mach number but includes pressure altitude biasing. Wing position is scheduled to the optimum sweep angle for developing maximum maneuvering performance. In addition to serving an automatic wing positioning function, the programer also defines the forward sweep limit that cannot be penetrated using any of the other electrical (manual or bomb) modes. The forward sweep limiter prevents electrical mispositioning of the wings from a wing structure standpoint.

Pilot selection of the AUTO mode or automatic transfer from the manual mode causes the AUTO flag to appear in the wing sweep indicator. Once in the AUTO mode, the four-way wing sweep mode switch can be in the center position without changing the command mode.

MANUAL MODE

The manual wing sweep mode is commanded by selecting AFT or FWD from the neutral position of the mode switch, driving the wings in the commanded direction to any wing sweep position aft of the automatic program. The switch is spring loaded to return to the center position. Manual command mode exists unless the wing sweep program is intercepted, at which point transfer to the AUTO mode is automatic. Indication of the existing mode is provided by the AUTO and MAN flags in the wing sweep indicator.

BOMB MODE

Bomb mode is selected by moving the wing sweep mode switch to the down (BOMB) position. With the switch in the BOMB position the following occurs:

- The wing SWEEP indicator shows a MAN flag.

WING SWEEP CONTROLS AND INDICATORS

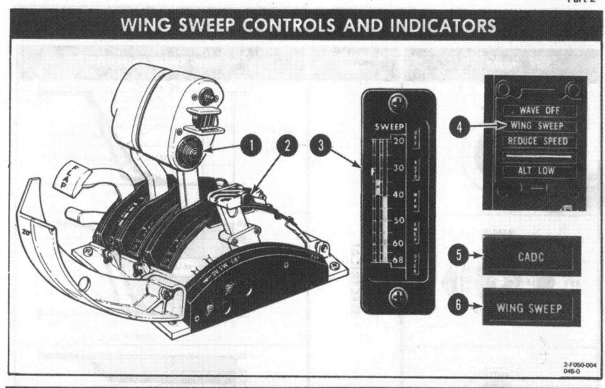

NOMENCLATURE	FUNCTION
① Wing sweep mode switch	AUTO - Wing sweep angles are determined by CADC according to wing sweep program. Detented switch position. BOMB - Wings are positioned at 55° or further aft if commanded by the CADC program. Detented switch position. AFT/FWD - The pilot can select AFT or FWD wing positions within limits imposed by the wing sweep program. Switch is spring-loaded to the center position. When the forward limit is intercepted, the mode is transferred to AUTO.
② Emergency WING SWEEP handle	Provides a mechanical means of wing sweep control that overrides the CADC program commands. Wing sweep angles between 20° and 68° are unrestricted except for flap interlocks. Oversweep to 75° is provided with weight-on-wheels, horizontal stabilizer authority restricter in reduced range, and air bag pressure dumped.
③ Wing SWEEP indicator	Displays (from right to left) actual wing sweep position, commanded position and wing sweep program position, which is the maximum forward angle at present airspeed and attitudes. Indicator windows show the operating mode.
④ WING SWEEP warning light	Indicates failure of both wing sweep channels and/or disengagement of spider detent. Wing sweep positioning is required using the emergency wing sweep handle.
⑤ CADC caution light	Indicates hardware failure and/or that certain computations of the air data computer are unreliable. Illumination of WING SWEEP warning and/or advisory light determines pilot action.
⑥ WING SWEEP advisory	Indicates failure of a single channel in the system. Illumination of both WING SWEEP advisory and CADC caution light indicates failure of one channel in CADC.

Figure 1-41. Wing Sweep Controls and Indicators

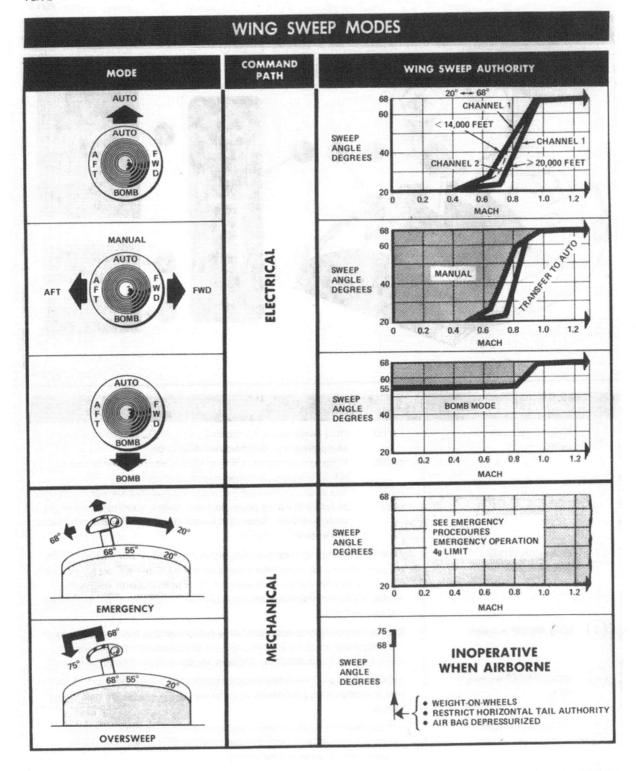

Figure 1-42. Wing Sweep Modes

- If the wing sweep is less than 55°, wings will drive to 55°.

- If wing sweep is greater than 55°, wings will not move.

- If maneuver flaps are extended, they will retract and wings will sweep to 55°.

- If speed is greater than Mach 0.35, glove vanes will extend.

As the aircraft accelerates and the AUTO wing sweep schedule is intercepted, the wings will follow the AUTO schedule even though the switch remains in BOMB mode. Upon decelerating, the wings will sweep forward to 55° and stop.

EMERGENCY MODE

During normal mode operation of the wing sweep system, the wing sweep control drive servo drives the hydraulic valve command input through a spider detent mechanism. The emergency handle under a transparent guard is moved in parallel with the servo output. The emergency mode provides an emergency method of controlling wing sweep. It bypasses the normal command path of the fly-by-wire system (CADC and control drive servo loop).

To select emergency mode, the handle must be extended vertically, and the guard should be moved out of the way before the handle is operated. Vertical extension of the emergency handle provides for better accessibility and leverage. The detent is not disengaged by raising the handle vertically. An initial fore or aft force of up to 30 pounds breakout and 13 pounds maximum is necessary for operation.

The spider detent is reengaged if the handle is repositioned to the detent (servo) position.

In aircraft BUNO 159825 and subsequent, the emergency WING SWEEP handle incorporates locks at approximately 4° increments between 20° and 68°. These locks are provided to eliminate random wing movement in the emergency mode should electrical system transients be experienced. When the locks are engaged, wing movement is inhibited provided that wings match handle position. The wing sweep locks eliminate the need for the installation or activation of wing sweep servo cutout switches. Locks are engaged by raising the handle 1 inch from the stowed position. In order to bypass the locks and select a wing position, the handle is raised an additional 1 inch (2 inches from stowed) and moved to the desired position. The handle is spring loaded to return to the lock position when released. The handle can be raised from 20° to 68° and oversweep, but can only be returned to the stowed position at 20° and oversweep. This feature is intended to prevent inadvertent engagement of the AUTO mode, commanding the wings to spread causing possible damage to the aircraft or injury to personnel in a confined area. The handle is spring-loaded toward the stowed position, but requires depressing the release button on the inboard side of the lever in order to return the handle to the stowed position.

CAUTION

Except for wing flap (main and auxiliary) and oversweep interlocks in the control box, the emergency mode does not prevent pilot mispositioning the wings, from a structural standpoint.

Since the wing sweep program acts as a forward limiter only for the normal modes of operation, the pilot must follow the following schedule in the emergency mode:

```
0.4 IMN . . . . . . . . . . . . . . . . 20°
0.7 IMN . . . . . . . . . . . . . . 25°
0.8 IMN . . . . . . . . . . . . . . 50°
0.9 IMN . . . . . . . . . . . . . . 60°
1.0 IMN . . . . . . . . . . . . . . 68°
```

When operating in the emergency mode, pulling the WING SWEEP DRIVE NO. 1 and WG SWP DR NO. 2/ MANUV FLAP circuit breakers on the pilot's left knee panel assures that the electrical command path cannot interfere with the emergency mode.

OVERSWEEP MODE (75°)

The wing oversweep mode allows sweeping the wings aft of 68° to 75° during ondeck operation only, thereby reducing the overall width of the aircraft for deck spotting. At the 75° position, the wing trailing edge is over the horizontal tail surface.

With the wings at 68°, oversweep can be initiated by raising the emergency WING SWEEP handle to its full extension and holding. Raising the handle releases air pressure from the wing seal air bags and activates the horizontal tail authority system, restricting the surface deflections to 18° trailing edge up and 12° trailing edge down. During motion of the horizontal stabilizers, the HZ TAIL AUTH caution light is illuminated. When the horizontal tail authority restriction is accomplished (approximately 15 seconds), the HZ TAIL AUTH caution light will go off and the OVER flag on the wing sweep indicator will be visible. This advises the pilot that the oversweep interlocks are free, allowing movement of the emergency WING SWEEP handle to 75° and stow. The EMER and OVER on the wing sweep indicator will be visible.

CAUTION

Failure of the oversweep interlocks while trying to achieve oversweep may result in damage to either the wingtip and horizontal tail trailing edge of the maneuver flap actuator, or both.

The reverse process takes place when sweeping forward from oversweep. However, there is no need to hold the emergency handle in the raised position at 68°. Motion out of oversweep is completed (wing seal airbag pressure established and horizontal tail authority restriction removed) when both the OVER flag and the HZ TAIL AUTH caution lights are off. Six seconds later the WING SWEEP advisory light will illuminate. Upon engagement of the spider detent by further unsweeping the emergency handle, MASTER RESET pushbutton is pressed to clear both WING SWEEP advisory and warning lights, thus activating the electrical command circuits of the WING SWEEP system.

CAUTION

When coming out of oversweep and a 68° wing position is desired, the wings should be moved forward to approximately 60° and then back to 68°.

Wing Sweep Interlocks

Automatic limiting of wing sweep authority is provided under normal in-flight control modes to prevent mispositioning of the wings at conditions that could result in the penetration of structural boundaries. Wing sweep interlocks within the CADC are shown in figure 1-43. Wing sweep is also electrically inhibited at normal accelerations less than −0.5g.

FLAP AND SLAT WING SWEEP CONTROL BOX

Electromechanical (auxiliary flaps, oversweep enable) and mechanical (main flap) interlocks in the control box limit aft wing sweep commands at 21°15', 50°, and 68°. Interlocks in the control box are shown in figure 1-42. These interlocks, which serve as a backup to the electronic interlocks in the CADC, are imposed on both the normal and the emergency inputs to the control box and assure noninterference between movable surfaces and the fuselage.

Wing Sweep System Test

CONTINUOUS MONITOR

The command and execution of the wing sweep system is continually monitored by a failure detection system. The failure detection system in the CADC governs the change from wing sweep channel 1 to channel 2 or the disabling of wing sweep channel 1 or 2 by switching the respective control drive servo off. A single channel failure in the wing sweep electrical command path is indicated by illumination of the WING SWEEP advisory light followed by normal operation on the remaining channel. Failure of the remaining channel is indicated by illumination of the WING SWEEP warning light, after which wing sweep control must be exercised through the emergency WING SWEEP handle. Transient failures in the CADC can be reset by pressing the MASTER RESET pushbutton, which recycles the failure detection system.

CAUTION

If WING SWEEP advisory light illuminates again, do not attempt to recycle a failure through the MASTER RESET pushbutton more than twice.

PREFLIGHT CHECK

A preflight check of the wing sweep system to assure proper operation of the electrical command circuits without moving the wings should be accomplished after starting engines while the wings are in oversweep (75°).

1. Set wing sweep mode switch to AUTO.

Note

In aircraft BUNO 159637 and subsequent the CADC caution light will illuminate and test will not run if AUTO is not selected on the wing sweep switch.

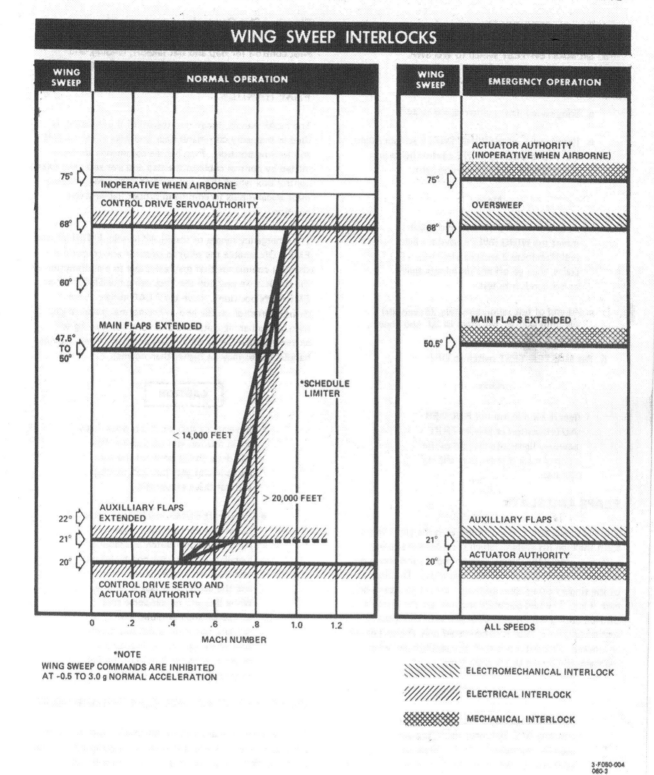

Figure 1-43. Wing Sweep Interlocks

2. Press MASTER RESET pushbutton.

3. Set MASTER TEST switch to WG SWP.

4. Monitor test by observing:

    a. Wing sweep limit pointer drives to 44°.

    b. Illumination of the WING SWEEP advisory light, FLAP and GLOVE VANE caution lights and the REDUCE SPEED warning light.

        Note

        In aircraft BUNO 159637 and subsequent the WING SWEEP advisory light will illuminate 3 seconds after test starts, then go off and illuminate again at 8 seconds into test.

    c. At end of test (approximately 25 seconds) the limit pointer will drive to 20° and above lights will go off.

5. Set MASTER TEST switch to OFF.

    Note

    Ignore illumination of RUDDER AUTH caution or MACH TRIM advisory lights and motion of the control stick if they occur during the test.

FLAPS AND SLATS

The flaps and slats, in conjunction with the glove vanes, form the high lift system, which provides the aircraft with augmented lift during the two modes of operation: takeoff or landing, and maneuvering flight. The flaps are of the single slotted type, sectioned into three panels on each wing. The two outboard sections are the main flaps, utilized during both modes of operation. The inboard section (auxiliary flap) is commanded only during takeoff or landing. The slats consist of two sections per wing mechanically linked to the main flaps.

Note

In aircraft 159637 and prior not incorporating AFC 364, only main flaps are used for maneuvering flight. Slats are used during takeoff and landing only.

Flap and Slat Controls

Pilot controls for flap and slat takeoff, landing, and maneuvering modes are illustrated in figure 1-44.

FLAP HANDLE

The FLAP handle, located outboard of the throttles, is used to manually command flaps and slats to the takeoff and landing position. Flap handle commands are transmitted by control cable to the flap and slat and wing sweep control box where they are summed with CADC electromechanical inputs to command proper flap and slat position.

The emergency ranges of the FLAP handle, EMER UP and EMER DN, enable the pilot to override any electromechanical commands that may exist due to a malfunction of the CADC. To position the flaps using the EMER UP or EMER DN positions, move the FLAP handle in the desired direction to the end of the normal travel range; then, move handle outboard and continue moving to extreme EMER UP or EMER DN position. While moving handle, forces may be higher than normal.

CAUTION

- Reversal of flap direction while flaps are in motion will damage the flexible drive shafts between the main flap and slat gear box and the flap and slat drive sequencer.

- In aircraft BUNO 160379 and subsequent and aircraft incorporating AFC 500, a slip clutch assembly is installed between the combined system forward flap hydraulic motor and the center gear box assembly. While this will relieve some stall torque on the hydraulic motor, extremely fast reversals of flap direction while flaps are in motion may result in eventual failure of the flap and slat flexible drive shaft.

MANEUVER FLAP AND SLAT THUMBWHEEL

The maneuver flap and slat thumbwheel is located on the left side of the stick grip and is spring loaded to the center position. With LDG GEAR and FLAP handles up,

NAVAIR 01-F14AAA-1

FLAP AND SLAT CONTROLS AND INDICATORS

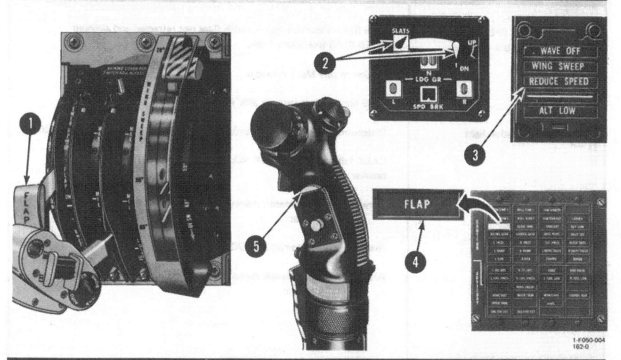

NOMENCLATURE		FUNCTION	
①	FLAP handle	UP	- Normal retraction of main and auxiliary flaps.
		DN	- Normal extension of main and auxiliary flaps.
		EMER UP	- Emergency retraction of main flaps to full up overriding any electromechanical command faults.
		EMER DN	- Emergency extension of main flaps to full down (35°) overriding any electromechanical command faults.
②	Flap and Slats indicator	▨	- Power off; maneuver slats extended.
		◢	- Slats extended (17°) { Slat position is an electrical pickoff of right slat position only
		◣	- Slats retracted (0°)
			- Flaps full up (0°) { Flap position is a pickoff from hydraulic motor for main flaps only
			- Maneuver flaps down (10°)
			- Flaps full down (35°)

Figure 1-44. Flap and Slat Controls and Indicators (Sheet 1 of 2)

NOMENCLATURE	FUNCTION
③ REDUCE SPEED warning light	Main flap comparator failures with flaps not retracted and airspeed >225 KIAS (see figure 1-39).
	Maximum safe Mach exceeded.
	Total temperature exceeds 388°F.
④ FLAP caution light	Disagreement between main and/or AUX flap position or asymmetry lockout.
	CADC failure. WG SWP DR NO. 2/MANUV FLAP (LE2) circuit breaker pulled.
⑤ Maneuver flap and slat thumbwheel	Forward - Commands maneuver flaps and slats and glove vanes to retract.
	Neutral - Automatic CADC program.
	Aft - Commands maneuver flasp and slats and glove vanes to extend.

Figure 1-44. Flap and Slat Controls and Indicators (Sheet 2 of 2)

automatic CADC flap, slat, and glove vane positioning can be overridden with pilot thumbwheel inputs to partially or fully extend or retract the maneuvering flaps, slats, and/or glove vanes; however, the next time angle of attack crosses an extension or retraction threshold, the automatic command will again take precedence, unless manually overridden again. Manual thumbwheel command is a proportional command.

Note

In aircraft prior to BUNO 159825 not incorporating AFC 364, the thumbwheel commands maneuver flaps and glove vanes only. It provides the only means of obtaining maneuver flaps since automatic CADC positioning is not incorporated.

MAIN FLAPS

The main flaps on each wing consist of two sections, simultaneously driven by four mechanical actuators geared to a common flap drive shaft. Each wing incorporates a flap asymmetry sensor and flap over-travel switches for both the extension and retraction cycles.

Cove doors, spoilers, eyebrow doors, and gusses operate with the flaps to form a slot to optimize airflow over the deflected flap. The cove doors are secondary surfaces along the underside of the wing forward of the flap. (See figure 1-45.) As the flaps pass 25° deflection, a negative command received from the AFCS depresses the spoilers to -4 1/2° to meet with the cove doors. Because the spoilers do not span the entire wing as do the flaps, gusses inboard and ouboard of the spoilers perform the flapdown function of the spoilers. With the flaps retracted, the eyebrow doors, which are the forward upper surface of the flaps, are spring-loaded in the up position to close the gap between the trailing edge of the spoiler or guss and the leading edge of the flaps. Mechanical linkage retracts the eyebrow door when the flaps are lowered, to provide a smooth contour over the upper surface of the deflected flap.

AUXILIARY FLAPS

The auxiliary flaps are inboard of the main flaps and are powered by the combined hydraulic system. The actuator is designed to mechanically lock the auxiliary flaps when in the up position. In the event of high dynamic pressure conditions, a bypass valve within each control valve opens causing the auxiliary flap to be blown back, thus avoiding possible structural damage. During loss of electrical power, the control valve is spring loaded to retract, retracting the auxiliary flaps within 1 minute. The auxiliary flaps use cove doors, eyebrow doors, and gusses identical in purpose and operation with those associated with the main flaps.

SLATS

The slats on each wing are divided into two sections, both of which are driven simultaneously by a single slat drive shaft. The slats are supported and guided by seven curved tracks.

Flap and Slat Operation

NORMAL OPERATION

The main flap and slat portion of the high-lift system is positioned with a dual redundant hydromechanical servo loop in response to the FLAP handle command. The auxiliary flap is a two-position control surface powered by the combined hydraulic system. With the FLAP handle exceeding 5° deflection, the auxiliary flaps fully extend. Conversely, they retract for a FLAP handle position equal to or less than 5°. The torque of the flap and slat drive hydraulic motor is transmitted by flexible drive shafts to each wing.

DEGRADED OPERATION

In the event of a combined hydraulic system failure, outboard spoiler module fluid is automatically directed to a backup hydraulic motor to lower main flaps and slats only. In the event of main flap asymmetry greater than 3°, slat asymmetry greater than 4° or flap surface overtravel, the flap and slat system is disabled. Flaps and slats will remain in the position they were in when failure or malfunction occurred. The auxiliary flaps are automatically commanded to retract. There is no asymmetry protection for the auxiliary flaps.

AUTOMATIC RETRACTION

In aircraft prior to BUNO 161167 not incorporating AYC 660P1 automatic flap and slat retraction from the landing and takeoff position is provided for airspeeds greater than 225 knots for any FLAP handle position within the normal range of 0 to 35° to prevent main flap panel overloading. Overload protection for the auxiliary flaps is provided by a blowback feature. Since air loads on the auxiliary flaps are low, they may only blow back partially and wing sweep may be restricted to 22°. At the time of flap and slat automatic retraction, the auxiliary flap stays commanded to extend, tripping the flap system failure detection logic, which illuminates the FLAP caution light.

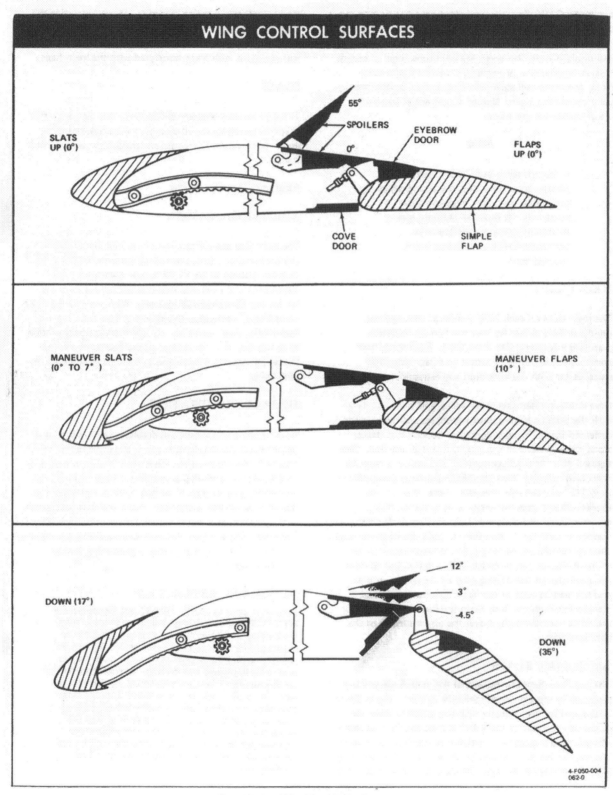

Figure 1-45. Wing Control Surfaces

FLAP WING INTERLOCKS

The main flap and auxiliary flap commands are interlocked electrically and mechanically with the wing sweep to prevent flap fuselage interference. An electrical interlock in the CADC and a mechanical command in the wing sweep control box prevent wing sweep aft of 22° with auxiliary flaps extended. In a similar manner, upon extension of the main flaps, the wings are electrically and mechanically limited to wing sweep angles less than 50°. The FLAP handle is mechanically prevented from moving to the down position if wing position is aft of 50°. If flaps are lowered with wings between 21° and 50°, main flaps will extend but auxiliary flaps will remain retracted.

CAUTION

- If flaps are extended with wings between 21° and 50°, auxiliary flap extension is inhibited and a large down pitch trim change will occur.

- Pulling the FLAP/SLAT CONTR SHUTOFF circuit breaker could eliminate mechanical or electrical main and auxiliary flap interlocks and may allow the wings to be swept with the flaps partially or fully down in the wing sweep emergency mode.

MANEUVER FLAP AND SLAT MODE

The main flaps can be extended to 10° with the slats extended to 7° within the altitude and Mach envelope shown in figure 1-46.

Maneuver flaps and slats are automatically extended and retracted by the CADC as a function of angle of attack and Mach number (figure 1-47.) The schedule commands full maneuver flaps and slats as soon as the slatted wing maneuvering efficiency exceeds that of the clean wing. The pilot may use the thumbwheel to manually override the automatic CADC command. Maneuver flaps are not available above Mach 1.

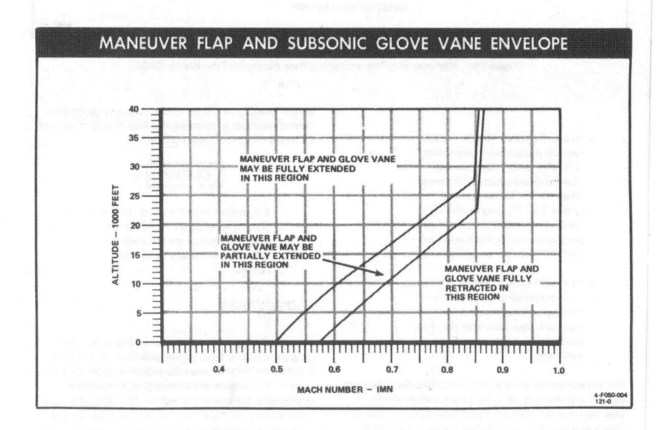

Figure 1-46. Maneuver Flap and Subsonic Glove Vane Envelope

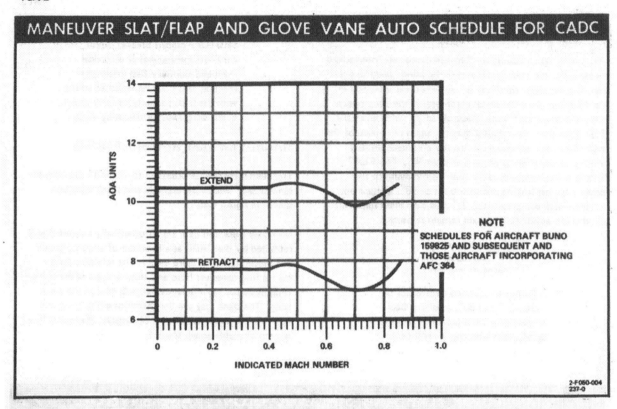

Figure 1-47. Maneuver Slat/Flap and Glove Vane Automatic Schedule for CADC

Note

- In aircraft BUNO 159825 and subsequent and aircraft incorporating AFC 364, CADC maneuver flap commands are automatically reset when the flap handle is placed down greater than 2°, wing sweep BOMB mode is selected or maneuver flaps are commanded to less than 1° by the CADC due to dynamic pressure.

- In aircraft BUNO 159825 and prior not incorporating AFC 364, the thumbwheel commands maneuver flaps and glove vanes only and there is no automatic maneuver flap extension.

The angle-of-attack input to the CADC from the alpha computer is inhibited and will retract the maneuver devices if they are extended when the LDG GEAR handle is lowered. This is to ensure that the maneuver devices are retracted before lowering the FLAP handle. Maneuver devices extended condition is indicated by a SLATS barberpole and an intermediate (10°) flap position.

If maneuver devices are not retracted prior to lowering the FLAP handle, a rapid reversal of the flaps will occur with possible damage to the flap system.

GLOVE VANES

An extendable glove vane is in the leading edge of each wing glove (figure 1-48). When extended, the triangular-shaped glove vanes increase the aircraft wing area and also serve to reduce the excessive longitudinal stability encountered in the supersonic regime. The right glove vane is powered by the flight hydraulic system while the left glove vane is powered by the combined hydraulic system.

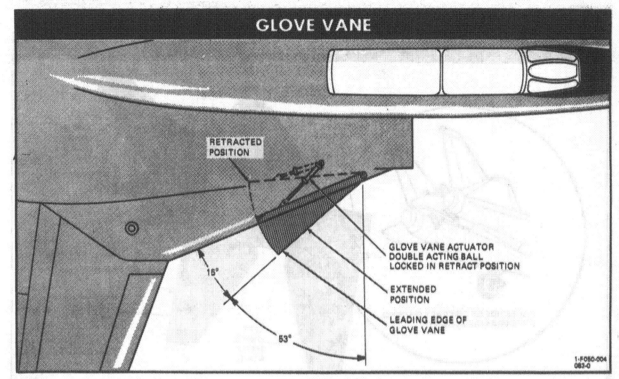

Figure 1-48. Glove Vane

The glove vanes automatically start extending with the maneuver flaps and slats with wings aft of 25°. Extension angle increases with greater wing sweep until, at 35° wing sweep the glove vanes are fully extended in conjunction with maneuver flaps and slats.

From 1.0 to 1.4 IMN, glove vanes will extend proportionally in response to pilot thumbwheel inputs.

Automatic extension of the glove vanes is effected within the CADC, starting at 1.35 IMN to reach full extension (15°) at 1.45 IMN. Above 1.5 IMN, the glove vanes remain full out (15°) overriding any pilot command through the thumbwheel. Through a feedback loop, the position of each glove vane is compared with the commanded position and, should a disparity exist, both glove vanes are automatically retracted and the GLOVE VANE caution light illuminates. Transient failures may be reset by depressing the MASTER RESET pushbutton.

Selection of wing sweep BOMB mode automatically extends glove vanes for M > 0.35 and retracts glove vanes for M ≤ 0.35. Manual glove vane commands are automatically reset to retract when 1.0 IMN is passed in either direction.

SPEED BRAKES

The speed brakes consist of three individual surfaces; one upper and two lower panels, on the aft fuselage between the engine nacelles. (See figure 1-49.) As a drag control device, the speed brakes may be infinitely modulated on the extension cycle but the retraction cycle is a single step. Operating time for full deflection is approximately 2 seconds. Hydraulic power is supplied by the combined hydraulic system (nonisolation circuit) and electrical power is through the essential no. 2 dc bus with circuit overload protection on the pilot's right circuit breaker panel (SPD BK/P-ROLL TRIM ENABLE) (RA2).

Speed Brake Operation

Pilot control of the speed brakes is effected by use of the three-position speed brake switch on the inboard side of the right throttle grip. (See figure 1-50.) Automatic retraction of the speed brakes occurs with placement of either or both throttles at MIL or in-flight loss of combined hydraulic pressure or electrical power.

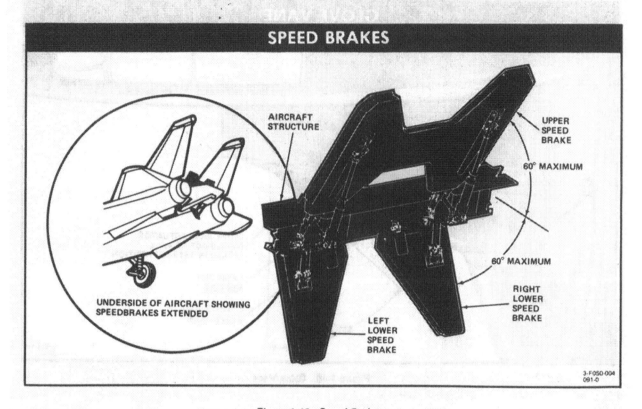

Figure 1-49. Speed Brakes

To avoid fuel impingement on the fuselage boattail and nozzles, fuel dump operations are prevented with the speed brakes extended.

Note

- With loss of combined hydraulic pressure, speed brakes will float.
- The speed brake switch is electrically bypassed during a combined hydraulic system failure, enabling the pilot to dump fuel when the speed brakes are floating or modulating. The electrical bypass is accomplished whenever the combined pressure falls below 500 psi.
- Do not extend the speed brakes in flight within 1 minute (nominal) after terminating fuel dump operations to allow residual fuel in the dump mast to drain.
- A throttle must be held in MIL (or greater) for approximately 3 seconds in order for the automatic function to completely retract the speed brake. Anything less will cause partial retraction.

The speed brakes will start to blowback (close) at approximately 400 knots and will continue toward the closed position as airspeed increases to prevent structural damage. A reduction in airspeed will not automatically cause the speed brakes to extend to the originally commanded position.

FLIGHT CONTROL SYSTEMS

Flight control is acheived through an irreversible, hydraulic power system operated by a control stick and rudder pedals. Aircraft pitch is controlled by symmetrical deflection of the horizontal stabilizers. Roll control is effected by differential stabilizer deflections and augmented by spoilers at wing sweep positions less than 57°. Directional control is provided by dual rudders. During power approach maneuvers, the aircraft flight path can be controlled through symmetric spoiler displacement by the pilot selecting direct lift control (DLC). Control surface indicators are shown in figure 1-51.

The horizontal stabilizer and rudders are powered by the flight and combined hydraulic systems and controlled by

SPEED BRAKE CONTROL AND INDICATOR

NOMENCLATURE	FUNCTION
① Speed brake switch	EXT - Momentary position used for partial or full extension. When released, switch returns to center (hold) position.
	RET - Normal position of switch. Retracts and maintains speed brakes closed.
② SPEED BRAKE indicator	▦ - Partial extension (hold).
	◼ - Full extension (60°).
	[IN] - Full retracted position.
	▨ - Speed brakes power is off.

Note

Automatic retraction of speed brakes occurs when either or both throttles are at MIL.

Figure 1-50. Speed Brake Control and Indicator

Section I
Part 2

NAVAIR 01-F14AAA-1

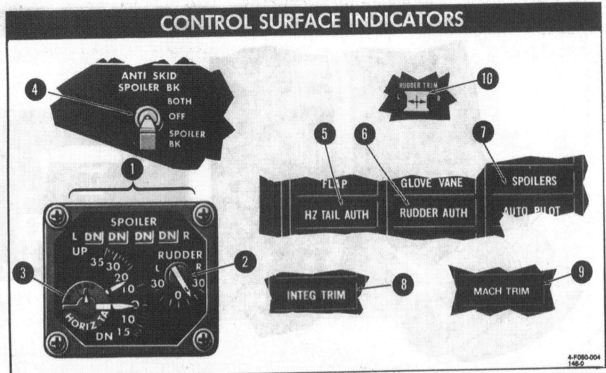

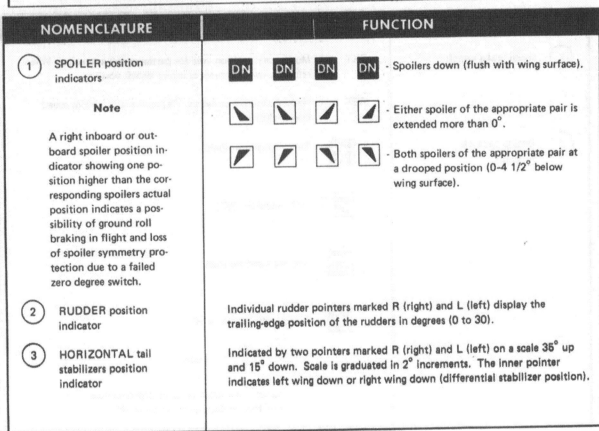

Figure 1-51. Control Surface Indicators (Sheet 1 of 2)

NOMENCLATURE	FUNCTION
(4) ANTI SKID SPOILER BK switch	BOTH - Antiskid. Spoiler brakes operate with weight on wheels and throttles at IDLE. OFF - Antiskid and spoiler inoperate with weight on wheels. SPOILER BK - Spoiler brakes operate with weight on wheels and both throttles at IDLE. Antiskid is deactivated.
(5) HZ TAIL AUTH caution light	Failure of lateral tail authority actuator to follow schedule or CADC failure.
(6) RUDDER AUTH caution light	Disagreement between command and position, failure of rudder authority actuators to follow schedule, or CADC failure.
(7) SPOILER caution light	Spoiler system failure, causing a set of spoilers to be locked down.
(8) INTEG TRIM advisory light	Discrepancy between input command signal and actuator position or an electrical power loss within the computer.
(9) MACH TRIM advisory light	Failure of Mach trim actuator to follow schedule. **Note** Transient failures involving HZ TAIL AUTH, RUDDER AUTH, or SPOILER caution lights and INTEG TRIM and MACH TRIM advisory lights can be reset by pressing the MASTER RESET pushbutton.
(10) Rudder trim switch	Controls the electromechanical actuator that varies the neutral position of the mechanical linkage for rudder trim.

Figure 1-51. Control Surface Indicators (Sheet 2 of 2)

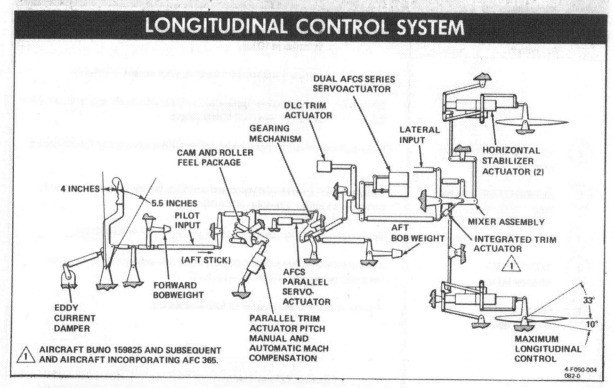

Figure 1-52. Longitudinal Control System

pushrods and bellcranks. A third independent flight control hydraulic power source is provided by the backup module. Sufficient control for safe return and field landing is provided to the stabilizer and rudder actuators by the backup flight control module should both the combined and flight hydraulic systems fail. Spoiler control is effected by an electrohydraulic, fly-by-wire system and powered by the combined hydraulic system and outboard spoiler module.

The automatic flight control system (AFCS) includes a stability augmentation system (SAS), an autopilot and auxiliary control functions for spoiler control, rudder authority control, lateral stick authority control, and Mach trim compensation. Aircraft BUNO 159825 and subsequent and aircraft incorporating AFC 400, include provisions for an aileron-rudder interconnect (ARI) which is not presently operational.

Longitudinal Control

Longitudinal control (figure 1-52) is provided by symmetric deflection of independently actuated horizontal stabilizer slabs. Control stick motion is transmitted to the stabilizer power actuators by pushrods and bellcranks to dual tandem actuators independently powered by the flight and combined hydraulic systems. The power actuators control the stabilizers symmetrically for longitudinal control and differentially for lateral control. This is accomplished by mechanically summing pitch and roll commands at the pitch-roll mixer assembly. Nonlinear stick-to-stabilizer gearing provides appropriate stick sensitivity for responsive and smooth control. Longitudinal system authority is shown in figure 1-53.

LONGITUDINAL FEEL

Artificial feel devices in the control system provide the pilot with force cues and feedback. A springloaded cam and roller assembly produces breakout force when the stick is displaced from neutral trim and provides increasing stick forces proportional to control stick displacement. Control stick forces, proportional to normal acceleration (g forces) and pitch acceleration, are produced by fore and aft bobweights. Aircraft overstress from abrupt stick inputs is prevented by an eddy current damper which resists large, rapid control deflections.

LONGITUDINAL SYSTEM AUTHORITY

ACTUATION	COCKPIT CONTROL MODE	MOTION	STABILIZER SURFACE AUTHORITY	RATE	PARALLEL TRIM AUTHORITY	AVERAGE RATE
CONTROL STICK	MANUAL	4 INCHES FORWARD 5.5 INCHES AFT	12° TED 33° TEU	36° PER SECOND	9° TED 18° TEU	1° PER SECOND
AFCS	SERIES (SAS)	NONE	±3°	20° PER SECOND	—	—
AFCS	PARALLEL (ACL ONLY)	4 INCHES FORWARD 5.5 INCHES AFT	10° TED 33° TEU	36° PER SECOND	9° TED 18° TEU	0.1° PER SECOND
MANEUVER FLAP (ITS) AND DLC THUMBWHEEL	SERIES	±45° DLC THUMBWHEEL MODE	8.4° TED MAXIMUM	36° PER SECOND	—	—
MANEUVER FLAP (ITS) AND DLC THUMBWHEEL	SERIES	±45° MANEUVER FLAP MODE	±3°	3° PER SECOND	—	—

Figure 1-53. Longitudinal System Authority

LONGITUDINAL TRIM

Longitudinal trim is provided by varying the neutral position of the cam and roller feel assembly with an electromechanical screwjack actuator. The manual pitch trim button on the stick is a five-position switch, spring loaded to the center (off) position. (See figure 1-54.) The fore and aft switch positions produce corresponding nose-down and nose-up trim, respectively. The manual trim switch is deactivated when the autopilot is engaged.

MACH TRIM

Mach trim control is provided by the AFCS and is continuously engaged to provide automatic Mach trim compensation during transonic and supersonic flight. A failure of Mach trim compensation is indicated by the MACH TRIM advisory light. Transient failures can be reset by depressing the MASTER RESET pushbutton.

The manual and AFCS automatic trim and Mach trim actuator is installed in parallel with the flight control system. Trim actuation produces a corresponding stick and control surface movement.

Integrated Trim System (ITS)

In aircraft BUNO 159825 and subsequent and those aircraft incorporating AFC 365, an integrated trim system (ITS) is incorporated to eliminate longitudinal trim changes due to the extension and retraction of flaps and speed brakes. Disagreement of command position removes power from the motor and illuminates the INTEG TRIM advisory light. Transient failures can be reset by pressing the MASTER RESET pushbutton. ITS schedules are shown in figure 1-55.

CONTROL STICK AND TRIM

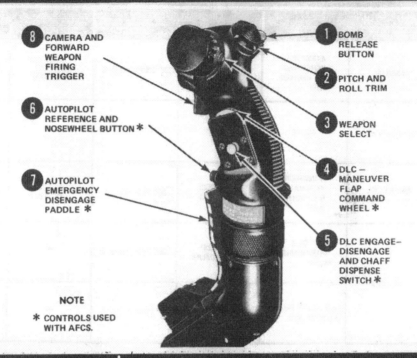

NOMENCLATURE	FUNCTION
① Bomb release button	Pilot control for release of stores.
② Pitch and roll trim button	Spring-loaded to (center) off position. Up and down positions control pitch trim and left and right positions control roll trim. Manual trim is inoperative during autopilot operation.
③ Weapon Selector switch	SP or PH - Selects Sparrow or Phoenix missiles. SW - Selects Sidewinder missiles. GUN - Selects gun. OFF - Inhibits all weapon selection by the pilot.
④ Maneuver flap, slat, and DLC command thumbwheel	Spring-loaded to a neutral position. With DLC engaged: Forward rotation extends spoilers; aft rotation retracts spoilers. With gear and flaps up: Forward rotation retracts maneuvering flaps/slats; aft rotation extends maneuvering flaps/slats.
⑤ DLC engage, disengage, and chaff switch	Momentary depression of the switch with flaps down, throttle less than MIL, and no failures in spoiler system engages DCL. With flaps up, switch will dispense chaff or flares. DLC is disengaged by momentarily pressing the switch, raising the flaps, or advancing either throttle to MIL.
⑥ Autopilot reference and nosewheel steering pushbutton	With weight on wheels, nosewheel steering can be engaged by depressing switch momentarily. Weight off wheels and autopilot engaged; switch engages compatible autopilot modes.

Figure 1-54. Control Stick and Trim (Sheet 1 of 2)

NOMENCLATURE	FUNCTION
7 Autopilot emergency disengage paddle	Disengages all autopilot modes and DLC, if selected, and releases all autopilot switches. Momentarily disengages (as long as switch is held) pitch and roll SAS servos; switches remain ON. Yaw SAS, spoilers, and authority controls are not disengaged.
8 Camera and forward weapon firing trigger	Pilot control of gun camera or forward weapons.

Figure 1-54. Control Stick and Trim (Sheet 2 of 2)

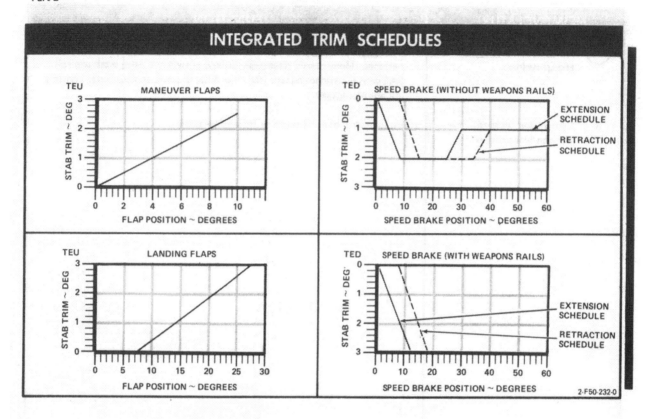

Figure 1-55. Integrated Trim Schedules

CAUTION

When the AIM-54 weapon rail pallet(s) is installed, the speed brake compensation schedule in the integrated trim computer changes. If less than four AIM-54 missiles are carried on the weapon rails, the ITS may overcompensate for the speed brake trim change. In the worst case (low altitude, between 0.7 and 0.8 IMN, pitch SAS off, and weapon rails without AIM-54 missiles) the ITS can cause an incremental 2-g nosedown trim change when the speed brake is extended. Under these conditions with the pitch SAS engaged, maximum trim change is reduced to approximately 1 g.

PREFLIGHT

The ITS is automatically energized with hydraulic and electrical power applied. It can be checked by operating flaps or speed brakes and observing a change in indicated stabilizer position.

Lateral Control

Lateral control (figure 1-56) is effected by differential displacement of the horizontal stabilizers and augmented by wing spoilers at wing sweep positions of less than 57°. A ±1/2-inch stick deadband is provided to preclude spoiler actuation with small lateral stick commands. The spoilers are interlocked in the flush-down position at wing sweeps greater than 57° (mechanical downlocks are installed at 62°) and roll control is provided entirely by differential stabilizer. Lateral stick commands are transmitted by pushrods and bellcranks to the independent

NAVAIR 01-F14AAA-1

LATERAL CONTROL SYSTEM

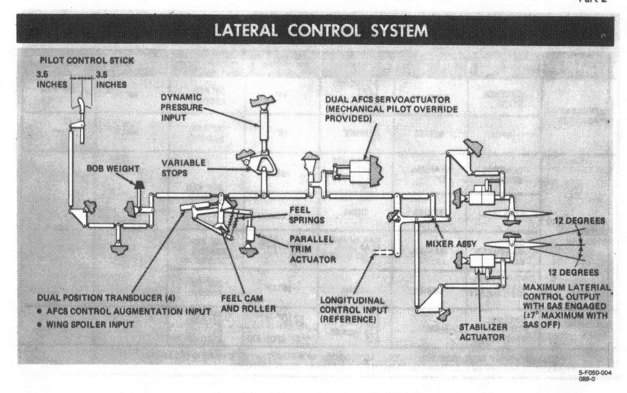

Figure 1-56. Lateral Control System

stabilizer power actuators and electrically to the spoiler actuators. Lateral system authority is tabulated in figure 1-57.

LATERAL FEEL

An artificial feel system provides the pilot with force cues and feedback. The lateral feel mechanism is a spring roller cam assembly with a neutral stick position detent and a constant stick deflection force gradient.

LATERAL TRIM

Lateral trim is by differential deflection of the horizontal stabilizers. The wing spoilers are not actuated for lateral trim control. Trim is provided by adjusting the neutral position of the spring roller cam feel assembly with an electromechanical screwjack. Left or right deflection of the roll trim button on the stick grip produces corresponding stick movement and left or right wing down trim, respectively. The normal stick grip trim switch is inoperative when the autopilot is engaged.

Note

With lateral trim set at other than 0°, maximum spoiler deflection is reduced in the direction of applied trim.

LATERAL CONTROL STOPS

To limit the torsional fuselage loads, variable lateral control authroity stops are installed. The lateral stick stops vary according to dynamic pressure airloads from full stick authority at low Q, to one-half stick-throw limits at high-Q conditions. Failure of the lateral stick stops is indicated by the HZ TAIL AUTH caution light. Transient failures can be reset with the MASTER RESET pushbutton. Failure of the stops in the one half stick position does limit low-Q rolling performance, however, ample roll control is available for all landing conditions and configurations. Failure in the open condition with SAS on, requires the pilot to manually limit stick deflection at higher speeds to avoid exceeding fuselage torsional load limits, as lateral stops do not limit SAS authority.

LATERAL SYSTEM AUTHORITY

CONTROL SURFACE	COCKPIT CONTROL			SURFACE		PARALLEL TRIM	
	ACTUATION	MODE	MOTION	AUTHORITY	RATE	AUTHORITY	RATE
DIFFERENTIAL STABILIZER	CONTROL STICK	MANUAL	3.5 INCHES LEFT 3.5 INCHES RIGHT	±7°	36° PER SECOND	±3°	3/8° PER SECOND
	AFCS	SERIES	NONE	±5°	33° PER SECOND	—	—
INBOARD AND OUTBOARD SPOILERS	CONTROL STICK	MANUAL 57°	3.5 INCHES LEFT 3.5 INCHES RIGHT	±55°	250° PER SECOND	NONE	NONE
	AFCS (ACL)	SERIES	NONE	15° MAXIMUM			
	DLC MANEUVER FLAP	MANUAL	MANEUVER FLAP COMMAND WHEEL ±45°	+9°, -7-1/2° ABOUT +3° NEUTRAL POSITION	125° PER SECOND (MINIMUM)		
	GROUND ROLL BRAKING ARMED, WEIGHT ON WHEELS	SERIES	NONE	55° UP	250° PER SECOND		
*LATERAL STOPS	CONTROL STICK RESTRICTED	MANUAL	1.75 INCHES LEFT 1.75 INCHES RIGHT	±3-1/2° DIFF. STABILIZER **28° SPOILER	36° PER SECOND 250° PER SECOND	—	—

*PROGRAMMED BY CADC (HORIZONTAL TAIL AUTHORITY) AS A FUNCTION OF DYNAMIC PRESSURE.
**MAXIMUM SAS OFF DEFLECTION LIMITS WITH FULL LATERAL STOPS ENGAGED.

Figure 1-57. Lateral System Authority

Spoiler Control

Four spoiler control surfaces (figure 1-58) on the upper surface of each wing augment roll control power, provide direct lift control, and implement aerodynamic ground roll braking. The inboard pairs are powered by the combined hydraulic system. The outboard pairs are powered by the outboard spoiler module. Hydraulic actuation of the servo actuators is controlled by electric servo valves at the actuator and commanded by control stick displacement. The aircraft has two spoiler gearing curves called cruise and power approach. Cruise spoiler gearing, is the schedule that spoilers follow in the clean configuration and is shown in figure 1-59. Power approach is the schedule that spoilers follow with the flaps down greater than 25° and is shown in figure 1-59 (DLC on). With direct lift control on, in the power approach mode, all spoilers are uprigged from the normal -4.5° position to the +3° position. Lateral stick inputs result in the spoilers extending on one side in the direction of stick displacement and depressing toward the -4.5° stowed position on the other side. This is the primary reason for better roll response in the landing configuration with DLC engaged.

As mentioned earlier, lateral trim is provided by adjusting the neutral position of the stick. This movement of the neutral position has an effect on the amount of spoiler deflection available. That is, as lateral trim is applied away from the 0 trim position, maximum spoiler deflection is reduced in the same direction. (Right trim - less right wing spoilers deflection.)

Full lateral trim in the same direction as lateral stick displacement will still provide approximately 25° of spoiler deflection to counteract an asymmetric flap and slat condition. (See figure 1-59.) This is sufficient to control full-flap asymmetry with symmetrically down slats.

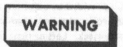

Full slat asymmetry (17°) can result in an out-of-control situation at 15 units AOA or greater, even with 55° of spoilers available.

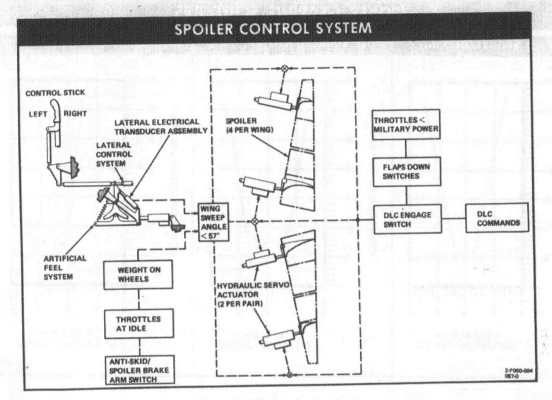

Figure 1-58. Spoiler Control System

GROUND ROLL BRAKING

Aerodynamic ground-roll braking is provided by symmetric deflection of all spoilers to +55°. Ground-roll braking is controlled by the ANTI SKID SPOILER BK switch on the pilot's left vertical console. The three-position switch allows optimal selection of BOTH (spoiler brake and wheel antiskid braking), SPOILER BK (spoiler brake only), or OFF where neither spoilers nor antiskid is armed. With SPOILER BK or BOTH selected, two conditions are required to actuate the spoilers:

- Weight on wheels
- Both throttles at idle

Failure to satisfy any one of the above conditions will cause the spoilers to return to the down position.

Note

During initial spoiler brake operation, it is normal for the indicators in the SPOILER window to momentarily flip-flop.

SPOILER FAILURE

Simultaneous deflection of corresponding spoiler pairs greater than 18°, or a stick displacement of 1 inch to oppose a symmetric spoiler greater than 18° will signal a spoiler failure, remove all electrical commands to the affected spoiler set, hydraulically power the failed spoiler pairs to the -4-1/2° position, and illuminate the SPOILER caution light. Transient spoiler failures can be reset by pressing the MASTER RESET pushbutton. With either SPOILER BK or BOTH selected and weight on wheels, the symmetric spoiler failure logic is disarmed to permit symmetrical 55° spoiler deflection for ground-roll braking.

Failure of the AFCS roll computer fails the inboard spoiler pairs. An AFCS pitch computer failure renders the outboard spoiler pairs inoperative.

In aircraft 160920 and subsequent and aircraft incorporating AFC 573, a spoiler failure override (SPOILER FLR ORIDE) panel on the pilot's right inboard console allows the pilot to manually override the automatic shutdown function. If a stuck up spoiler failure occurs, selecting the appropriate INBD or OUTBD spoiler switch ORIDE position (figure 1-60) and then depressing MASTER RESET allows the remaining spoilers in that set to operate.

Varying ac voltage during an electrical short condition may yield one or more of the following characteristics:

- Loss of one or both channels of computer operation
- Differential tail and rudder hardovers for up to 3 seconds

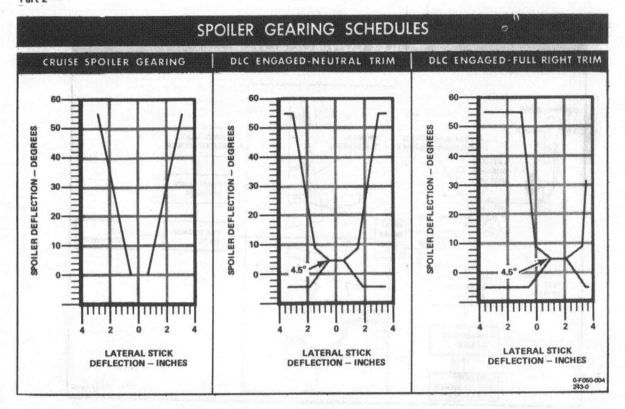

Figure 1-59. Spoiler Gearing Schedules

- Inoperative spoilers and SAS outputs even when voltage was reset to nominal 115 volts
- Inability to reset computers through master reset and reengage pitch and roll switches

Note

On deck, when the flap handle is cycled to the UP position, the outboard spoiler module is shut down. This may cause the outboard spoilers to indicate a droop or down position. If this occurs, position the flap handle to DN and move the control stick laterally. This will result in a proper spoiler indication.

SPOILER TEST

The spoiler test provides a continuity and logic test for the spoiler 18° and stick switches. The test is conducted with the flaps down and spoilers up. Select STICK SW position on the MASTER TEST switch (SPOILERS caution light illuminates, spoilers droop). Move the stick left and right of center one inch. Observe GO light illuminate on the MASTER TEST panel after moving the stick in each direction. Return the MASTER TEST switch to off and depress the MASTER REST pushbutton to reset the spoilers.

Yaw Control

Yaw control (figure 1-61) is effected by twin rudders, one on each vertical tail. The rudder pedals adjust through a 10-inch range in 1-inch increments with the adjust control on the lower center pedestal, forward of the control stick.

Yaw commands are transmitted mechanically from the rudder pedals to the rudder power actuators by pushrods and bellcranks. Tandem power actuators are powered independently by the flight and combined hydraulic systems. Yaw system authority is tabulated in figure 1-62.

NAVAIR 01-F14AAA-1

Section I
Part 2

NOMENCLATURE	FUNCTION	
① INBD spoiler override switch	ORIDE -	Overrides inboard spoiler symmetry protection logic. If an inboard spoiler fails more than 18° up, allows remaining inboard spoilers to operate after MASTER RESET is depressed.
	NORM - (guarded position)	Safety guard down. Allows inboard spoiler symmetry protection. If an inboard spoiler fails up, all inboard spoilers are commanded to the droop position and the SPOILERS light illuminates.
② OUTBD spoiler override switch	ORIDE -	Overrides outboard spoiler symmetry protection logic. If an outboard spoiler fails more than 18° up, allows remaining outboard spoilers to operate after MASTER RESET is depressed.
	NORM - (guarded position)	Safety guard down. Allows outboard spoiler symmetry protection. If an outboard spoiler fails up, all outboard spoilers are commanded to the droop position and the SPOILERS light illuminates.

Figure 1-60. Spoiler Failure Override Panel

YAW CONTROL SYSTEM

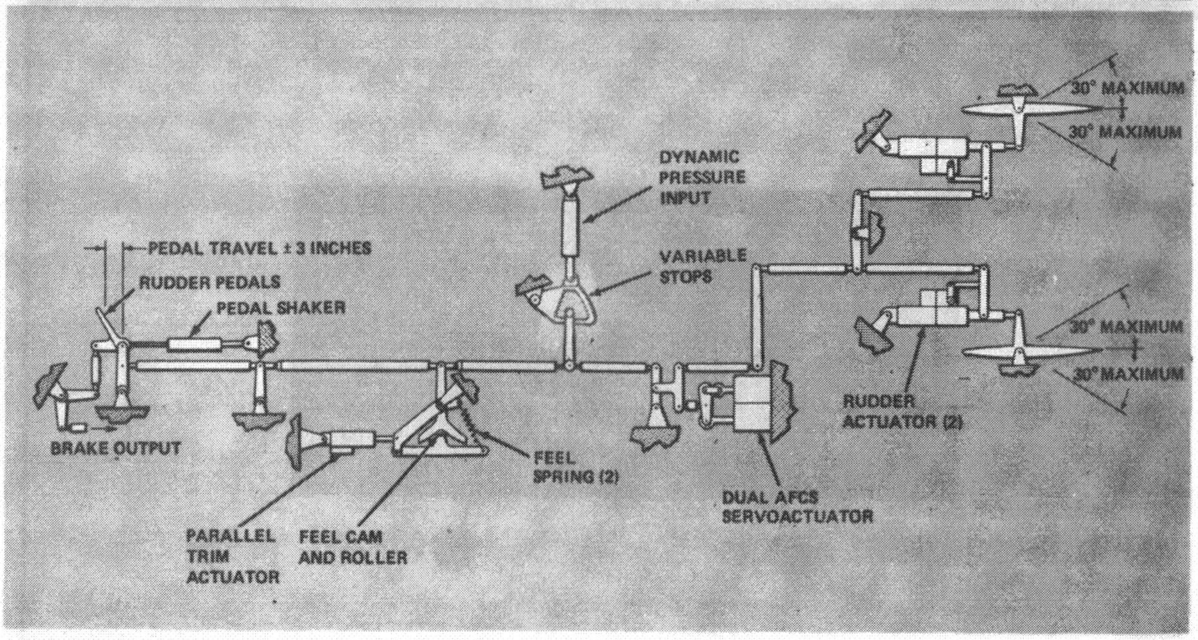

Figure 1-61. Yaw Control System

YAW SYSTEM AUTHORITY

ACTUATION	COCKPIT CONTROL		RUDDER SURFACE		PARALLEL TRIM	
	MODE	MOTION	AUTHORITY	RATE	AUTHORITY	RATE
RUDDER PEDALS	MANUAL (UNRESTRICTED)	3 INCHES LEFT 3 INCHES RIGHT	±30° MAXIMUM	106° PER SECOND	7°	1.13° PER SECOND
	*MANUAL (RESTRICTED)	1 INCH LEFT 1 INCH RIGHT	±9.5° MINIMUM	106° PER SECOND		
AFCS	SERIES	NONE	**±19°	80° PER SECOND	—	—

*STOPS PROGRAMMED BY CADC (RUDDER AUTHORITY) AS A FUNCTION OF DYNAMIC PRESSURE.
**IN AIRCRAFT BUNO 159637 AND PRIOR WITHOUT AFC 400, YAW AUTHORITY IS ±9.5°.

4-F050-004
153-0

Figure 1-62. Yaw System Authority

RUDDER FEEL

Artificial feel is provided with a spring roller-cam mechanism similar to the longitudinal and lateral feel systems.

Rudder force with pedal deflection is nonlinear with a relatively steep gradient about the neutral detent and gradually decreasing with increased pedal travel.

RUDDER TRIM

Rudder trim is effected by varying the neutral position of the feel assembly with an electromechanical screwjack actuator. Rudder trim control is actuated by a three-position switch on the left console outboard of the throttle quadrant. Left (L) and right (R) lateral switch movement commands left and right rudder trim respectively. The switch is spring loaded to the center off position. Trim actuation produces an associated movement of the rudder pedals, rudders, and rudder indicator.

RUDDER AUTHORITY STOPS

Rudder authority control stops limit rudder throws in the high-Q flight environment. Rudder deflection limits are scheduled by the CADC, commencing at about 250 KIAS. Above approximately 400 KIAS, the stops are fully engaged, restricting manual rudder deflection to 9.5°. Disagreement between command and position removes power from the motor and illuminates the RUDDER AUTH caution light.

CAUTION

A CADC failure may drive the rudder authority stops to 9.5°. This condition should be determined prior to making a single engine or crosswind landing. With the 9.5° stops in, rudder control may be insufficient to maintain directional control with single engine afterburner operation or during crosswind conditions. Nosewheel steering authority is greatly reduced with the 9.5° stops engaged.

RUDDER PEDAL SHAKER

On aircraft BUNO 159825 and subsequent and those aircraft incorporating AFC 400, a rudder pedal shaker is installed to warn the pilot of excessive angle of attack during approach. No cockpit controls or visual indications are associated with the warning devices; the shaker is automatically armed with the flaps at 25° or more and is activated when the aircraft exceeds 21±4 units AOA, as sensed by the ARI alpha probe and computer. Circuit protection is provided by a 28-volt dc ALPH COMPUTER/PEDAL SHAKER circuit breaker (7C8).

Direct Lift Control (DLC)

During landing approaches the spoilers and horizontal stabilizers can be controlled simultaneously to provide vertical glidepath correction without changing engine power setting or angle of attack.

Before DLC can be engaged, the following conditions are required:

- Flaps down greater than 25°
- Throttles less than MIL power
- Spoilers operational
- Pitch and roll AFCS computers operational

DLC is engaged with the control stick DLC switch and commanded by the thumbwheel. The thumbwheel is spring-loaded to a neutral position. Forward rotation of the wheel extends spoilers and aft rotation retracts them proportionally to the degree of thumbwheel rotation. DLC is provided by the pitch and roll computers. Upon engagement of DLC, the roll and pitch computers extend all the spoilers 3° above the 0° position. The pitch computer displaces the trailing edges of the horizontal stabilizers 6° down from their trim position. If the thumbwheel control is rotated fully forward, the spoilers extend to their 12° position and the stabilizer trailing edges are displaced 8° down. This increases the rate of descent. If the thumbwheel control is rotated fully aft, the spoilers retract to their -4.5° position and the stabilizer trailing edges return to the trim position. This decreases the rate of descent.

AUTOMATIC FLIGHT CONTROL SYSTEM (AFCS)

The automatic flight control system (AFCS) (page FO-14) augments the aircraft's natural damping characteristics and provides automatic commands for control of attitude, altitude, heading and approach modes selected by the pilot. All AFCS functions are integrated into the primary flight control system. Additionally, in aircraft BUNO 159825 and subsequent and aircraft incorporating AFC 400, an ARI is installed. The ARI is an auxiliary function of the stability augmentation system. A built-in-test (BIT) capability is provided to exercise inflight monitoring and to conduct an automatic operational readiness test for preflight checks. AFCS rates and authorities are tabulated in figure 1-63.

AFCS RATES AND AUTHORITIES

AXIS	ACTUATOR		SURFACE	AUTHORITY	SURFACE RATE
PITCH	DUAL SERIES	SAS	STABILIZER	±3°	20° PER SECOND
		ITS	STABILIZER	±3°	3° PER SECOND
	PARALLEL (ACL ONLY)		STABILIZER	10° TED 33° TEU	36° PER SECOND
	PARALLEL TRIM		STABILIZER	10° TED 18° TEU	0.1° PER SECOND
ROLL	DUAL SERIES		DIFFERENTIAL STABILIZER	±5°	33° PER SECOND
			SPOILERS (ACL)	15° MAXIMUM	250° PER SECOND
YAW	DUAL SERIES		RUDDER	*±19°	80° PER SECOND

*IN AIRCRAFT BUNO 159637 AND PRIOR AIRCRAFT WITHOUT AFC 400, YAW AUTHORITY IS ±9.5°.

Figure 1-63. AFCS Rates and Authorities

Stability Augmentation System (SAS)

Stability augmentation is provided for all three aircraft axes (pitch, roll, and yaw). Control surface commands are generated by the roll, pitch, and yaw computers in response to inputs from AFCS sensors.

Computer outputs are fed through the flight control system dual series actuators which drive the control system mechanical linkages in series (no stick motion) to produce surface motion. Stability augmentation is controlled by the three STAB AUG switches on the upper half of the AFCS control panel (figure 1-64). SAS is engaged by placing these switches to the ON position during normal post start procedures. The PITCH and ROLL STAB AUG switches are solenoid held and the YAW STAB AUG switch is a manually operated toggle switch. All three STAB AUG switches should normally remain in the ON position throughout the flight. Pitch and roll SAS axes consist of two redundant channels that provide failsafe operation.

Yaw SAS consists of three electronic channels to provide failsafe operational capability.

Note

In aircraft BUNO 159637 and prior, full lateral wing sweeps on the deck or rapid lateral stick inputs may illuminate either ROLL STAB 1 or ROLL STAB 2 caution lights or both and disengage the roll SAS. Cycle the ROLL STAB AUG switch to regain roll SAS.

Voltage Monitor Control Unit (VMCU)

In aircraft BUNO 160681 and subsequent and aircraft incorporating AFC 564, the voltage monitor control unit (VMCU) has been designed to detect abnormal voltage transients in the aircraft electrical supply system and reduce the voltage to a low enough level to allow the

NAVAIR 01-F14AAA-1

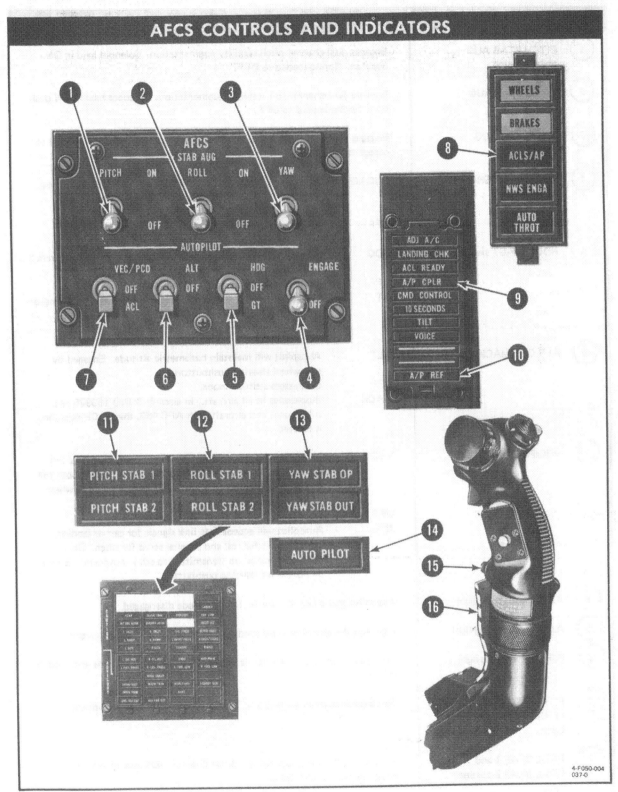

Figure 1-64. AFCS Controls and Indicators (Sheet 1 of 3)

NOMENCLATURE	FUNCTION
(1) PITCH STAB AUG engage switch	Engages dual-channel pitch stability augmentation. Solenoid held in ON position. Spring-loaded to OFF.
(2) ROLL STAB AUG engage switch	Engages duel-channel roll stability augmentation. Solenoid held in ON position. Spring-loaded to OFF.
(3) YAW STAB AUG engage switch	Engages three-channel yaw stability augmentation and RARI. Required to engage the autopilot.
(4) AUTOPILOT ENGAGE-OFF switch	ENGAGE - Engages autopilot. PITCH, ROLL, and YAW SAS must be engaged. No warmup required. Engages attitude hold. Requires weight off wheels. OFF - Disengages autopilot.
(5) HDG-OFF-GT switch	HDG - Autopilot will lock on constant aircraft heading when aircraft is less than ±5° roll. OFF - Disengages heading hold and ground track. GT - Selects autopilot ground tracking computed at time of engagement using INS data. Engaged by nosewheel steering pushbutton.
(6) ALT-OFF-MACH switch	ALT - Autopilot will maintain barometric altitude. Engaged by nosewheel steering pushbutton. OFF - Disengages altitude mode. MACH - Inoperative in all aircraft. In aircraft BUNO 159825 and subsequent and aircraft with AFC 488, the MACH position is deleted.
(7) VEC/PCD-OFF-ACL	VEC/PCD - Autopilot roll axis commands steer aircraft using data link signals for vectoring. If the PCD discrete is present both roll and pitch axis commands are used. Engaged by nosewheel steering pushbutton. OFF - Disengages VEC/PCD and ACL modes. ACL - Autopilot will accept data link signals for carrier landing, using spoilers for roll and parallel servo for pitch. Only pitch commands are transmitted to stick movement. Engaged by nosewheel steering pushbutton.
(8) ACLS/AP caution light	Autopilot and automatic carrier landing mode disengaged.
(9) A/P CPLR warning light	Indicates the autopilot is coupled to automatic carrier landing system.
(10) A/P REF warning light	Autopilot mode is selected but is not engaged. (Except attitude and heading hold.)
(11) PITCH STAB 1 and PITCH STAB 2 caution lights	Illuminated channel inoperative. Single channel 50% loss of authority.
(12) ROLL STAB 1 and ROLL STAB 2 caution lights	Illuminated channel inoperative. Single channel, 50% loss of authority. Indicates ROLL SAS failure.

Figure 1-64. AFCS Controls and Indicators (Sheet 2 of 3)

NAVAIR 01-F14AAA-1

Section I
Part 2

NOMENCLATURE	FUNCTION
⑬ YAW STAB OP and YAW STAB OUT caution lights	OP - One channel inoperative. OUT - Two channels inoperative. Yaw stabilization inoperative. Applicable to YAW SAS.
⑭ AUTO PILOT caution light	Indicates failure of one or more of pilot relief modes.
⑮ Autopilot reference and nosewheel steering pushbutton	Engages the ALT, GT, ACL or VEC/PCD autopilot mode selected. Autopilot must be engaged and compatible autopilot modes selected. Requires weight off wheels.
⑯ Autopilot emergency disengage paddle	Disengages all autopilot modes and DLC, if selected, and releases all autopilot switches. On aircraft BUNO 159858 and earlier without AFC 400, disengages pitch and roll SAS servos as long as switch is held. On aircraft 159859 and subsequent and aircraft with AFC 400, pitch and roll SAS servos remain disengaged when paddle is released. In aircraft BUNO 161157 and subsequent and aircraft incorporating IAFC 623/IAVC 2342, reverts throttle system from AUTO or BOOST mode to MANUAL mode only while depressed and with weight on wheels.

Figure 1-64. AFCS Controls and Indicators (Sheet 3 of 3)

AFCS to detect it and disengage the SAS. If the VMCU detects a low voltage condition of 97 volts or less in the ac, phase A supply, the unit will activate a relay to drop the voltage to zero, and maintain that level for a minimum of 1.25 seconds plus the duration of the detected low-voltage condition. Activation of the VMCU due to a low-voltage condition results in the simultaneous illumination of the following lights for the duration of the transient:

- PITCH STAB 1 and 2
- ROLL STAB 1 and 2
- YAW STAB OP and OUT
- SPOILERS
- HZ TAIL AUTH
- RUDDER AUTH
- AUTO PILOT
- MACH TRIM
- R INLET

While the VMCU is activated, all spoilers will be commanded to -4.5°, all SAS and AFCS functions will be turned off, and the PITCH and ROLL STAB AUG switches drop to OFF. When normal supply voltage is sensed, the VMCU is deactivated and normal power is restored to the phase A bus, all lights will go out, except the RUDDER AUTH light at low speeds and the HZ TAIL AUTH light at high speeds, and SAS will be available for the yaw axis since the YAW STAB AUG switch remains ON. SAS operation of the remaining axis will be available when the PITCH and ROLL STAB AUG switches are set to ON and the illuminated light(s) will reset with the MASTER RESET pushbutton.

The VMCU is checked prior to flight during the POST-START procedure. The voltage drop obtained when the emergency generator is turned off is sufficient to activate the VMCU and illuminate the affected caution and advisory lights.

Autopilot

The autopilot is controlled by four switches on the lower half of the AFCS control panel (figure 1-64) and the autopilot reference and nosewheel steering pushbutton on the stick grip. With all three SAS axes engaged, autopilot operation is commanded by placing the ENGAGE/OFF switch to the ENGAGE position. No warmup period is required. The autopilot may be engaged with the aircraft in any attitude. If, however, aircraft attitude exceeds ±30° in pitch and ±60° in roll, the autopilot will automatically return the aircraft to these limits. Normally IMU is the prime reference and AHRS a backup.

AFCS SERIES ACTUATOR

The AFCS series actuator is a dual-channel servo actuator that is controlled and commanded by the AFCS computers to provide a low authority input that can be mechanically overridden by the pilot. Each servo of the dual actuator is monitored to provide failure detection and automatic shutdown of a malfunctioning actuator channel. PITCH STAB and ROLL STAB caution lights notify the pilot of a malfunction. The remaining functional channel will continue to provide half authority for the pitch and roll axes. Autopilot modes may be engagable but will have reduced authority.

The YAW STAB OP caution light indicates one of the three yaw channels is inoperative. Full authority is retained.

Note

Taxiing with one engine shut down and the HYD TRANSFER PUMP off, may illuminate the pitch, roll, and/or yaw caution lights.

AFCS PARALLEL ACTUATOR

The AFCS parallel actuator is a single-channel electro-hydraulic servo actuator controlled and commanded by the pitch AFCS computer. It is used only in the longitudinal control system for automatic carrier landings. Since the actuator can command full surface authority, implementation is through a mechanical disconnect link that disconnects the actuator from the control system when a control stick force of approximately 25 pounds is applied. When the ACL mode is not engaged, the actuator is disconnected from the system by an internal hydraulically controlled clutch.

AUTOMATIC PITCH TRIM

Automatic pitch trim is used in all autopilot pitch modes to trim the aircraft in order to minimize pitch transients when disengaging autopilot functions. The pitch servo position is monitored to drive the aircraft pitch trim motor at one tenth manual trim rate. The pilot's manual trim button on the control stick is inoperative during all autopilot operation.

AFCS EMERGENCY DISENGAGE

Operation of the autopilot emergency disengage paddle on the control stick (figure 1-64), disengages the autopilot and DLC. All autopilot switches will return to OFF. The pitch and roll SAS servos are disengaged while the paddle is held depressed. On aircraft BUNO 159858 and prior not incorporating AFC 400, the pitch and roll SAS engage switches will remain ON. Release of the paddle will result in reengagement of the pitch and roll SAS servos if all monitors reflect normal operation. On aircraft BUNO 159859 and subsequent and aircraft with AFC 400, depressing the paddle disengages the pitch and roll SAS servos and causes the pitch and roll SAS switches to move to the OFF position. The yaw SAS channels are not affected by operation of the autopilot emergency disengage paddle, and will remain ON.

Note

In aircraft BUNO 159825 and subsequent and aircraft incorporating AFC 488, the AUTO PILOT light may or may not illuminate when the autopilot is disengaged with the autopilot emergency disengage paddle.

Pilot Relief and Guidance Modes

CONTROL STICK STEERING

With the autopilot engaged, the aircraft may be maneuvered using control stick steering. In control stick steering mode, the AFCS automatically synchronizes to the new attitude.

ATTITUDE HOLD

Attitude hold is selected by setting the AUTOPILOT ENGAGE switch to the ENGAGE position. To change attitude, use control stick sterring. Reengagement is achieved by releasing pressure on the stick. The autopilot will hold pitch attitudes up to $\pm 30°$ and bank angles up to $\pm 60°$. IMU failure will cause mode disagreement and the engage switch will return off. The mode may be reengaged using AHRS as a reference.

HEADING HOLD

Heading hold is engaged by setting the HDG-OFF-GT switch to the HDG position. After maneuvering the aircraft to the desired reference heading, release the control stick at a bank angle of less than $\pm 5°$. The autopilot will then hold the aircraft on the desired heading. Heading reference is obtained from the attitude heading and reference (AHRS) system via the CSDC.

GROUND TRACK

To engage ground track, set the HDG-OFF-GT switch to GT. When the A/P REF warning light (on left side of the vertical display indicator) illuminates, press the nosewheel steering pushbutton on the control stick grip. When the A/P REF warning light goes off, the mode is engaged.

Disengagement will occur if more than 1-1/2 pounds lateral stick force is applied and will be indicated by the A/P REF light. The ground track mode may be reengaged by releasing the stick force and pressing the nosewheel steering pushbutton.

Ground track steering computations are performed by the AWG-9 computer, based on inputs from the CSDC, IMU, and AHRS. The computer output, in the form of ground-track error signals, is processed in the CSDC which generates steering commands to the autopilot roll axis. Bank angles are limited to $\pm 30°$. Failure of INS or AHRS will cause loss of ground tracking steering.

CAUTION

When the AFCS is in ground track mode, the AWG-9 BIT sequence 2 must not be performed. Performing sequence 2 will cause the AFCS AUTO PILOT caution light to illuminate, which may cause the ground track mode to disengage. After BIT sequence 2 is completed, the AUTO PILOT caution light can be extinguished by cycling the HDG-OFF-GT switch from GT to OFF to GT. Ground track mode is then selected by depressing the autopilot reference and nosewheel steering pushbutton on the control stick.

ALTITUDE HOLD

Altitude hold mode is engaged by setting the ALT-OFF-MACH switch to ALT. When the A/P REF warning light illuminates, press the nosewheel steering pushbutton when at the desired altitude. This will engage the altitude hold mode and the A/P REF warning light will go off. Applying 10 pounds longitudinal stick force will cause the A/P REF warning light to illuminate. The mode may be reengaged by depressing the nosewheel steering pushbutton on the stick grip, when at the desired altitude, and observing that the A/P REF warning light goes off. Altitude hold should not be engaged during any maneuvers requiring large, rapid, pitch trim changes due to limited servo authority and slow auto trim rate. Disengagement of altitude hold is accomplished by applying 10 pounds or more longitudinal stick force, or by placing the ALT-OFF-MACH switch in the OFF position.

Note

Do not actuate inflight refueling probe with altitude hold engaged due to large transients in pitot-static systems sensed by the CADC.

DATA LINK VECTOR — PRECISION COURSE CORRECTION (PCD)

(Operational in aircraft BUNO 159825 and subsequent and aircraft incorporating AFC 488)

This mode is engaged by placing the VEC/PCD-OFF-ACL switch in the VEC/PCD position, and pressing the nosewheel sterring pushbutton. Mode engagement is evidenced by the A/P REF warning light going off. Disengagement of the mode is accomplished by application of stick forces of 7-1/2 pounds lateral or 10 pounds longitudinal, or disengagement by placing the VEC/PCD-OFF-ACL switch in the OFF position. If the switch is left in the VEC/PCD position, the A/P REF warning light will illuminate and the mode may be reengaged by depressing the Autopilot reference and nosewheel steering pushbutton.

Determination of whether data link or precision course direction signals are present is made in the AFCS pitch and roll computers in response to inputs from the data link converter and CSDC. If the data link vector discrete is present, the autopilot roll axis will respond to data link heading commands and bank angle authority will be limited to ±30°.

When the PCD discrete is present, the autopilot roll and pitch axes will respond to data link commands.

AUTOMATIC CARRIER LANDING (ACL)

(Aircraft BUNO 159825 and subsequent and aircraft incorporating AFC 488)

To engage the ACL mode, place the VEC/PCD-OFF-ACL switch in the ACL position and press the nosewheel steering pushbutton. Mode engagement is evidenced by the A/P REF light going out. Disengagement of the ACL mode causes complete autopilot disengagement. All autopilot switches return to OFF and the AFCS reverts to SAS modes only. The AUTO PILOT caution light and the ACLS/AP caution light on the windshield frame are illuminated. They may be extinguished by pressing the MASTER RESET pushbutton.

The pilot can disengage from the ACL mode (and autopilot) by application of 7-1/2 pounds of lateral or 10 punds of longitudinal force, by disengaging the AUTOPILOT ENGAGE switch, or by pressing the autopilot emergency disengage paddle.

Note

Routine downgrading from an ACL pass by pressing the autopilot emergency disengage paddle is not recommended since DLC will also be disengaged.

The ACLS mode (and autopilot) will be disengaged automatically by loss of data link A/P discrete, no message for more than 2 seconds or a wave off discrete from the boat.

ACL commands control the aircraft through the autopilot by spoiler commands in roll and by parallel servo in pitch. Pitch commands move the control stick.

CAUTION

If a pitch parallel actuator is disconnected, the A/P CPLR warning light will go out when ACL is selected on the AFCS control panel but the aircraft will not respond to SPN-42 commands.

AFCS Test

An independent test program and associated hardware is in each AFCS axis computer.

AFCS OBC

Preflight indication of AFCS performance can be obtained during post start OBC. All SAS switches should be engaged. After selecting OBC on the MASTER TEST switch, the autopilot switch should be placed to ENGAGE. A complete test sequence encompassing all functions specifically associated with the selected axis (axes) will commence upon test initiation and will cycle to OFF after completion. The PITCH STAB 1 and 2, ROLL STAB 1 and 2, YAW

STAB OP, and YAW STAB OUT caution lights illuminate and serve as an AFCS BIT running indication. All other AFCS caution lights will illuminate momentarily during BIT testing.

In aircraft BUNO 159825 and subsequent and aircraft incorporating AFC 488, large longitudinal control stick deflections should be observed during OBC if the pitch parallel actuator is functioning properly. An OBC with the flaps down requires a longitudinal trim of 3° or more noseup; an OBC with the flaps up requires not less than 0° noseup. A disconnected pitch parallel actuator during OBC is indicated by illumination of the AUTO PILOT caution light, a PC acronym, and the absence of large control stick deflections.

AFCS IN-FLIGHT BIT

AFCS in-flight BIT provides automatic failure initiated testing for the AFCS pitch and roll axes only. In-flight BIT is not provided for the yaw axis. If yaw axis failures occur, the appropriate caution light will illuminate and flight envelope restrictions must be observed as applicable. In-flight testing in the pitch and roll axes is automatically initiated when an AFCS computer failure is detected. When this failure is detected, the affected axes will disengage, both SAS caution lights for the affected areas will illuminate, and BIT will automatically run for 9 to 14 seconds, depending upon the axis affected. Upon completion of BIT, the failed channel light remains illuminated and the other channel light goes out. Weapon replaceable assembly (WRA) failure indications can be observed in flight by the RIO on the TID by selecting the SPL category on the computer address panel and depressing the OBC button.

LANDING GEAR SYSTEMS

The aircraft has fully retractable tricycle landing gear operated by combined hydraulic pressure in the normal mode of operation and a stored source of pressurized nitrogen for emergency extension. The landing gear retract forward so that airloads and gravity assist on emergency extension. Air-oil shock struts with oil metering pins reduce landing loads transmitted to the airframe, and the struts are fully extended with the gear in the wells. All landing gear doors remain open with the gear extended.

The landing gear handle mechanically positions the landing gear valve for normal operation. Pulling the handle mechanically selects emergency extension of the gear using the pneumatic backup source. Both modes of gear operation can be accomplished without electrical power except for the gear position indication which requires essential no. 2 bus dc power. Gear downlock actuators incorporate internal mechanical finger locks which maintain the downlock inserted position in the absence of hydraulic pressure.

Design limit landing sink speed for the aircraft is 25.3 feet per second (nominal landing sink speed is about 11 feet per second). Normal and emergency controls and displays associated with operation of the landing gear are shown in figure 1-65.

Main Landing Gear

Each main landing gear shock strut consists of an upper outer cylinder and a lower internal piston, which has a maximum stroke of 25 inches. A hard step (31,000 pounds required for further compression) in the strut air curve provides a consistent 4-inch stroke remaining in the ground static condition. A side brace link is mechanically extended from the inboard side of the strut outer cylinder to engage in a nacelle fitting and thus provides additional side load support for ground operations.

The path of the wheel assembly is controlled by the drag brace as it folds (jack-knives upwards) during gear retraction and unfolds during extension. The fully extended shock strut and jack-knifed drag brace retracts forward and rotates the wheel assembly 90° to lie flat in the wheel well. Inboard and outboard and aft main gear doors are individually actuated closed in sequence to provide fairing for the retracted gear. An uplock hook on the shock strut engages a roller in the wheel to hold the gear in the retracted position. The main landing gear actuator on the inboard side of the shock strut retracts and extends the gear assembly.

The gear downlock actuator, mounted at the drag brace knee pin, extends to prevent unlocking (jack-knifed) of the drag brace. Hydraulic pressure must be supplied to the downlock actuator in order to retract it against the spring action of the integral locking mechanism. A paint stripe across the drag brace knee pin provides an external visual indication of the drag brace locked condition. A ground lock device clamps onto the downlock actuator rod for safetying the main gear.

Maximum strut extension and wheel alignment are controlled by torque arms that incorporate cam-operated microswitches to detect a weight-on-wheels condition (greater than 5 inches of strut compression). The single split-type wheel assembly incorporates thermal fuse blow plugs, and a pressure relief device to prevent overinflation of the tire.

LANDING GEAR CONTROLS AND INDICATORS

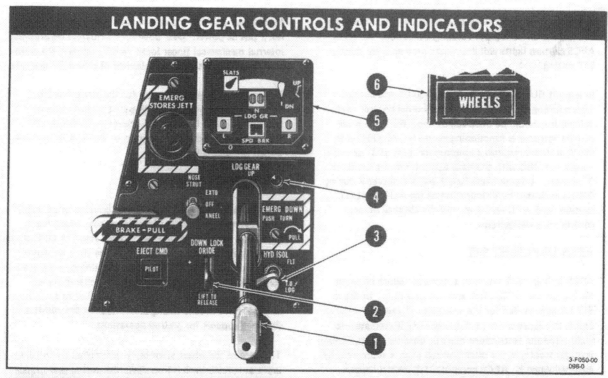

NOMENCLATURE		FUNCTION
① LDG GEAR handle	Normal -	Up and down overcenter action provides normal retraction and extension by the combined hydraulic system.
	Emergency -	Down-push-turn clockwise pull action provides emergency extension of all gear by a compressed nitrogen charge.
② DOWN LOCK ORIDE lever	Down -	Weight-on-wheels indication, prevents gear handle being retracted without pilot override (raising lever).
	Up -	Weight-off-wheels indication, does not inhibit pilot raising gear handle. Automatic operation by electrical solenoid.
③ HYD ISOL switch	FLT -	Combined system hydraulic pressure is shut off to the landing gear, nosewheel steering, wheel brakes, and tail hook operation.
	T.O./LDG -	Switch is automatically placed in this position with gear handle down. Combined hydraulic pressure is available to all components.
④ Landing gear transition light	On whenever gear and door positions (including main landing gear sidebrace actuators) do not correspond to handle position. Out when gear and doors are locked in position selected by handle.	
⑤ LDG GR indicator	-	Landing gear down and locked (except main landing gear sidebrace actuator).
	-	Unsafe gear or power off indication.
	UP -	Landing gear retracted and doors closed.
⑥ WHEELS warning light	Light flashes with flaps greater than 10° deflection and either or both throttles less than approximately 85% rpm, and all landing gear not down and locked. Approach lights and indexer will not illuminate unless all gear are down and locked.	

Figure 1-65. Landing Gear Controls and Indicators

> **CAUTION**
>
> Illumination of indexer lights does not indicate that the main landing gear are clear of the runway. Raising the gear before a positive rate of climb is established will result in blown main tires.

Nose Landing Gear

The dual wheel nose landing gear has a shock strut consisting of an outer cylinder and a lower internal piston, which has a maximum stroke of 18 inches. During normal ground operations, the strut is fully extended. Pilot control is provided to kneel the strut (4 inches stroke remaining) for catapult operations. During retraction the fully extended nose strut is rotated forward by the retract actuator into the well and enclosed by two forward and two aft doors. The forward doors are operated by a separate actuator which also engages the gear uplock whereas the two aft doors are mechanically linked to the shock strut. An uplock hook actuator engages a roller on the lower piston to hold the gear and doors in the retracted position. During extension, the telescoping drag brace compresses so that a downlock actuator mechanically locks the inner and outer barrel to form a rigid member for transmission of loads to the airframe.

> **Note**
> - In aircraft BUNO 160930 and prior without AFC 601, there is no foolproof visual check of the nose landing gear locked down status. Neither the downlock mechanism, which is concealed in the fuselage, nor insertion of the ground lock pin will provide a positive indication of gear locked status. Inflight, the pilot must normally rely on his indicator. In aircraft with AFC 655 incorporated, visual determination of nose landing gear unlocked status is assisted by paint stripes on the drag brace oleo. If white is visible between the red stripes, the nose gear is not locked.
> - In aircraft BUNO 161133 and subsequent and aircraft incorporating AFC 601, an additional sequencing switch in series with the existing down and locked switch provides the pilot with a positive indication of nose gear position.

Maximum strut extension and wheel steering angle are controlled by torque arms interconnecting the steering collar and the lower piston. The split-type wheel assembly incorporates tire pressure relief device to prevent overinflation of the tire. Additional hardware on the nose landing gear include the launch bar, holdback fitting, approach lights, nosewheel steering actuator, and taxi light. The wheel axles incorporate recessed holes for attachment of a universal tow bar with maximum steering angle of ±120°.

> **CAUTION**
>
> Restrict nosewheel deflection to ±90° to prevent structural damage to nose gear steering unit.

Landing Gear Normal Operation

The landing gear handle is mechanically connected to the landing gear valve that directs combined hydraulic fluid into the gear up and down lines and provides a path for return flow. In the down position, the handle mechanically sets the hydraulic isolation switch to provide hydraulic pressure for gear operation. The handle is electromechanically locked in the down position with weight-on-wheels to prevent inadvertent gear retraction. Pilot override of the solenoid operated handle lock can be effected by lifting the downlock lever next to the gear handle. Vertical movement of the gear handle causes a corresponding up and down selection of the landing gear with the combined hydraulic system pressurized. Three flip-flop indicators provide a position display for each of the landing gear and a gear transition light on the control panel illuminates any time the gear position and handle do not correspond. In addition, a WHEELS warning light alerts the pilot if the landing gear is not down with flaps deflected greater than 10° and either or both throttles set for less than approximately 85% rpm.

> **CAUTION**
>
> - Wait for gear to transition (15 seconds with normal hydraulic pressure) before recycling the handle or gear door damage may result.
>
> - Maximum landing gear tire speed is 190 KIAS.

Placement of the landing gear handle to the UP position actuates the landing gear valve which ports hydraulic pressure to the downlock actuators, gear retract actuators and in sequence to the door and uplock actuators. The gear shock strut and door uplocks are hydraulically operated into a mechanical overcenter position. An UP indication is displayed on the gear position indicators when the gear are in the uplock and all doors closed.

Placement of the LDG GEAR handle to the DN position actuates the gear control module to port hydraulic pressure to the door uplocks, door actuators and the strut uplocks. The landing gear are hydraulically extended and assisted by gravity and air loads. A gear down symbol (wheel) is displayed on the gear position indicators when the gear downlocks are in the locked position. The gear transition light will go out when the main gear side brace links are engaged.

Note

With the main gear downlock inserted but the side brace link not engaged, landing sink speed is restricted to 8 feet per second. Minimize yaw and sideslips on touchdown and rollout.

Emergency Gear Extension

Although emergency gear extension can be initiated with the landing gear control handle in any position, it is preferable that the LDG GEAR handle be placed in the DN position before actuating the emergency extension system.

CAUTION

The landing gear handle must be held in the fully extended emergency position for a minimum of 1 second to ensure complete actuation of the air release valve. Approximately 55 pounds pull force is required to fully actuate the emergency nitrogen bottle. The pulling motion should be rapid and continuous to ensure the air release valve goes completely overcenter to the locked position. The landing gear handle will be loose in its housing as an indication of complete extension of the handle. An incomplete handle motion could cause partial porting of gaseous fluid, initiating the emergency dump sequence. Interruption of handle motion without completing the overcentering action of the valve could cause the extending gears to contact and damage the strut doors.

The emergency landing gear nitrogen bottle is located in the nosewheel well. Normal preflight bottle pressure is 3,000 psi at 70°F. Minimum bottle pressure for accomplishing emergency extension to the down and locked position is 1,800 psi.

Pneumatic pressure is directed by separate lines to power open the gear door actuators in sequence, release the gear uplock actuators, pressurize the nose gear actuator to extend the gear (main gear free fall), and pressurize the downlock actuators. A normal gear down indication is achieved upon emergency gear extension. Following emergency gear extension, nosewheel steering is disabled. Once the landing gear is extended by emergency means, it cannot be retracted while airborne and must be reset by maintenance personnel.

Note

- Emergency extension of the landing gear shall be logged in the Yellow Sheet (OPNAV Form 3760-2).

- To facilitate in-flight refueling probe extension when the landing gear has been blown down, raise landing gear handle to give priority to the refueling probe system.

WHEEL BRAKE SYSTEM

The wheel brake system provides power boost hydraulic control of the multiple disk-type main wheel brakes using pressurized fluid in the landing gear down line from the combined hydraulic system. Individual or collective wheel brake control can be modulated by depression of the rudder toe pedals or collective, unmodulated brake control is available with the parking brake. An antiskid system is provided to operate electrohydraulically in conjunction with the normal wheel braking mode. Wheel brake controls are shown in figure 1-66.

Brake pedal and parking brake control motions are mechanically transmitted to the power brake module together with the anti-skid valve. Separate hydraulic lines transmit normal and emergency fluid pressure from the power brake module to the left and right wheel brake assemblies. At each brake assembly the normal and emergency lines input to the brake shuttle valve which applies brakes as a function of normal or emergency line fluid pressure. Two wear indicator pins on the brake piston housing measure lining wear for preflight inspection. For new brakes, these pins extend approximately 1/4 inch above the piston housing. When the pin is flush with the piston housing with the parking brake applied, the brake assembly is worn to the point of replacement.

Four thermal relief plugs are mounted in each main wheel assembly to relieve tire pressure and thus avert a blowout due to hot brakes if the local wheel temperature exceeds 350°F.

NAVAIR 01-F14AAA-1

Section I
Part 2

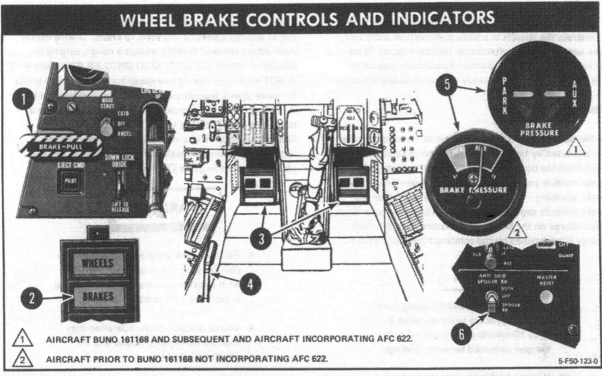

WHEEL BRAKE CONTROLS AND INDICATORS

⚠1 AIRCRAFT BUNO 161168 AND SUBSEQUENT AND AIRCRAFT INCORPORATING AFC 622.
⚠2 AIRCRAFT PRIOR TO BUNO 161168 NOT INCORPORATING AFC 622.

NOMENCLATURE	FUNCTION
① Parking brake handle	Forward - Parking brake released. Modulated braking action available with brake pedal depression. Aft - Parking brake set. No modulation of control, locks both main wheel brakes.
② BRAKES warning light	Indicates antiskid failure or failure of priority valve in the brake power module to switch to combined hydraulic side, or when the parking brake handle is pulled.
③ Brake Pedals	Press top of rudder pedals to command normal or auxiliary braking.
④ Hand pump	Recharges auxiliary and parking brake accumulators with gear handle down. With gear handle up, provides emergency extension of refueling probe.
⑤ BRAKE PRESSURE gage	Provides pilot indication of brake accumulator pressure remaining which is indicative of auxiliary and emergency brake cycles remaining.
⑥ ANTI SKID SPOILER BK switch	BOTH - Antiskid and spoiler brakes operative with weight on wheels and both throttles in IDLE. OFF - Antiskid and spoiler brakes inoperative. SPOILER BK - Spoiler brakes operate with both throttles IDLE. Antiskid is deactivated.

Figure 1-66. Wheel Brake Controls and Indicators

1-117

The capacities of the wheel brake assemblies are sufficient to restrain the aircraft in a static condition on a dry surface with zone no. 1 afterburner (nozzle position 2) set on both engines. The minimum hydroplaning speed for the main tires on a wet runway is approximately 90 knots.

Normal Braking

In the normal mode of operation wheel brake application is modulated by brake pedal depression using pressurized fluid from the combined hydraulic system through the brake module and through the normal brake line to the brake assembly. In the normal mode of operation the brake pressure gage indication should continue to indicate a full charge on the brake accumulators since this fluid energy is maintained by the combined hydraulic system.

CAUTION

- After heavy or repeated braking, such as during practice landings, allow a 5- to 10-minute cooling period with the gear extended between landings.

- If heavy braking is used during landing or taxiing followed by application of the parking brake, normal brake operation may not be available following release of the parking brake if the brakes are still hot. Check for normal brake operation after releasing the parking brake and prior to commencing taxiing.

Antiskid

The antiskid system operates electrohydraulically in conjunction with the normal mode of wheel brake operation to deliver maximum wheel brakes upon pilot command without causing a skid. Essential no. 2 bus dc power for antiskid operation is supplied through the ANTI SKID/R AICS LK UP PWR circuit breaker (7E1) and controlled by the ANTI SKID SPOILER BK switch (figure 1-66). When energized, approximately 200 milliseconds are required for antiskid system warmup. Individual wheel rotational velocity is sensed by skid detectors mounted in the wheel hubs and transmitted to the skid control box. The control box detects changes in wheel deceleration and reduces fluid pressure in the normal brake lines to both wheels simultaneously to prevent a skid.

With the antiskid system armed in flight, the touchdown circuit in the control box prevents braking until weight is on both main gear and the wheels have spun up, regardless of brake pedal application. The antiskid system is inoperative at ground speeds of less than 15 knots. During maximum effort antiskid braking, expect a rough, surging deceleration. When the ANTI SKID SPOILER BK switch is in BOTH position during low-speed taxi (less than 10 knots for more than a few seconds), subsequent acceleration of the aircraft through approximately 15 knots will cause a temporary loss of brakes lasting from 2 to 10 seconds. Should this happen, use of the brakes can be regained instantly by turning antiskid OFF. To preclude this possibility, antiskid must be OFF during taxi.

WARNING

- Failure of the weight-on-wheels switch results in continuous release signal with antiskid selected. Normal braking is available with antiskid off.

- Erratic antiskid brake operation may indicate overheated brakes and impending brake lock.

- If the antiskid system remains operative below 15 knots, place the ANTI SKID SPOILER BK switch in the OFF position otherwise the aircraft cannot be stopped using normal braking.

The antiskid system is inoperative when the wheel brakes are in the auxiliary or parking modes of operation since the emergency brake lines bypass the brake valve. If an electrical failure occurs in the antiskid system or if hydraulic pressure is withheld from either brake for greater than 1.2 seconds by the control box, the system automatically becomes inoperative and illuminates the BRAKES warning light with the ANTI SKID SPOILER BK switch in the BOTH position.

ANTISKID GROUND TEST

During ground operation, a self-test of the antiskid system can be initiated on the face of the control box with the system energized, parking brake handle released, and the aircraft in a ground static condition. Before taxiing (chocks in place), but after releasing the parking brake and while the pilot presses the toe pedal brakes, the plane captain should press the antiskid test pushbutton on the control box in the nosewheel well. Approximately 10 seconds is required for self-test, which checks the operational status of the control box, brake valve, and wheel sensors. Any discrepancies detected will be displayed by the BIT flags on the face of the control box (figure 1-67).

A valid BIT test requires that three criteria be met; the BIT flags on the face of the control box must check good, the pilot must feel both brakes release during BIT test, and the BRAKES warning light must not remain illuminated. A flash of the BRAKES light coinciding with brake pedal thumps during the antiskid BIT check is acceptable.

WARNING

Before initiating antiskid self-test by pressing the antiskid pushbutton on the control box, ensure that the aircraft chocks are in place. Initiation of antiskid self-test will release aircraft brakes.

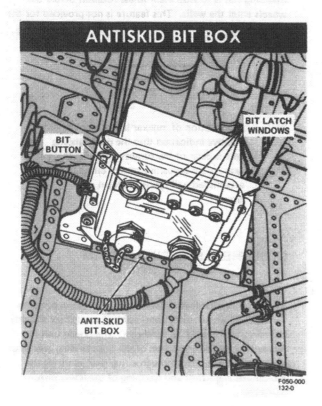

Figure 1-67. Antiskid BIT Box

Auxiliary Brake

Transfer of brake operation to the auxiliary mode is automatic without the requirement for pilot action upon the loss of combined hydraulic system pressure. Two brake accumulators provide pressure for auxiliary and parking modes of operation when combined hydraulic system pressure is not available. Accumulators deliver 3,000 psi when fully charged by the combined hydraulic system or hydraulic handpump (with gear handle down only). When the combined hydraulic system pressure decreases below 1,425 psi, shuttle valve in the power brake module shift the brake system to the auxiliary mode. With accumulator pressure greater than 1,150 psi, accumulator fluid pressure is directed through the emergency brake lines to the emergency side of the shuttle valves on the main landing gear with the wheel brake cylinders operating in response to brake pedal depression as in the normal mode.

In aircraft prior to BUNO 161168 without AFC 622, approximately 6 to 8 full dual brake pedal applications, are available in the auxiliary mode when starting with fully charged accumulators. A full charge in the accumulators is indicated on the front cockpit-mounted (center pedestal) brake cycles-remaining gage by full clockwise displacement of a single needle at the top of the green (AUX) band. Full capacity operation of the brake accumulators in the emergency modes of operation is predicated on the system serviced with a nitrogen precharge of 950±50 psi without combined hydraulic system pressurized.

In aircraft BUNO 161168 and subsequent and aircraft incorporating AFC 622, approximately 8 to 10 full dual brake applications are available in the auxiliary mode. A dual pneumatic brake pressure gage on the front cockpit-mounted (center pedestal) shows auxiliary and park accumulator pressures on individual dials. The green band of dial indicates pneumatic pressure between 3,000 psi at the top of the band to 2,150 psi; the red band (emergency mode) indicates pneumatic pressures between 2,150 psi and 1,900 psi at the bottom of the band. Approximately 2 brake applications are available in the emergency mode (red band).

When the brakes are cycled in the auxiliary mode with an attendent reduction in accumulator pressure, the AUX brake pressure gage needle moves from the green band to the red emergency band. Under these conditions braking action by the pedals is no longer available.

Additional braking can only be achieved by pulling the parking brake handle aft. If the shuttle valve in the power brake modules does not return to the normal position with combined hydraulic pressure greater than 2,000 psi, the BRAKES caution light will illuminate when a brake pedal is depressed. In this instance the wheel brake accumulators can be recharged only by the hydraulic handpump with the landing gear handle down. Pilot manual isolation or system automatic isolation of the combined hydraulic system cuts off the supply of combined hydraulic pressure to the power brake module so that depression of the brake pedals will cause depletion of the brakes accumulator charge.

The BRAKES warning light will illuminate whenever auxiliary brake pressure is applied to the brakes via the brake pedals, indicating the combined hydraulic system pressure is not available to the brakes and cautioning the pilot to monitor brake application with the auxiliary brake pressure indicator.

In aircraft BUNO 161416 and subsequent and aircraft incorporating AFC 636, a postlight is installed above the BRAKE PRESSURE gage to illuminate the dial.

Note

The postlight requires electrical power. Brake riders on carrier night respot must use a flashlight to check the cockpit BRAKE pressure gage.

Parking Brake

The parking brake mode provides a means for collective locking of the wheel brakes to maintain a ground static position during normal operations or during emergency conditions. Aft movement of the parking handle provides for unmodulated porting of accumulator fluid pressure through emergency lines to the shuttle valve at the wheel brake assembly. In the parking brake mode, the rudder pedals have no effect on wheel brake operation. Pushing the parking brake handle forward releases wheel brake pressure and the power brake module reverts to the normal and auxiliary braking mode. When auxiliary mode braking action is no longer available by depression of the rudder pedals, sufficient accumulator fluid pressure remains for a minimum of one parking brake application.

CAUTION

To avoid fusing the brake disks after heavy braking, do not use parking brake until the brake disks have cooled.

In the absence of a pressurized combined hydraulic system, the wheel brake accumulators can only be recharged by the pilot's hydraulic handpump with the landing gear handle in the down position.

Wheel Antirotation

During the initial phase of the landing gear retraction cycle, pressurized fluid from the gear up lines is directed to the power brake module to displace the normal and auxiliary metering valves to stop main wheel rotation before the wheels enter the wells. This feature is not provided for the nosewheels.

CAUTION

Illumination of indexer lights is not a positive indication that the main landing gear is clear of the runway. Raising the gear before a positive rate of climb is established will result in blown main tires.

NOSEWHEEL STEERING SYSTEM

The electrohydraulic nosewheel steering system provides for on deck aircraft directional control, nosewheel shimmy damping, and nosewheel centering. The power unit is located on the lower portion of the nose landing gear strut outer cylinder, which, through a ring gear, controls the directional alignment and damping of the lower piston assembly.

Combined hydraulic system pressure is the motive power used for steering and centering. Electrical power is supplied from the essential dc bus with circuit protection by the

NOSEWHEEL STEER/AFCS circuit breaker (RB2) on the pilot's right knee panel. Hydraulic pressure is derived from the gear-down line such that steering control is disabled subsequent to emergency extension of the landing gear.

Note

If nosewheel steering is inoperative, the emergency gear extension air release valve may be tripped, which will prevent gear retraction.

Nosewheel steering control during ground operations is energized by momentarily pressing the autopilot reference and nosewheel steering pushbutton on the lower forward side of pilot's stick grip. (See figure 1-68.) The system cannot be engaged without weight on wheels. The system will remain engaged until weight is off wheels, electrical power is interrupted or the pushbutton switch is pressed again. Engagement of nosewheel steering is indicated by illumination of the NWS ENGA caution light.

With the system engaged, nosewheel steering is controlled by rudder pedal position. Centering is unaffected by directional trim displacement. Maximum steering authority is 70° either side of neutral and the nosewheel can swivel a maximum of 120° about the centered position. With greater weight on the nosewheel (wings forward, high gross weight, etc) the steering torque can only turn nosewheel +5° with the aircraft static. However, only a slight forward movement will provide the pilot with full power steering authority. In a full pedal-deflection turn using nosewheel steering, the aircraft pivots about a point between the main gear such that the inboard main wheel rolls backward. Under this condition application of either main wheel brake will only serve to increase the radius of turn. Because of the outboard location of the engines, the application of thrust in tight turns should be made on the outboard engine to efficiently complement the turning movement of the nose gear.

Nosewheel Centering

The nosewheel is automatically centered during gear retraction before the nosewheel enters the wheel well. During gear retraction with weight off wheels, hydraulic pressure from the combined system bypasses the steering unit shut-off valve to center the nosewheel independent of rudder pedal movement. If the nosewheel is cocked beyond 15° either side of center after takeoff, the nosewheel is automatically prevented from retracting and the LAUNCH BAR advisory light illuminates.

During carrier arrestment the nosewheel is centered with weight on wheels and hook down when both throttles are retarded to IDLE to prevent castoring during rollback. After arrestment and roll back, the nosewheel will remain centered until nosewheel steering is engaged.

Shimmy Damping

Shimmy damping is provided in the steering actuator. Increased shimmy damping action is obtained with NWS disengaged.

CAUTION

If excessive nosewheel shimmy is encountered, disengage nosewheel steering.

NOSE GEAR CATAPULT SYSTEM

Catapult connection components on the nose landing gear shock strut piston provide nose gear catapult capability. A launch bar attached to the forward face of the nose gear steering collar guides the aircraft onto the catapult track and serves as the tow link which engages the catapult shuttle. A holdback fitting secures the holdback restraint prior to launch. The two position nose strut uses the stored energy catapult principle to impart a positive pitch rotational moment to the aircraft at shuttle release, thus providing for a hands off launch flyaway technique.

Nose Strut Kneel

Prior to catapult hookup, the nose strut is compressed 14 inches. Control of the nose strut kneel function is provided by the NOSE STRUT switch on the landing gear control panel. (See figure 1-69.) The three position (EXTD, OFF, KNEEL) toggle switch is spring loaded to return to the detented center OFF position. The position of the strut remains in the last commanded position independent of electrical or hydraulic power interruptions. In both cases, the transfer control valve source of electrical power is the essential no. 2 bus and combined hydraulic system fluid is used as the transfer medium. With external electrical power on the aircraft, the combined hydraulic

Section I
Part 2

NAVAIR 01-F14AAA-1

NOSEWHEEL STEERING CONTROLS

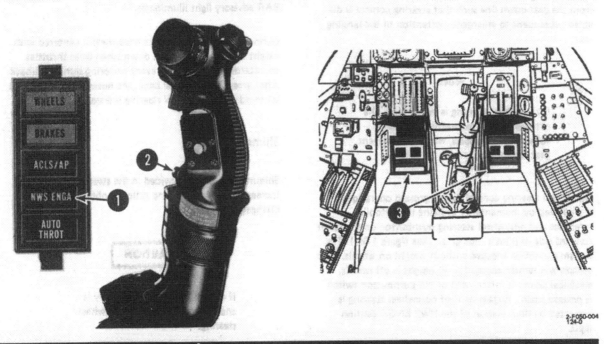

NOMENCLATURE	FUNCTION
① NWS ENGA caution light	Illumination when nosewheel steering engaged and will respond as a function of rudder pedal displacement. Nosewheel steering automatically centers with hook down. Nosewheel centering requires throttles at IDLE and weight-on-wheels with hook down.
② Autopilot reference and nosewheel steering pushbutton	Press to engage and disengage nosewheel steering. Requires weight-on-wheels.
③ Rudder pedals	Controls nosewheel steering position with system engaged.

Figure 1-68. Nosewheel Steering Controls

1-122

NAVAIR 01-F14AAA-1

Section I
Part 2

LAUNCH BAR CONTROLS

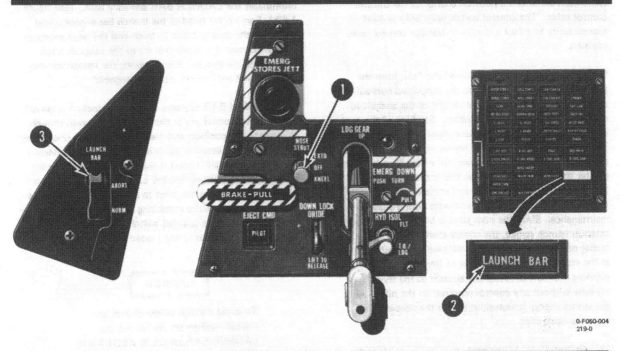

NOMENCLATURE		FUNCTION	
①	NOSE STRUT switch	EXTD	Hydraulic pressure causes strut to extend. Combined hydraulic system must be pressurized before switch is activated on external power. Launch bar is lifted into the up-lock position by torque arms as strut extends 14 inches.
		OFF	Spring-loaded return position.
		KNEEL	Nose strut transfer control valve releases pressure in the shock strut, which strokes 14 inches. Combined hydraulic system must be pressurized before switch is active on external power. Launch bar uplock can be released to allow bar to lower to deck by turning nosewheel ±10°.
②	LAUNCH BAR advisory light	Illuminates under the following conditions: Weight-on-wheels, aircraft kneeled, throttle less than MIL (goes out when throttles are advanced to MIL to provide lights out criterion for catapult launch). • Launch bar not up and locked (normal operation) Weight-off-wheels (inhibits nosegear retraction). • Launch bar not up and locked • Nosewheel not within ±15° of center	
③	LAUNCH BAR	ABORT	Enables pilot to disengage the launch bar from the catapult while remaining at MIL power and in the kneel position. Must be held in ABORT position.
		NORM	Spring-loaded to this position.

Figure 1-69. Launch Bar Controls

system must be pressurized (> 500 psi) before the control switch can command a position change of the transfer control valve. The control switch need only be held momentarily to effect a change in transfer control valve position.

Selection of KNEEL releases hydraulic fluid from the shock strut transfer cylinder to the combined hydraulic system return line, causing the weight of the aircraft to compress the shock strut 14 inches. Stroking of the nose strut causes the aircraft to rotate about the main wheels. The aircraft may be taxied or towed in the strut kneeled position except for the nuisance trip of the launch bar at greater than 10° steering angle; this is the position used for taxiing onto the catapult, and enhances accessibility to the forward fuselage compartments during ground maintenance. Since the nose strut is bottomed during the catapult launch stroke, the energy stored in the last 4 inches of strut piston stroke is released upon shuttle release at the end of the catapult stroke to impart a noseup pitching moment to rotate the aircraft to the fly-away attitude without any control required by the pilot. All the stored energy is expended before the nosewheels leave the deck edge.

Note

Under certain launch conditions (high wind over deck and light aircraft gross weights) the nose strut will not be fully compressed during the catapult stroke. Subsequent nose rotation following shuttle release will be at a less than normal rate. Aircraft launch bulletins for the aircraft are written to ensure that catapult launch pressures are sufficient to provide safe launch pitch rates and fly-away capability.

Full extension of the nose strut after launch and weight off wheels provide a redundant and automatic transfer of the control valve to the extend position. With weight off wheels, the NOSE STRUT switch is inoperative.

Launch Bar

The launch bar is attached to the nose gear and serves as the tow link for catapulting the aircraft. (See figure 1-70.) With the nose strut extended, the launch bar is held in the retracted position. The launch bar can be lowered by kneeling the aircraft and turning the nosewheel greater than ±10° from the centered position. The launch bar can also be lowered by the deck crew with no pilot action after the aircraft has been kneeled. A proximity sensing switch on the uplock detects the latch out of the locked position and illuminates the LAUNCH BAR advisory light. (See figure 1-69.) Ears on the head of the launch bar engage under the lip of the catapult lead in track and the head serves as a guide to steer the nosewheel on to the catapult track and engage the shuttle. For an abort, the launch bar cannot be raised until the shuttle is disengaged.

The LAUNCH BAR advisory light is interlocked to go off when both throttles are at the MIL position even though the launch bar position and mechanism remains unchanged; this action is effected to establish a "lights out" criterion for launch, the light circuit is disabled with nose gear up and locked. A pilot-controlled LAUNCH BAR switch is installed that enables the pilot to disengage the launch bar from the catapult while remaining at MIL power and in the kneel position. This guarded switch is on the pilot's left vertical console and is spring-loaded at the NORM.

CAUTION

To avoid damage to the launch bar retract mechanism, do not set the LAUNCH BAR switch to ABORT with the nosewheel deflected off center.

Note

The LAUNCH BAR switch is spring-loaded and must be held in the ABORT position until the catapult officer signals to lower the launch bar.

After the catapult launch stroke, extension of the strut mechanically cams the launch bar up to the retracted and locked position. If the launch bar is not engaged in the uplock with weight off wheels, the LAUNCH BAR advisory light will illuminate and nose gear retraction will be electrically inhibited.

Holdback Fitting

The holdback fitting is provided on the nose strut for insertion of the holdback bar. A ground crewman must manually attach the bar before the aircraft is taxied into the catapult lead track. The holdback bar is reusable and provides for repeated releases at a tow force of 76,000 pounds. Force greater than this on launch causes the holdback bar to release the aircraft holdback fitting.

NAVAIR 01-F14AAA-1

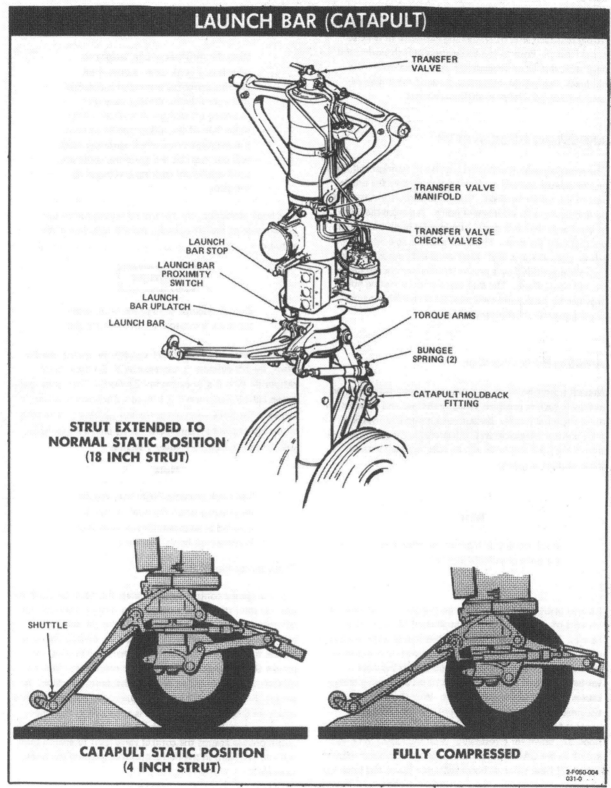

Figure 1-70. Launch Bar (Catapult)

High-power engine turnup operations can use the holdback fitting to attach aircraft restraining hardware to deck secured fittings. Prior to the application of high power, the nose strut should be kneeled and slack taken out of the holdback mechanism, otherwise dynamic loads may exceed mechanism design strength conditions.

ARRESTING HOOK SYSTEM

The arresting hook installation consists of a stinger tail hook and associated control mechanism mounted to the underside of the center fuselage. The hook shank is free to pivot up and down at its attachment point. A pneumatic dashpot preloads the hook down to minimize hook bounce on contact with the deck. The hook shank is free to pivot left or right within a ±26° sway angle with positive centering action provided by a pneumatic damper housed inside the tail hook shank. The trail angle of the arresting hook provides for hookpoint-deck contact even with the nose landing gear strut fully compressed.

Arresting Hook Operation

Normal operation of the arresting hook requires combined hydraulic system pressure, dashpot charged and dc essential no. 2 electrical power. Because of a redundant means of pilot control (electrical and mechanical), emergency extension of the arresting hook can be accomplished without these sources of power.

Note

Hook retraction requires electrical and combined hydraulic power.

Normal operation (figure 1-71) on the pilot's hook control consists of a straight down-up movement of the HOOK handle. This action actuates switches that provide electrical command signals to the hook control valve. For lowering the hook, the uplock is released and the lift cylinder is vented. Dashpot pressure of 800±10 psi assisted by gravity causes the arresting hook to lower. With the hook down, the pneumatically charged dashpot provides holddown contact force to minimize hook bounce and maintain the hook 37° down for arrestment. With the HOOK BYPASS switch in the CARRIER position, the approach and indexer lights will flash when airborne with gear down and hook up.

Note

With the throttles at idle, weight on wheels and hook down a nosewheel steering centering command is automatically provided to prevent nose gear castoring on arrestment rollback. Pilot deflection of the rudder pedals while in the automatic nosewheel centering mode will not override the centering command until nosewheel steering is engaged by the pilot.

For hook retraction, the control valve pressurizes the retract side of the lift cylinder and the lock side of the actuator.

CAUTION

Do not attempt to raise the hook when the hook is engaged in the arresting gear.

When the arresting hook roller engages the uplock mechanism, the lift cylinder is depressurized. On deck, hook retraction time is approximately 3 seconds. The hook transition light is illuminated as long as a discrepancy exists between the hook and cockpit handle positions. The transition light remains illuminated during on-deck extension, which requires approximately 1 second.

Note

The hook transition light may remain illuminated when the hook handle is lowered at airspeeds greater than 300 knots due to hook blowback.

Emergency Hook Extension

The emergency control system lowers the hook by mechanically (cable) tripping the uplock and venting the hook lift cylinder pressure. With the handle in the DN position, emergency hook extension is initiated by pulling the hook handle full aft (approximately 4 inches) and turning the handle 90° counterclockwise. After emergency hook extension, restow handle to avoid inadvertent retraction. In aircraft BUNO 160672 and subsequent and aircraft incorporating AFC 570, when the hook is extended in the emergency mode, a leaf spring wiper on the uplock release mechanism presses against the control valve lever to ensure that the valve remains in the dump position, keeping the hook extended.

NAVAIR 01-F14AAA-1

ARRESTING HOOK CONTROLS

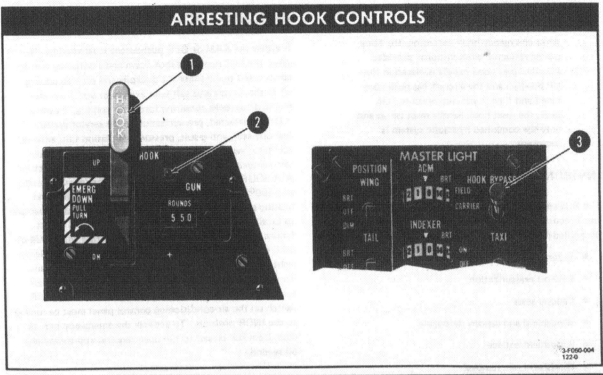

NOMENCLATURE	FUNCTION	
① Arresting HOOK handle	UP	Electrically energizes hydraulic retract actuator to raise hook into uplock.
	DN	Electrically releases hydraulic uplock actuator and allows hook to extend by dashpot pressure and gravity.
	EMERG	(Pull-twist) mechanically releases uplock actuator and allows hook to extend by gravity and dashpot pressure.
② Hook transition light	Illuminates whenever arresting hook position does not correspond with handle position. Light will not go out in down position until hook is in full trail angle.	
③ HOOK BYPASS switch	FIELD	Used for nonarrested landings. Bypasses the flashing feature of the approach lights and indexer when landing gear is down and hook retracted.
	CARRIER	Used for arrested landings. Approach lights and indexer flash when landing gear is down and the hook retracted.

Figure 1-71. Arresting Hook Controls

Note

After emergency hook extension, the hook can be retracted while airborne provided that the reset hook handle is placed in the UP position and the aircraft has both combined and flight hydrualic pressure. On deck, the reset hook handle must be up and only the combined hydraulic system is necessary.

ENVIRONMENTAL CONTROL SYSTEM (ECS)

The ECS regulates the environment of flightcrew and electronic equipment. The system provides temperature-controlled pressure-regulated air for the following systems:

- External drop tank pressurization
- Cockpit pressurization
- Canopy seals
- Windshield and canopy defogging
- Windshield anti-ice
- Windshield rain removal
- Anti-g suit inflation
- Pressure and exposure suit vent air
- Wing air bag seals
- Ammunition drum and gun gas purging
- Electronic equipment cooling and pressurization
- Temperature control of liquid coolant supplied to AN/AWG-9 control system, AN/AIM-54 Phoenix missile.

This system description also covers the chemical rain-repellent system, which is an environmentally related system not using conditioned air.

ECS Air Sources
BLEED AIR

The normal source of ECS air is 16th-stage bleed air from from both engines. Through a series of manifolds and valves, this air is cooled and mixed to reduce temperature and pressure to usable levels. The primary valves are the two engine bleed air shutoff valves, the dual pressure regulating and shutoff valve, and the turbine compressor modulating and shutoff valve, which are all controlled by the AIR SOURCE selector pushbuttons: L ENG, R ENG and BOTH ENG. (See figure 1-72.)

RAM AIR SOURCE

If either the RAM or OFF pushbutton is selected by the pilot, the ECS system is shut down and emergency ram air can be used to ventilate the cockpits and provide cooling air to the service and suit heat exchanger and those electronic subsystems requiring forced air cooling. However, if OFF is selected, pressurization to the service systems (canopy seal, anti-g-suit, pressure/ventilation suit, external fuel tank, wing air-bag seal) and 400°F air supply to the rain removal, defog, and heating systems is lost. Selecting AIR SOURCE RAM will provide air to the service systems and 400°F manifold air to the rain removal, defog, and heating systems. This action however, may result in damage to local structure, components, and wire should a duct failure exist. Interconnects inhibit gun firing with RAM or OFF selected. The emergency ram air door is on the lower right side of the fuselage, inboard of the right glove vane. To activate the ram air door, either the OFF or RAM air source pushbutton must be depressed and the RAM AIR switch on the air-conditioning control panel must be moved to the INCR position. To activate the emergency ram air door from full closed to full open requires approximately 50 seconds.

WARNING

Selection of the AIR SOURCE pushbutton to RAM with a failure of the ECS duct system will continue to circulate 400°F air throughout the system surrounding aircraft components and may cause a fire.

CAUTION

Before opening the ram air door, reduce airspeed to 350 KIAS or 1.5 Mach, whichever is lower, to prevent ram air temperatures above 110°F from entering the system. After ram air flow is stabilized, airspeed may be varied as required for crew comfort or to increase flow to electronic equipment.

For maximum cockpit ram airflow, the cockpit pressurization must be dumped. Pressing either L ENG, R ENG or BOTH ENG pushbuttons, automatically closes the ram air door if it is open.

AIR CONDITIONING AND PRESSURIZATION CONTROLS AND INDICATORS

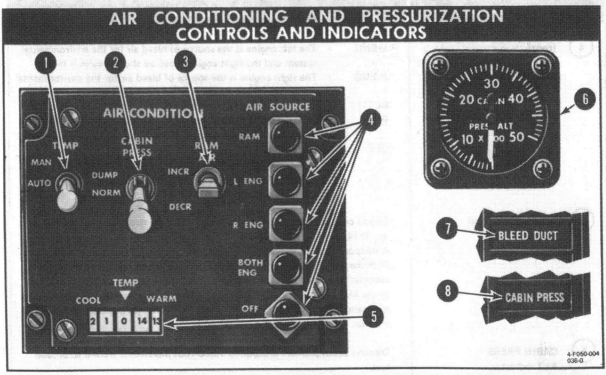

NOMENCLATURE		FUNCTION	
1	TEMP mode selector switch	AUTO	- Cockpit and pressure suit temperature is automatically maintained at that comfort level selected on the temperature control selector.
		MAN	- Cockpit temperature must be manually selected as airspeed and altitude change to maintain a desired temperature.
2	CABIN PRESS switch Lever-lock switch which must be lifted to be moved to DUMP.	NORM	- Cockpit pressure will be maintained at an altitude of 8000 feet up to 23,000 feet, above which the regulator maintains a 5-spi pressure differential.
		DUMP	- The cockpit safety valve is opened, depressurizing the cockpit.
3	RAM AIR switch	INCR DECR	- Manually modulates the ram air door and regulates the amount of ram air supplied to the cabin and electronics bay after the AIR SOURCE pushbutton is selected to RAM or OFF. (Approximately 50 seconds.)
4	AIR SOURCE selector pushbuttons	RAM	- Closes the turbine inlet, regulating and shutoff valve. Inhibits gun firing. The ram air control switch is enabled. Combined ram air and regulated 400°F bleed air are available to the cockpits and air cooled electronic equipment for temperature control.
			When either BOTH ENG, L ENG or R ENG are selected, the ram air door automatically closes.

Figure 1-72. Air Conditioning and Pressurization Controls and Indicators (Sheet 1 of 2)

NOMENCLATURE	FUNCTION
④ (cont)	L ENG - The left engine is the source of bleed air for the environmental system and the right engine bleed air shutoff valve is closed. R ENG - The right engine is the source of bleed air for the environmental system and the left engine bleed air shutoff valve is closed. BOTH ENG - The right and left engine bleed air shutoff valves are open and both supply bleed air to the environmental control system. This is the normal position. Automatically closes ram air door. OFF - Both the left and right engine bleed air shutoff valves and the dual pressure regulator valve are closed. Inhibits gun firing. Pressurization and air conditioning are not available. Enables the RAM air switch.
⑤ TEMP thumbwheel control	Selects cockpit and suit air temperature. It can be rotated through a 300° arc (0-14) with mechanical stops at each end placarded COOL and WARM. A midposition temperature (7) is approximately 70°F in the automatic mode. With the TEMP mode selector switch in AUTO the temperature selected is automatically maintained by the modulating temperature control valves. In the MAN position, the TEMP control thumbwheel must be repositioned to maintain cockpit and suit air temperature with changes in airspeed and altitude.
⑥ CABIN PRESS ALT indicator	Displays cabin pressure altitude in 1,000-foot increments from 0 to 50,000 feet.
⑦ BLEED DUCT caution light	Indicates overheating along the high temperature (1,000°F maximum) bleed air duct routing forward of the engine fire wall up to the primary heat exchange and the 400°F modulating valve.
⑧ CABIN PRESS caution light (RIO's cockpit)	Indicates cabin pressure has dropped below 5-psi pressure differential or cockpit altitude is above 27,000 feet.

Figure 1-72. Air Conditioning and Pressurization Controls and Indicators (Sheet 2 of 2)

EXTERNAL AIR

The adapter for connecting a ground air-conditioning unit is under the fuselage, aft of the nosewheel well. An additional provision for connecting an external source of servo air is in this same area.

Because the standard ground air-conditioning unit is not capable of cooling the cockpits, avionics, and AWG-9/AIM-54 simultaneously, a deverter valve is installed just downstream of the external air inlet. Cockpit control of the diverter valve is provided by the GND CLG switch on the flight officer's right console. Either OBC/CABIN or AWG-9/AIM-54 switch position may be selected.

If the OBC/CABIN position is selected, ground cooling air will go to the cockpit, all the forced air-cooled avionics, and a controlled amount of air will be available to cool the AWG-9 liquid loop at a reduced heat load. If AWG-9/AIM-54 position is selected, a ground cooling air will go to the AWG-9/AIM-54 liquid cooling loop and all essential forced-air-cooled avionics.

External electrical power is automatically inhibited from BDIG, MDIG, APX-76, CADC, and the CDSC if external air conditioning is not connected to the aircraft.

Pressure switches on both sides of the ground air diverter valve (page FO-15) interrupt electrical power to specific forced air cooled equipment, depending upon position selected by the GND CLG switch or shutoff electrical power if air pressure is insufficient.

Note

> The equipment not cooled by ground cooling air when the GND CLG switch is in the AWG-9/AIM-54 position are AN/ALQ-100, MDIG, CADC, VDIG, AN/APX-76 and in aircraft BUNO 159895 and earlier, TACAN.

Cockpit Air Conditioning

ECS manifolding consists of:

- The high-temperature (bleed air) manifold
- The 400° manifold
- The cold-air manifold

High-temperature engine bleed air is routed through the primary heat exchanger. The cooled output of this heat exchanger is split and a portion is mixed with hot, engine bleed air to a temperature of approximately 400°F; the remainder is further cooled by the turbine compressor. Here the air is compressed, run through the secondary heat exchanger, and then expanded in the turbine section, resulting in cold air that is mixed with 400° air to obtain any temperature desired. The primary and secondary heat exchangers are between the left and right engine inlets and the fuselage. At speeds above 0.25 Mach, ram air across the heat exchangers is used for cooling. During ground operations and at airspeeds less than 0.25 Mach, airflow across the heat exchanger is augmented by air-powered turbine fans.

Note

> Ground operation of the aircraft with either one or both engines in idle power reduces ECS cooling capacity. This can result in the illumination of the RIO's COOLING AIR and/or AWG-9 COND advisory light. If the cooling capacity is inadequate at idle power, the throttle should be advanced to turn off the lights. It should not be necessary to advance the throttle beyond the point at which the engine exhaust nozzle closes (monitor exhaust nozzle position indicator). To increase airflow to forced air cooled equipment, place CANOPY DEFOG - CABIN AIR control lever in CABIN DEFOG position.

The third heat exchanger is the service-air-to-air heat exchanger. This normally uses cold air from the cold air manifold as a heat sink but can use emergency ram air if the cold air manifold is not operating. Air from the service heat exchanger is used by exposure suit, g-suit, canopy seal, and for pressurization of radar wave guide, antenna, transmitter, WCS and I missile liquid cooling loop tanks and TV sight unit.

TEMPERATURE MANAGEMENT

The pilot can control cockpit temperature by selecting either a manual (MAN) mode or automatic (AUTO) mode with the TEMP mode selector switch. (See figure 1-69.) In the AUTO mode, temperature (60°F to 80°F) is selected by the pilot with the TEMP thumbwheel control. This desired temperature is maintained by a cabin temperature sensor in the forward left side of the cockpit. In the MAN mode, the TEMP thumbwheel control manually positions the cockpit hot-air-modulating valve to maintain cockpit and suit air temperatures. If cockpit inlet airflow temperature (in either AUTO or MAN) exceeds 250°F, a cockpit overtemperature switch closes the hot-air-modulating valve.

The conditioned air entering the cockpit is divided forward and aft, with 50% of the air going to each cockpit. A CANOPY DEFOG-CABIN AIR lever on the right console in each cockpit individually controls the percentage of airflow through the cockpit diffusers and the canopy defog nozzles. When the lever is in the CABIN AIR position (full aft), 70% of the air is directed through the cockpit diffusers and 30% through the canopy defog nozzles. In the CANOPY DEFOG position, 100% of the air is directed through the canopy defog nozzles.

PRESSURE SUIT AND VENTILATION AIR

The pressure suit and ventilation air system provides temperature controlled, pressure regulated engine bleed air to both pressure suits or to the ventilated seat cushions when pressure suits are not worn. The pilot can select either manual or automatic temperature control mode and regulate a desired suit or seat cushion temperature. Each crewman can individually control the volume of air flow through the pressure suit or seat cushion by the VENT AIRFLOW thumbwheel on his left inboard console.

ANTI-g SUIT

Each anti-g suit is connected to the aircraft pressurization system by an anti-g suit hose, which delivers pressurized air to the suit control valve and then to the suit through a composite disconnect. Below 1.5 g, the suit remains deflated. A spring-balanced anti-g valve automatically opens when g forces exceed 1.5 g. Operation of the anti-g suit valve may be checked by depressing the test button marked "G" VALVE on each crewman's left console.

Electronic Equipment Cooling

Ambient cooled electronic equipment in the electronic bays is cooled by the air exhausted from the cockpits. Equipment incapable of being cooled by free convection is cooled from the cold air manifold.

A schematic of the AWG-9 and AIM-54 and electronic equipment is shown on page FO-16. Controls and lights are shown in figure 1-73.

LIQUID COOLING

AWG-9 equipment is cooled by liquid coolant (page FO-16). The heat is rejected in the ram air heat exchanger. This is accomplished by circulating coolant fluid through the electronics and ram air heat exchanger and/or the AWG-9 and AIM-54 heat exchanger. The cooling loop is also used for automatic warmup of the AWG-9 using bleed air.

The AWG-9 liquid cooling loop incorporates a separate ram-air-liquid heat exchanger. A ram-air door is located under the right glove, forward of the primary heat exchanger inlet. There are no cockpit controls for this ram-air door. It is controlled by the AWG-9/AIM-54 controller and is independent of the air conditioning and pressurization system. The AWG-9 system ram-air heat exchanger automatically maintains the liquid temperature within operating limits.

WARNING

A potential fire hazard is present in the event of an AWG-9 coolant pump failure that can lead to the pump spraying flammable fluid into the ECS compartment. A PM acronym will appear on the TID in the AWG-9 CM buffer when the AWG-9 coolant pressure, flow, or temperature sensors from flycatcher 7-00166 are activated. This PM acronym should alert the flight crew to the need for immediate action to secure the AWG-9 coolant pump.

Figure 1-73 shows the controls and lights associated with the AWG-9 and AIM-54 cooling loop. The AWG-9 cooling loop is activated by the LIQ COOLING switch on the RIO left outboard console. In the AWG-9 and AIM-54 switch position, both the AWG-9 radar and AIM-54 missile cooling loops are activated for airborne operation. If the AWG-9 position is selected, only the radar cooling pump is activated. A temperature sensor in the AWG-9 and AIM-54 heat exchanger outlet lights the AWG-9 COND advisory light when the liquid temperature goes above 104°F. In addition, a pressure switch in the AWG-9 pump illuminates the AWG-9 COND advisory light when pump output pressure is too low.

In aircraft BUNO 160887 and subsequent and aircraft incorporating AFC 580, an overtemperature switch is added to the AWG-9 coolant pump. When the coolant

NAVAIR 01-F14AAA-1

Section I
Part 2

AWG-9 AND AIM-54 LIQUID COOLING CONTROLS AND LIGHTS

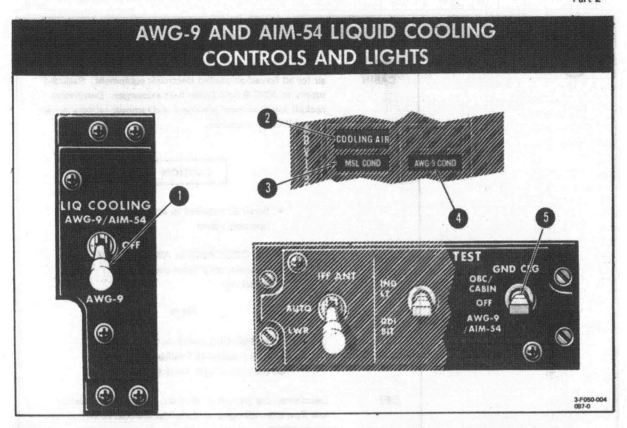

NOMENCLATURE	FUNCTION
① LIQ COOLING switch	AWG-9/AIM-54 - Activates the AWG-9 and AIM-54 cooling pump motors. OFF - Deactivates the AWG-9 and AIM-54 cooling pumps. AWG-9 - Activates the AWG-9 cooling pump for ground and airborne thermal conditioning.
② COOLING AIR advisory light	Illuminates after a delay of 25 to 40 seconds when an overtemperature condition exists in the electronic forced air cooling system.
③ MSL COND advisory light	Illuminates when temperature of the coolant flow to the missiles is above $115° \pm 3°F$; also indicates pump output pressure is below 60 ± 5 psi. With WCS switch in STBY or XMIT and Phoenix fairing installed, light will illuminate when LIQ COOLING switch is in the OFF or AWG-9 position. In aircraft BUNO 160917 and subsequent and aircraft incorporating AFC 597 and AAC 733, the light illuminates when the overtemperature switch shuts down the AIM-54 coolant pump.
④ AWG-9 COND advisory light	Illuminates when coolant exiting the heat exchanger reaches $104°F$ or pump output pressure is below 60 psi. In aircraft BUNO 160887 and subsequent and aircraft incorporating AFC 580, the light illuminates when the overtemperature switch shuts down the AWG-9 coolant pump.

Figure 1-73. AWG-9 and AIM-54 Liquid Cooling Controls and Lights (Sheet 1 of 2)

NOMENCLATURE	FUNCTION	
⑤ GND CLG switch	OBC/CABIN	- Positions ground cooling diverter to allow external cooling air for all forced-air-cooled electronic equipment. Reduced supply to AWG-9 liquid loop heat exchanger. Deactivates cockpit low-flow-override signal and removes primary power from AWG-9 transmitter.

CAUTION

- Servo air required to actuate servo operated valves.
- Use OBC/CABIN or AWG-9/AIM-54 positions only when engines are shutdown.

Note

With GND CLG switch in OBC/CABIN AWG-9 position BIT failures will occur on the ground with weight on wheels.

	OFF	- Deactivates the ground air diverter. Activates the cockpit low flow override signal and shall be selected when engines are operating.
	AWG-9/AIM-54	- The ground cooling diverter is positioned to allow external cooling air for the AWG-9 and AIM-54 liquid cooling system and all essential forced-air-cooled equipment. Deactivates the cockpit low-flow override signal.

Figure 1-73. AWG-9 and AIM-54 Liquid Cooling Controls and Lights (Sheet 2 of 2)

temperature rises to 230°±5°F; the thermal switch closes, shutting down the pump to prevent pump failure, and illuminates the AWG-9 COND advisory light.

> **WARNING**
>
> Failure to turn off the AWG-9 or AIM-54 coolant pumps after illumination of an AWG-9 COND light, or failure to respond properly to other indications of an overheating AWG-9 coolant pump could lead to a failure of the pump causing the flammable coolant fluid to spray into the compartment with subsequent potential for fire. Other indications of an overheating pump may be automatic shutdown of the AWG-9 with concurrent XM acronym, AWG-9 failure due to low coolant pressure, coolant temperature high, or low coolant flow with an associate PM acronym on the TID and/or popping of the AWG-9 coolant pump three phase circuit breakers.

During ground operation, with electrical power, external air conditioning and servo air available to the aircraft and the GND CLG switch in the AWG-9 and AIM-54 position, the ground cooling diverter valve directs the external conditioned air to the AWG-9 and AIM-54 systems and to essential forced-air-cooled avionics. When the OBC/CABIN position is selected, external conditioned air is directed to the cockpit, and all forced air cooled avionics, and a limited amount of air is available to cool the AWG-9 liquid loop. With the GND CLG switch in the OBC/CABIN position, the AWG-9 radar transmitter is inhibited from ground operation with engines running on the ground, select the OFF position of the ground cooling switch.

COCKPIT AIR PRIORITY FUNCTION

The cockpit air priority function is operational during all engine-on operations. (See FO-15.) It provides the cockpit with priority over the AWG-9 liquid cooling loop in the event there is a shortage of conditioned air. On engine power, the GND CLG switch (figure 1-73) should always be in the OFF position and the canopy locked to enable the cockpit air priority function.

Up to 8,000 feet, a small positive differential pressure (inside to outside) is maintained within the cockpit to provide a controllable back pressure and, thereby, limit flow. If the flow limits cannot be met because of excessive leakage of air (causing excessive flow) from the cockpit, the cockpit priority system progressively closes the cold air valve to the AWG-9 liquid heat exchanger to make more air available to the cockpit.

There is no indication to the flightcrew that the cockpit priority action is taking place unless it progresses to the point that an AWG-9 COND advisory light illuminates. Even then, it is only one of several problems that could have triggered the light.

AIM-54 MISSILE COOLING

The AIM-54 cooling (page FO-16) is required to dissipate heat generated in the AIM-54 missile electronic systems. This is accomplished by circulating 18 gallons of coolant fluid per minute through one to six missiles (3 gal/min per missile) at a controlled inlet fluid temperature of 70°F to the missiles. Heat picked up in the missile is rejected to ECS air in the missile AWG-9 and AIM-54 heat exchanger. The coolant fluid is also used for missile warmup.

If the temperature of the coolant delivered to the missiles goes above 115°F, the MSL COND advisory light illuminates. A pressure switch in the AIM-54 cooling loop pump will illuminate the MSL COND advisory light when pump output pressure is low.

In aircraft BUNO 160917 and subsequent and aircraft incorporating AFC 597 and AAC 733, and overtemperature switch is added to the AIM-54 coolant pump. When the coolant temperature rises to 230°±5°F, the thermal switch closes, shutting down the pump to prevent pump failure, and illuminating the MSL COND advisory light.

Pressurization

COCKPIT PRESSURIZATION

From sea level to 8,000 feet altitude the cockpit is unpressurized. Between altitudes of 8,000 feet to 23,000 feet the system maintains a constant cockpit pressure altitude of 8,000 feet. At altitudes above 23,000 feet, the cockpit pressure regulator maintains a constant 5-psi pressure differential greater than ambient pressures. An illustration of the cabin pressure schedule is shown in figure 1-74.

A cockpit pressure altimeter (figure 1-71) is provided for the pilot. In the rear cockpit, the RIO has a low-pressure caution light on the CAUTION and ADVISORY light panel. This low-pressure caution light placarded

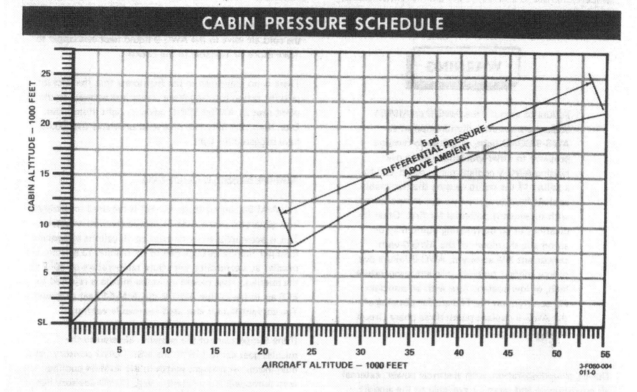

Figure 1-74. Cabin Pressure Schedule

CABIN PRESS, illuminates when cockpit pressure drops below 5 psi absolute pressure or cockpit altitude is above 27,000 feet.

If the cockpit pressure regulator malfunctions, the cockpit safety valve will open to prevent a cockpit pressure differential from exceeding a positive 5.5-psi or a negative differential of 0.25 psi. The cockpit pressure regulator and the safety valve are pneumatically operated and function independently through separate pressure sensing lines.

Cockpit pressurization can be dumped by the pilot by selecting the DUMP position with the CABIN PRESS switch. When the DUMP position is selected the safety valve is immediately opened and the cockpit is depressurized.

CANOPY SEAL PRESSURIZATION

Pressurized air from the air-conditioning system is ducted through the cockpit to the canopy seal. The seal is automatically inflated when the canopy actuator is moved to the closed position. A check valve in the canopy pressure regulating valve prevents the loss of canopy seal pressurization if the conditioned air manifold is depressurized. Initial movement of the canopy actuator automatically deflates the seal.

Windshield Rain Removal and Anti-Ice

Compressor bleed air at approximately 390°F and at a pressure of 22 psi is directed over the outside of the windshield through a fixed area nozzle. This blast of hot air over the windshield will evaporate rain and ice and prevent its further accumulation. The system is activated by selecting the AIR (center) position with the WINDSHIELD lock lever switch on the EXT ENVIRONMENT panel (figure 1-75). A temperature overheat sensor at the base of the windshield protects the windshield from overheating. When the sensor detects overheating (300°F), a signal closes the pressure regulating valve and illuminates the WSHLD HOT advisory light on the pilot's CAUTION ADVISORY light panel.

NAVAIR 01-F14AAA-1

Section I
Part 2

WINDSHIELD RAIN REMOVAL AND DEFOG CONTROLS AND INDICATOR

NOMENCLATURE		FUNCTION
① WINDSHIELD rain removal and repel switch	RAIN REPEL	Momentary switch position. Lock lever switch must be pulled out and moved up to select rain repel. Provides a chemical spray and a blast of hot air over the exterior windshield. Should be used only during heavy rain. Currently not used.
	AIR	Provides a continuous blast of hot air (390°F) over the exterior windshield. Used for windshield anti-ice. Switch will spring return to this position from RAIN REPEL.
	OFF	Closes the shutoff valve after a 5-second delay. The system is deenergized.
② WSHD HOT advisory light		Light illuminates when windshield overheats (300°F).
③ CANOPY DEFOG-CABIN AIR lever (both cockpits)	CABIN AIR	70% of the conditioned air directed through the cockpit air diffusers and 30% is through the canopy air diffusers. This is normal position.
	CANOPY DEFOG	Air flow is directed through the canopy air diffusers only.
④ WSHLD DEFOG switch In Aircraft BUNO 160696 and earlier aircraft only.	MAX	Increases the flow of 400°F air mixed with cabin air, providing maximum defog capability.
	NORM	400°F air is added to cabin air for better defogging.
	MIN	Cabin air is blown on to the windscreen. Normal flight position of switch.

Figure 1-75. Windshield Rain Removal and Defog Controls and Indicator

Gun-Gas Purging

Gun-gas purging uses ECS turbine discharge air to ventilate the ammunition drum breech, ammunition and gun barrels. Immediately before gun firing, the gun control unit electrically opens the modulating valve, allowing conditioned air to enter the purge systems. A hydraulically operated gun-gas exhaust door at the bottom of the gun drum compartment is opened to expel the gun gas overboard. To ensure complete ventilation and purging of the gun breech, the cool air shutoff valve and the exhaust door remain open for approximately 35 seconds after the gun has ceased firing.

Degraded ECS Operation

There are various temperature and pressure safeguard systems that cause the ECS system to shut down if an unsafe situation is detected. A complete failure of the dual valve will cause it to shut down the pressurization and air-conditioning system. Should that fail to close, a pressure sensor will close both individual engine bleed air shutoff valves if an overpressure situation exists in the downstream duct (155 psi). A shutdown of the bleed air supply duct either automatically or pilot selected AIR SOURCE OFF pushbutton, will cause total ECS air shutdown.

CAUTION

Failure of the weight-on-wheels switch to the in-flight mode while on the deck will cause the loss of ECS engine compartment air ejector pumps, causing a subsequent failure of either or both generators prior to CSD thermal disconnect.

Note

- After an automatic shutdown of the system, the pilot should select either OFF or RAM AIR SOURCE to enable the emergency ram-air door and then hold RAM AIR switch to INCR for approximately 50 seconds to provide ram-air cooling to electronic equipment and to the cabin.

- Loss of electrical power with bleed air still operating will result in smoke entering the cockpit through the ECS system when the aircraft is on the deck. In flight, only cold air will be supplied to the cabin and suit. Icing of the water separator may occur, causing reduced flow to the cabin. Since the ECS panel is dependent on electrical power, selector pushbuttons will be inoperative.

If the 400° manifold reaches 475°, a 400° shutoff valve closes, stopping the flow of unconditioned engine bleed air to the 400° manifold.

If either compressor inlet or turbine inlet temperature becomes excessive, the regrigration unit will shutdown. Cockpit indications will be as follows:

- No cockpit airflow
- RIO's COOLING AIR advisory light illuminated
- RIO's AWG-9 COND advisory light illuminated
- RIO's MSL COND AIM-54 (aboard) advisory light illuminated.
- With COOLING AIR advisory light illuminated on deck, advancing throttles to approximately 75% may extinguish light. Light may be caused by under-servicing or excessive temperatures on deck. If light remains illuminated, suspect an ECS malfunction.
- Extended flight with AIR SOURCE OFF could cause an overheating condition of the CSDC and a subsequent loss of primary attitude and navigational indications, that is, VDI, HUD, NAV aides.

The pilot should press the AIR SOURCE RAM pushbutton and set the RAM AIR switch to INCR to open the ram-air door to provide forced-air cooling to the electronic equipment and to the cabin.

ECS duct failures may be indicated by diminishing cabin cooling airflow and/or cabin pressurization with or without COOLING AIR advisory light illumination. Duct failures may additionally be indicated by pressurization loss to the service systems, and air flow loss to rain removal, defog, and heating systems. This cannot be verified if the system is not in use. Selection of AIR SOURCE OFF and ram air increase is appropriate when any indication of duct failure exists. ECS malfunctions that are not caused by duct failure are usually indicated by loss of temperature control without a cabin or system air flow/pressurization degradation. The 400°F modulating valve will not cause illumination of the COOLING AIR light. Any duct failure in this area associated with the COOLING AIR light is strictly coincidental. However, the duct failure between the primary heat exchanger and the turbine compressor assembly (TCA), or between the secondary heat exchanger and the TCA, could cause degraded cooling air flow and a COOLING AIR light to illuminate.

Actuation of the overtemperature switch results in cycling of the 400° valve. During this period the heating capacity of the 400° manifold would be degraded.

OXYGEN SYSTEM

Breathing oxygen is provided each crewman by converting liquid oxygen to gaseous oxygen within either one or two 10-liter liquid oxygen converters. Emergency oxygen is available to both crewmen through a high-pressure gaseous oxygen bottle in each ejection seat pan.

Normal Oxygen System

The normal oxygen system provides crewmen with 100% pressure-regulated, temperature-controlled oxygen. The system is designed for use with a pressure-demand type regulator that may be mask-mounted or chest-mounted. It is also compatible with the full-pressure suit oxygen regulator. Normally, one 10-liter liquid oxygen converter is installed. Provisions are made for installation of a second converter, if required. The liquid oxygen converters are in the right side of the fuselage adjacent to and beneath the forward cockpit. An access door in the side of the fuselage is provided for servicing the system. Oxygen quantity duration chart is shown in figure 1-76.

An OXYGEN ON-OFF switch is located aft on the left console in each cockpit. In the ON position, oxygen is free to flow from the converter through the regulator to the mask. In the OFF position, a dual shutoff and relief valve at the control panel closes to shut off flow to the regulator.

OXYGEN QUANTITY INDICATOR

An oxygen quantity indicator is on the right side of the pilot's right knee panel. The indicator is calibrated in liters and indicates quantity remaining in increments of one liter. The indicator operates on 115V power from essential bus no. 2. If electrical power is lost, the indicator will indicate less than zero and the OFF flag on the lower face of the indicator will appear. The liquid oxygen indicator is tested through the built-in-test (BIT) system. When the MASTER TEST switch on the pilot's MASTER TEST panel is set to INST and depressed, the indicator pointer should read 2 liters, and the pilot's and RIO's OXY LOW caution lights illuminate.

OXYGEN CAUTION LIGHT

An OXY LOW caution light is on the pilot's and RIO's CAUTION and ADVISORY panels, forward on the right consoles. The amber light illuminates when oxygen remaining is less than 2 liters or pressure is low. When the light illuminates, a cross check with the oxygen quantity indicator will disclose whether the oxygen quantity is low or pressure is low. The light is powered by the essential dc bus no. 2 through the OXY/BINGO CAUTION circuit breaker (7F6).

Emergency Oxygen System

The oxygen bottle in the survival kit of each ejection seat provides limited supply of gaseous oxygen for use in ejection and in the event of failure or depletion of the normal system. The bottle is charged to 1,800 to 2,100 psi and is opened automatically on ejection. Flow from the emergency bottle is routed through a pressure reducer. It then follows the path of the normal oxygen system, flowing through the oxygen regulator to the face mask. The supply of oxygen available from the emergency bottle is adequate for approximately 15 minutes. The emergency oxygen gage is under the right rear of the seat cushion, in the survival kit of each ejection seat. The system is activated by pulling the green ring under the left side of the ejection seat cushion.

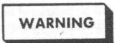

If normal oxygen system is contaminated, set OXYGEN switch to OFF before pulling emergency oxygen handle.

OXYGEN DURATION CHART

CABIN ALTITUDE	LITERS LIQUID OXYGEN								LESS THAN 1
	20	15	10	5	4	3	2	1	
35,000 FEET AND ABOVE	61.8	46.4	30.9	15.4	12.3	9.2	6.1	3.0	EMERGENCY DESCEND TO AN ALTITUDE NOT REQUIRING OXYGEN
30,000 FEET	45.2	33.9	22.6	11.3	9.0	6.7	4.5	2.2	
25,000 FEET	34.8	26.2	17.4	8.7	6.9	5.2	3.4	1.7	
20,000 FEET	26.3	19.7	13.1	6.5	5.2	3.9	2.6	1.3	
15,000 FEET	21.2	15.9	10.6	5.3	4.2	3.1	2.1	1.0	
10,000 FEET	17.0	12.7	8.5	4.2	3.4	2.5	1.7	0.8	
8000 FEET	15.5	11.6	7.7	3.8	3.1	2.3	1.5	0.7	
5000 FEET	13.5	10.1	6.7	3.3	2.7	2.0	1.3	0.6	
SEA LEVEL	10.9	8.1	5.4	2.7	2.1	1.6	1.0	0.5	

HOURS OF OXYGEN REMAINING – 100%

NOTE

- HOURS OF OXYGEN REMAINING ARE BASED ON TWO-MAN CONSUMPTION.
- DURATION DATA SHOULD BE USED AS A GUIDE ONLY SINCE OXYGEN CONSUMPTION VARIES WITH THE INDIVIDUAL.
- CONVERSION OF LIQUID OXYGEN TO GASEOUS OXYGEN IS 860 LITERS OF GASEOUS TO 1 LITER OF LIQUID OXYGEN.
- CONSUMPTION RATES ARE BASED ON MIL-D-19326D.

Figure 1-76. Oxygen Duration Chart

PITOT-STATIC SYSTEM

The pitot-static pressure system supplies impact (pitot) and atmospheric (static) pressure to the pilot's and RIO's flight instruments, to the central air data computer (CADC), and to the engine air inlet control system programmers. Some systems require static pressure only; others require static and pitot pressure. (See figure 1-77.)

The pitot-static system is composed of two separate systems with individual pitot-static probes, one on each side of the forward fuselage.

The left pitot pressure (P_T) probe supplies the pilot's airspeed and Mach indicator, the left AICS programmer, and the CADC with airspeed indications. The right pitot pressure (P_T) probe supplies the RIO's airspeed and Mach indicator and the right AICS programmer with airspeed indications. An electrical total pressure (P_T) input from the right AICS programmer is supplied to the CADC backup channel as airspeed indications for wing sweep. The left and right forward static ports (P_{S1}) are manifolded to provide static pressure to the pilot's airspeed and Mach indicator, servoed barometric altimeter, vertical velocity indicator, and the CADC. Static pressure from the right aft (P_{S2}) static ports supply the RIO's airspeed and Mach indicator, servoed barometric altimeter, and the right AICS programmer.

Note

With the inflight refueling probe extended, the pilot's and RIO's altimeter and airspeed and Mach indicators show erroneous readings because of changes in air flow around the pitot-static probes.

Pitot-Static Heat

Each pitot-static probe is equipped with electrical heating elements to prevent icing. Pitot-static heat is controlled by the pilot through the ENG/PROBE ANTI-ICE switch on the pilot's right console. In the AUTO position, pitot probe heat is available only with weight off wheels. The ORIDE position activates the probe heat elements independently of the weight-on-wheels switch and illuminates the INLET ICE caution light on the CAUTION and ADVISORY panel.

NAVAIR 01-F14AAA-1

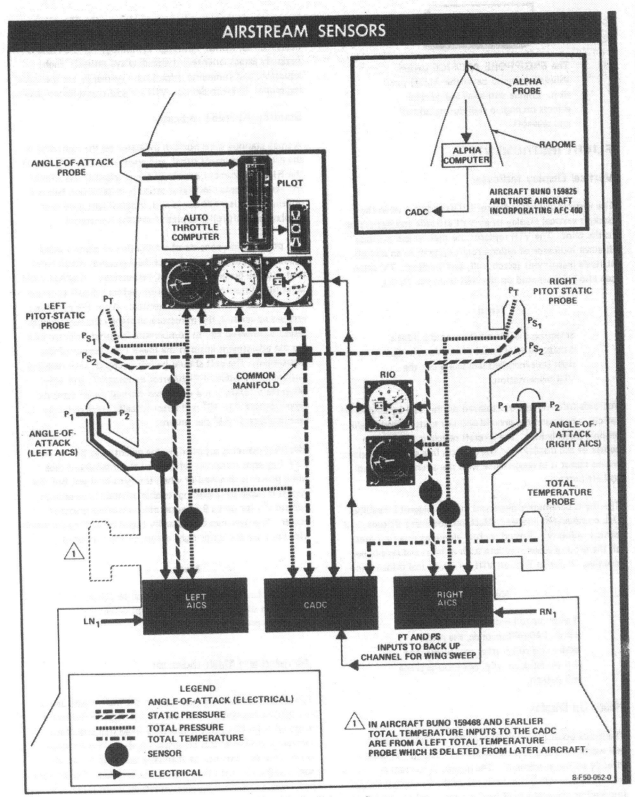

Figure 1-77. Airstream Sensors

> **WARNING**
>
> The ENG/PROBE ANTI-ICE switch should normally be in the AUTO position. Engine anti-icing has adverse effects on engine stall during takeoff and waveoff.

FLIGHT INSTRUMENTS

Vertical Display Indicator

The vertical display indicator (VDI) provides an in-the-cockpit vertical display of aircraft attitude and navigation to the pilot. The VDI replaces the mechanical attitude director indicator of older aircraft systems as an aircraft attitude instrument (pitch, roll, and heading). TV video can also be presented on the VDI from the TVSU.

> **Note**
>
> Selecting the TV position on the pilot's display control panel without a television sight unit installed will blank out the VDI presentation.

Attitude information is displayed on the VDI by an aircraft reticle, a horizon line, ground and sky texture, and a calligraphic pitch ladder. The aircraft reticle is fixed at the center of the display, and the horizon line and pitch ladder moves about it in accordance with the aircraft pitch and roll attitudes.

The flight parameters displayed include magnetic heading, D/L commanded airspeed (Mach number) and altitude, and vertical velocity. Ground texture elements superimposed on the ground plane simulate both motion and simple perspective. Refer to Section VIII for additional information.

> **Note**
>
> If pitch or roll data is not updated within 240 milliseconds, the pitch ladder and roll marker will be blanked and the horizon, sky, and ground plane will darken.

Heads-Up Display

The heads-up display (HUD) provides a combination of real-world cues and flight direction symbology, projected directly on the windshield. The display is focused at infinity, thereby creating the illusion that the symbols are superimposed on the real world (and so that visual cues received from outside the aircraft are not obscured). The pilot usually steers based on his interpretation of the visually observed real world. Although symbology for the HUD is basically attack oriented, it also displays attitude, flight situation, and command information primarily for the landing mode. Refer to Section VIII for additional information.

Standby Attitude Indicator

A small standby gyro horizon indicator on the right side of the pilot's instrument panel, and another on the left side of the RIO's instrument panel, are for emergency use should the vertical display indicator attitude information become unreliable. It is a self-contained, independent gyro that displays aircraft roll and pitch from the horizontal.

The presentation consists of a miniature airplane viewed against a rotating gray and black background, which represent sky and ground conditions, respectively. Caging should be accomplished at least 2 minutes before takeoff to allow the spin axis to orient to true vertical. After the gyro has erected to vertical, the miniature airplane reference may be raised or lowered $\pm 5°$ to compensate for pitch trim by turning the adjustment knob in the lower right corner of the instrument. The unit should be left uncaged, with recaging initiated only when in-flight error exceeds $10°$ and only when the aircraft is in a wing-level normal cruise attitude. Errors of less than $10°$ will automatically erect out data nominal rate of $2.5°$ per minute.

Electrical power is supplied by the essential ac buses. An OFF flag appears on the left side of the instrument face when power is removed or when the gyro is caged, but the gyro is capable of providing reliable attitude information (within $6°$) for up to 9 minutes after a complete loss of power. The gyro can be manually caged by pulling the pitch trim knob on the lower right corner of the instrument.

> **Note**
>
> The standby gyro should not be placed in the locked position except when removed from the aircraft.

Airspeed and Mach Indicator

The airspeed and Mach indicator on the instrument panel is a pitot-static system instrument that displays indicated airspeed from 80 to 850 knots on a fixed dial and Mach number from 0.4 to 2.5 on a rotating dial. Two independent index pointers can be manually set: one for an airspeed index and the other from a Mach index. The air data computer computes a maximum safe Mach number as a function of pitot-static pressure corrected for standard day temperature deviations. When a safe Mach is exceeded, the REDUCE SPEED warning light on the right side of the pilot's VDI illuminates.

Note

- When power failure occurs, the safe Mach marker drives to the 12 o'clock position.

- In aircraft BUNO 159636 and earlier, the pilot's airspeed indicator includes command Mach and command airspeed signals provided by the digital data link system.

Servoed Barometric Altimeter (AAU-19/A)

The counter-drum-pointer servoed barometric altimeter consists of a pressure altimeter combined with an ac-powered servomechanism. Altitude is displayed in digital form by a 10,000-foot counter, a 1,000-foot counter, and a 100-foot drum. Also, single pointer indicates hundreds of feet on a circular scale. A barometric pressure setting (baroset) knob is used to insert the local pressure in inches of Hg. The baroset knob has no effect on the digital output (mode C) of the CADC, which is always referenced to 29.92 inches Hg. The altimeter has a primary (servoed) mode and a standby (pressure) mode of operation, controlled by a spring-loaded, self-centering mode switch, placarded RESET and STBY. In the primary (servoed) mode, the altimeter displays altitude, corrected for position error, from the synchro output of the CADC. In the standby mode, the altimeter receives static pressure directly from the static system (uncorrected for position error) and operates as a standard pressure altimeter.

The primary mode is selected by positioning the mode switch to RESET for approximately three seconds, provided that ac power is on. During standby operation, a red STBY flag appears on the dial face. The altimeter will automatically switch to standby operation if electrical power is interrupted for more than 3 seconds, or if there is a system failure in the altimeter or altitude computer. Standby operation can also be selected by placing the mode switch to STBY. During standby operation, it is possible for the transponder to continue to transmit altitude information (corrected for position error) on mode C, while the altimeter is displaying altitude uncorrected for position error.

Note

- When CADC/OBC is initiated, the altimeter is put into STBY and must be reset when the test is completed.

- When operating on the STBY mode (barometric) altimeter position, error corrections must be added to instrument readings.

An ac-powered internal vibrator is automatically energized in the standby mode to minimize friction in the display mechanism.

When the local barometric pressure is set, the altimeter should agree within 75 feet at field elevation in both modes, and the primary or standby readings should agree within 75 feet. In addition, the allowable difference between primary mode readings of altimeters is 75 feet at all altitudes.

CAUTION

- If momentary locking of the baro-counters is experienced, during normal use of the baroset knob, do not force the setting. Application of force may cause internal gear disengagement and result in excessive altitude errors in both primary and standby modes. If locking occurs, rotate the knob a full turn in the opposite direction and approach the setting again with caution.

- To prevent stripping barometric setting gear, do not force settings below 28.10 inches Hg or above 31.00 inches Hg.

Vertical Velocity Indicator

The vertical velocity indicator on left side of pilot's instrument panel is a sealed case connected to a static pressure line through a calibrated leak. It indicates rate of climb or descent. Sudden or abrupt changes in attitude may cause erroneous indications due to sudden change of air flow over static probe.

Turn-and-Slip Indicator

The turn-and-slip indicator on the pilot's instrument panel gives information on rate of turn of aircraft around its vertical axis and turn coordination. The driving mechanism for the pointer is a dc meter movement receiving its input from a rate of turn transmitter. A needle-width deflection of pointer will initiate a 360° turn in 4 minutes. The inclinometer position of the instrument contains damping fluid and a ball that moves from center in an uncoordinated turn.

Note

In aircraft BUNO 161416 and subsequent and earlier aircraft modified by AFC 620, lighting for the pilot's turn-and-slip indicator is controlled by the INSTRUMENT thumbwheel.

In earlier aircraft not modified by AFC 620, turn-and-slip lighting in controlled by the ACM lighting thumbwheel.

Accelerometer

The accelerometer on the pilot's instrument panel is a direct-reading instrument that measures the accelerations of the aircraft along its vertical axis. The dial is graduated in g units from -5g to +10g. The normal reading of the instrument at rest is +1g. The instrument has three pointers of which one continuously indicates vertical acceleration of aircraft. One of the other pointers will stop and remain at maximum positive acceleration value attained, while the other will function in the same manner for negative acceleration values.

These two pointers remain at the highest values reached until reset by depressing a PUSH TO SET button on the lower left corner of instrument.

Note

Accelerometers may indicate over 1/2g low, if the pull-up rate is high.

Standby Compass

A conventional standby compass is above the pilot's instrument panel. It is a semifloat-type compass suspended in compass fluid.

Clock

A mechanical 8-day clock is on the instrument panel in each cockpit. It incorporates a 1-hour elapsed time capability. A winding and setting selector is in the lower left corner of the instrument face. The knob is turned in a clockwise direction to wind the clock, and pulled out to set the hour and minute hands. An elapsed time selector in the upper right corner controls the elapsed time mechanism. This mechanism starts, stops, and resets the sweep second and elapsed time hands.

ANGLE-OF-ATTACK SYSTEM

The angle-of-attack system measures the angle between the longitudinal axis of the aircraft and the relative wind. This is used for approach monitoring and to warn of an approaching stall. Optimum approach angle-of-attack is not affected by gross weight, bank angle, density altitude, or load configuration. (See figure 1-78 for angle-of-attack conversions.)

The system includes a probe type transmitter, approach lights, an indicator, and an indexer. The indexer and the indicator are electrically slaved to the sensor probe transmitter. In flight, the probe which is on left side of the fuselage, aligns itself with the relative airflow like a weather vane.

Probe anti-icing is provided by means of a 115-volt ac heating element along the probe and probe housing. The heating element is controlled by the ENG/PROBE ANTI-ICE switch on the pilot's right console. During ground operation probe heat is on with the landing gear handle down and the switch in the ORIDE position. With weight on wheels, the OFF and AUTO positions deactivate the probe heating element.

Note

In aircraft BUNO 159825 and subsequent and aircraft incorporating AFC 400, a second angle-of-attack system is used in conjunction with the ARI system and the rudder pedal shaker. Description of the system and its function is in this section, under flight controls.

Angle-of-Attack Test

A safety of flight check of the angle-of-attack indicator and other aircraft instruments can be performed while in flight or on the deck. When the INST position on the pilot's MASTER TEST switch is selected, the reference bar on the angle-of-attack indicator should indicate 18±0.5 units. A check of the indexer can be made by selecting the LTS position of the MASTER TEST switch.

NAVAIR 01-F14AAA-1

Section I
Part 2

ANGLE-OF-ATTACK CONVERSION

SEA LEVEL

DATE 1 JANUARY 1974
DATA BASIS: FLIGHT TEST

AIRCRAFT CONFIGURATION:
FLAPS AND SLATS, GEAR DOWN,
SPEED BRAKES EXTENDED,
FOUR AIM-7F

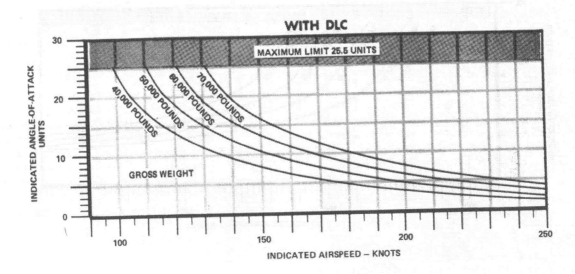

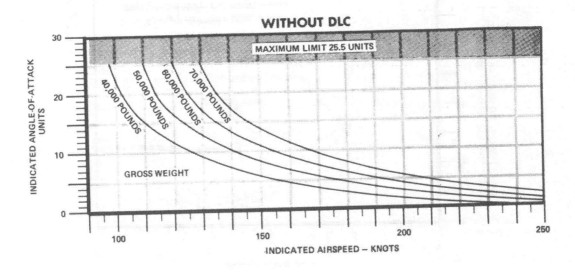

Figure 1-78. Angle-of-Attack Conversion (Sheet 1 of 2)

1-145

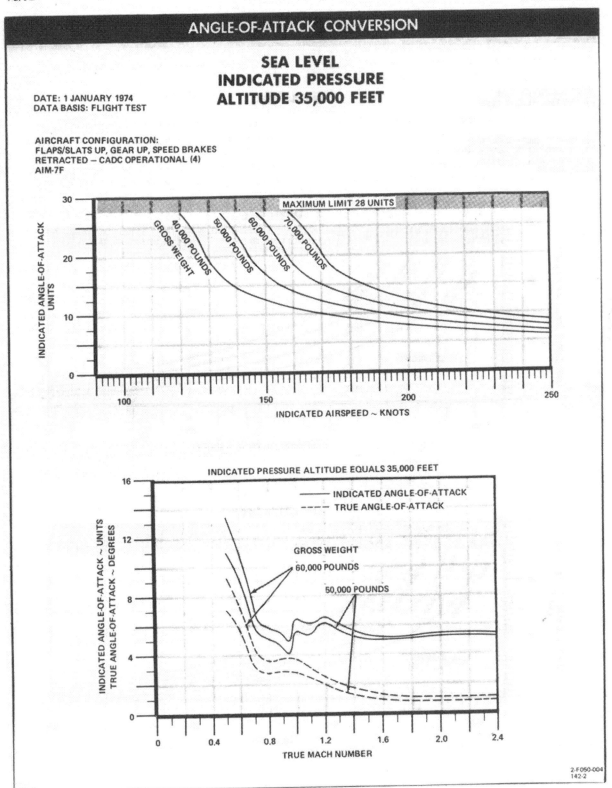

Figure 1-78. Angle-of-Attack Conversion (Sheet 2 of 2)

Angle-of-Attack Indicator

This indicator (figure 1-79) displays the aircraft angle of attack, provides a stall warning reference marker, a climb bug, cruise bug, and an angle of attack approach reference bar for landing approach.

Angle of attack is displayed by a vertical tape on a calibrated scale from 0 to 30 units, equivalent to a range of −10 to +40° of rotation of the probe. The approach reference bar is provided for approach (on speed) angle of attack at 15 units. The angle-of-attack indexer and approach lights will automatically follow the indicator.

Note

When the landing mode is selected the AOA symbol on the HUD display is faulty. Use only the angle-of-attack indicator readings.

The climb reference marker is set at 5.0 units, the cruise marker at 8.5 units, and the stall warning marker at 29 units. These reference markers are preset to the optimum angle-of-attack values and cannot be changed by the pilot.

Angle-of-Attack Indexer

The angle-of-attack indexer on the pilot's glare shield (figure 1-79) has two arrows and a circle illuminated by colored lamps to provide approach information. The relay operated contacts in the angle-of-attack indicator also control the angle-of-attack indexer. The upper arrow is for high angle of attack (green), the lower arrow is for low angle of attack (red), and the circle is for optimum angle-of-attack (amber). When both an arrow and a circle appear, an intermediate position is indicated.

The indexer lights function only when the landing gear is down. A flasher unit causes the indexer lights to pulsate when the arresting hook is up with the HOOK BY-PASS switch in the CARRIER position. The intensity of the indexer lights is controlled by the INDEXER thumbwheel control on the pilot's MASTER LIGHT panel.

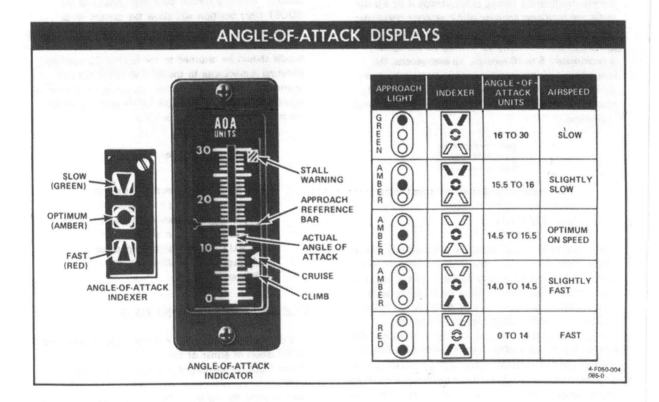

Figure 1-79. Angle-of-Attack Displays

Approach Lights

The approach lights consist of red, amber, and green indicator lights above the noise gear strut. The lights are actuated by the angle-of-attack indicator and provides qualitative angle-of-attack information to the LSO during landing approaches. A flasher unit in the angle-of-attack system will cause the approach lights to pulsate when the arresting hook is up with the landing gear down and the HOOK BY-PASS switch in the CARRIER position. When the FIELD position of the HOOK BY-PASS switch is selected the flasher unit is disabled.

A green approach light indicates a high angle of attack, slow airspeed; an amber light indicates optimum angle of attack; and a red approach light indicates a low angle of attack, fast airspeed.

CANOPY SYSTEM

The cockpit is enclosed by a one-piece, clamshell, rear-hinged canopy. Provisions are included to protect the pilot and RIO from lightning strikes by the installation of aluminum tape on the canopy above the heads of the crew. Normal opening and closing of the canopy is by a pneumatic and hydraulic actuator with a separate pneumatic actuator for locking and unlocking. The canopy can be opened to approximately 25° for ingress and egress in approximately 8 to 10 seconds. In emergencies, the canopy can be jettisoned from either crew position, or externally from either side of the forward fuselage. For rescue procedures, refer to Section V, Part 1, Emergency Procedures.

> **CAUTION**
>
> The maximum permitted taxi speed and head wind component with the canopy open is 60 knots.

The canopy system is controlled with the canopy control handle under the right forward canopy sill at each crew position. An external canopy control handle is on the left side of the fuselage directly below the boarding ladder. A CANOPY caution light on the CAUTION ADVISORY panel at each crew position illuminates when the canopy is not locked. Electrical power for the CANOPY caution light is supplied from the essential dc bus no. 2, through the CAN/LAD/CAUT/EJECT CMD IND circuit breakers (7C5). A 1-inch by 2-inch white stripe is painted on the canopy frame and sill above the aft cockpit canopy control handle panel. Alignment of this stripe provides an additional visual guide that the canopy is in a closed and locked position.

Pneumatic pressure for normal canopy operation is stored in a high-pressure, dry-nitrogen reservoir. Servicing is accomplished externally through the nosewheel well. Normal pressure should be serviced to 3,000 psi. A pressure gage in the nosewheel well should be checked during preflight. A fully charged nitrogen bottle provides approximately 10 complete cycles (open and close) of the canopy before the system is reduced to a minimum operating pressure of 225 psi. If pneumatic pressure drops below 225 psi, the canopy control module automatically prevents further depletion of the main reservoir and the canopy must be opened by the auxiliary mode.

Canopy Operation

EXTERNAL CANOPY CONTROLS

Access to the external canopy control is obtained through an access door on the left fuselage directly below the boarding ladder. Pulling the handle out and rotating it counterclockwise to the NORM CL position closes the canopy. Rotating further counterclockwise to the BOOST close position will allow the canopy to be closed under a high headwind or cold weather conditions. If BOOST position is used to close the canopy, the handle should be returned to the NORM CL position. Rotating it clockwise to the NORM OPEN position opens the canopy under normal operating conditions and rotating it further to the AUX OPEN position allows the canopy to be opened manually.

> **Note**
>
> The NORM OPEN position is not detented, therefore do not rotate the handle further clockwise unless the AUX OPEN position is desired. Using the AUX OPEN position unnecessarily will deplete the auxiliary uplock nitrogen bottle.

COCKPIT CANOPY CONTROLS

The canopy pneumatic and hydraulic system is operated by actuation of either of the cockpit control handles (figure 1-80), or the external control handle, which positions valves within the pneumatic control module to open or close the canopy. The canopy pneumatic and pyrotechnic systems are shown on page FO-17. Modes of operation available are: OPEN, AUX OPEN, HOLD, CLOSE, and BOOST.

NAVAIR 01-F14AAA-1

Section I
Part 2

COCKPIT CANOPY CONTROL HANDLE

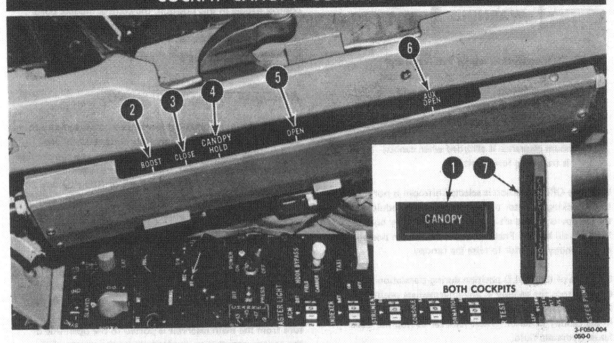

NOMENCLATURE	FUNCTION
1. CANOPY caution light	Advises flightcrew that the canopy is not in a down and locked position. Lights in both cockpits.
2. BOOST	Used to close the canopy in cold weather or when headwinds are greater than 30 to 60 knots.
3. CLOSE	Used under all flight conditions.
4. HOLD	Used to hold canopy in any position other than closed.
5. OPEN	Used to open the canopy under normal conditions.
6. AUX OPEN	Used to open canopy manually, which is required when nitrogen bottle pressure drops below 225 psi.
7. CANOPY JETTISON handle	Used to jettison canopy.

Figure 1-80. Cockpit Canopy Control Handle

1-149

> **WARNING**
>
> Flightcrews shall ensure that hands and foreign objects are clear of front cockpit handholds, top of ejection seats, and canopy sills to prevent personal injury and/or structural damage during canopy opening or closing sequence. Only minimum clearance is afforded when canopy is transiting fore or aft.

When the OPEN position is selected nitrogen is ported to the locking actuator through the control module and the canopy is moved aft disengaging the canopy hooks from the sill hooks. Pneumatic pressure is then ported to the canopy actuator to raise the canopy.

Selection of the HOLD position during translation of the canopy stops the canopy in any intermediate position between closed and open by pressurizing the lock valves in the canopy actuator. These lock valves stop the transfer of hydraulic fluid.

With the canopy in any intermediate (HOLD) position moving the handle slowly toward the OPEN position will allow the canopy to begin to close until the handle is finally in the OPEN position. This occurs because of the first motion of the handle moves the selector valve cam, which vents pressure from the lock valves and allows the canopy weight to transfer hydraulic fluid. Once the selector valve cam is completely moved to the OPEN position pressure is then applied to the open side of the canopy actuator. If the canopy handle is left in an intermediate position for an extended period, the canopy will slowly close.

The CLOSE position allows the canopy to close under normal conditions, (30-knot headwind) using its own weight without an expenditure of stored nitrogen. When the control handle is set to CLOSE, both sides of the canopy actuator are vented to the atmosphere, allowing the canopy to lower itself. The final closing motion actuates a pneumatic timer, which directs pressure from the control module to the locking actuator and the canopy is moved forward to engage the canopy hooks in the sill hooks.

To close the canopy under high headwind conditions (30 to 60 knots) or cold weather conditions, the BOOST position is used. The BOOST mode is activated by rotating the canopy control handle outboard past the CLOSE position stop and pushing the handle forward. With the control handle in this position, the control module ports additional regulated pneumatic pressure to the close side of the canopy actuator. If the BOOST position is used to close the canopy the handle should be returned to the CLOSE position.

AUXILIARY CANOPY OPENING

When the main pneumatic reservoir pressure is reduced to 225 psi, the canopy control module automatically prevents further depletion of reservoir pressure and the canopy must be opened manually. Actuation of the auxiliary mode can be effected from either the pilot's or RIO's canopy control handle or from the ground external canopy control. To open the canopy from the cockpit in this mode, the canopy control handle in the cockpit must be rotated outboard to move the handle past the OPEN position stop and then pulled aft to the AUX OPEN position. This activates a pneumatic valve, which admits regulated pneumatic pressure from an auxiliary nitrogen bottle to the locking actuator and moves the canopy aft out of the sill locks. When the canopy is unlocked, pneumatic pressure from the main reservoir is ported to the open side of the canopy actuator to counterbalance the weight of the canopy allowing the canopy to be manually opened or closed by the flightcrew.

Before leaving the cockpit, the control handle should be returned to the HOLD position. If left in the AUX OPEN position the canopy's own weight or a tailwind could force the canopy down with low pressure in the main reservoir. Once the auxiliary canopy unlock bottle is used, the canopy will not return to the normal mode of operation, and cannot be locked closed until the auxiliary pneumatic selector valve on the aft canopy deck is manually reset (lever in vertical position).

The auxiliary canopy unlock nitrogen bottle is on the turtleback behind the canopy hinge line. (Refer to page FO-17.) Servicing of the auxiliary bottle is through the small access panel immediately behind the canopy on this turtleback. A fully charged bottle will provide approximately 20 canopy cycling operations in the auxiliary open mode.

CANOPY JETTISON

The canopy can be jettisoned from either cockpit or from external controls on each side of the fuselage. An internal control handle in each cockpit (figure 1-80) is on the forward right side of each flightcrew's instrument panel, and is painted yellow and black for ease of identification. To activate the jettison control handle, pull handle in, squeezing the outer face of the handle, and then pull out.

The length of pull is approximately 1/2 to 3/4 inch and the handle comes free of the aircraft when actuated. Pulling either canopy jettison handle actuates an initiator that ignites the canopy separation charge and actuates the canopy gas generator. The canopy separation charge ignites the expanding shielded mild detonating cord lines, routed through the canopy sill hooks, breaking the sill hook frangible bolt: this allows the hooks to rotate upward, releasing the canopy. The canopy gas generator produces high-pressure gas that forces the canopy hydraulic actuator shaft upward, ballistically removing the canopy.

Ejection cannot be performed through the canopy; therefore, the canopy is jettisoned as part of the normal ejection sequence. A downward pull on the face curtain or an upward pull on the alternate ejection handle jettisons the canopy prior to ejection.

EXTERNAL CANOPY JETTISON HANDLES

There are two external emergency jettison handles located on the lower left and right fuselage below the pilot's cockpit, appropriately marked for rescue. Opening either access door and pulling the T handle fires an initiator that detonates the canopy separation charge and actuates the canopy gas generator. The sequence is the same as when the cockpit handles are pulled. The jettison control handles require squeezing the inner face of the handle and then pulling for actuation. The length of pull is approximately 1/2 to 3/4 inch and the T handle comes free on the aircraft when actuated. See Section V, Part 1, Emergency Procedures for canopy external jettisoning procedures.

EJECTION SYSTEM

The aircraft is equipped with an automatically sequenced command escape system incorporating two Martin-Baker MK GRU-7A rocket-assisted ejection seats. Both seats are identical in operation and differ only in rocket nozzle diameter to provide a divergent ejection trajectory. When either crew member initiated the command escape system, the canopy is ballistically jettisoned and each crew member is ejected in a preset time sequence. Ejection trajectories are canted laterally to provide additional separation of the seats. The RIO is ejected to the right and the pilot to the left.

A command ejection lever (figure 1-81) above the RIO's left outboard console allows the RIO to select either pilot or RIO control of the command ejection system. Each position has an internal locking detent. The handle is unlocked by lifting upward, and moved by a forward or aft motion. If the handle is released before reaching the aft position it is spring-loaded to return forward. It will automatically lock in the forward position, however a downward motion is required to positively lock it into the aft position. To select the MCO command ejection position, raise the handle and pull aft. An EJECT CMD flip-flop type indicator on the landing gear panel indicates the command mode selected. The RIO may eject individually when the command ejection lever is in the pilot control position. When the command ejection lever is in the MCO command position, the RIO can initiate ejection of both seats. Regardless of the position of the command ejection lever, an ejection initiated by the pilot will always eject both crewmen. Command ejection by either flight-crew member will eject the RIO first and the pilot 0.4 second later. Total time for ejection is 0.9 second.

The ejection system is capable of safe ejection on the deck at zero airspeed. Preflight procedures and location of safety pins are shown in Section III of this manual, ejection procedures are discussed in Section V. Ejection sequence is illustrated on page FO-18.

Ejection Seat Operation

FACE CURTAIN HANDLE

The automatic ejection sequence is initiated by pulling the face-curtain handle immediately aft and above the occupant's head. When the face curtain handle is pulled, the curtain is removed from its retainer and the system initiator is mechanically actuated starting the ejection sequence. The face-curtain protects the occupant's head and face from windblast as the seat leaves the aircraft. As the pilot's seat rises, the drogue gun and time-release mechanism sears are extracted by trip rods, and the emergency IFF transponder and ECM destruct switch is activated.

Note

The seat is fired after approximately 7 inches of face-curtain travel.

LOWER EJECTION HANDLE

The lower ejection handle is on the front of the seat bucket, between the occupant's legs. The sequence for ejection remains the same, since both the face-curtain handle and the lower ejection handle initiate seat firing by mechanically actuating the escape system initiator. The seat initiator is fired after 2 inches of alternate firing handle travel.

Section I
Part 2

NAVAIR 01-F14AAA-1

COMMAND EJECTION LEVER

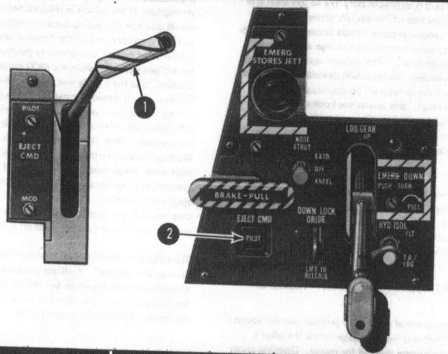

NOMENCLATURE		FUNCTION
(1) EJECT CMD lever (RIO cockpit)	PILOT	Ejection initiated by the pilot will eject himself and RIO - the RIO first. Ejection initiated by the RIO will eject only the RIO. Pilot eject command indicator - PILOT.
	MCO	Ejection initiated by the pilot will eject himself and RIO - the RIO first. Ejection initiated by the RIO will eject himself and pilot - the RIO first. Pilot eject command indicator - MCO.
(2) EJECT CMD (flip-flop) indicator	PILOT	Indicates command ejection lever is in PILOT position. Only the pilot can eject himself and RIO - RIO-initiated ejection will eject only himself.
	MCO	Indicates command ejection lever is in MCO position. Both pilot and RIO can eject both flightcrew members - RIO will eject first.

Figure 1-81. Command Ejection Lever

> **WARNING**
>
> Actuation of the lower ejection handle requires approximately 50 pounds of pull.

Note

If the face-curtain handle does not actuate the ejection seat, the face-curtain handle should be held while the lower ejection handle is actuated, to prevent the possibility of entanglement with the drogue chute when the drogue gun fires.

> **WARNING**
>
> The alternate firing handle lock may be overridden in the locked position by a pull force of approximately 80 to 100 pounds. Continued pulling of the handle after the lock has been overridden will initiate the ejection sequence provided the ground locks have been removed.

Ejection Initiation

When either ejection handle is pulled, the ejection pyrotechnic system is actuated and the following sequence of events occur:

- The seat shoulder harness power retraction reel is fired.
- The sill locks are released and the canopy is ballistically jettisoned.
- The seat safe and arm device is armed.

When canopy jettison is complete, the safe-and-arm device is fired by an 8-foot lanyard attached to the canopy. During a command sequence ejection the pyrotechnic train instantaneously triggers the RIO's seat and the pilot's seat is delayed 0.4 second. A 10-inch backup lanyard attached to the canopy activates an initiator with a 1.1 second time delay as a backup to fire the safe and arm device.

The powered inertia reel pulls the occupant back in the seat within 0.25 second. When the catapult gun fires, the expanding gases drive the catapult tubes upward and eject the seat from the aircraft. As the seat rises, the legs are pulled back into the leg pads and the restraint cords are pulled tight. As the seat continues to rise, tension built up in the leg restraint cords shear the rivets securing the cords to the deck. The legs remain secured to the seat by snubbers wedging against the restraint cord, preventing any flailing of the occupant's legs. The legs remain secured to the seat until the time-release mechanism actuates a plunger, freeing the legs prior to separation from the seat. While the seat is going up the rail, the following events occur:

- Oxygen hose, vent air, and anti-g lines are disconnected.
- The drogue-gun triprod is pulled.
- The time-release mechanism trip rod is pulled.
- The emergency oxygen bottle is activated.
- The rocket motor gas generator lanyard is pulled.

As the pilot's seat rises, the drogue gun and time-release mechanism sears are extracted by trip rods, and the emergency IFF transponder and ECM destruct switches are activated.

When fully extended, the rocket motor lanyard fires the gas generator, which fires the rocket motor under the seat.

SEAT CATAPULT AND ROCKET FIRING

One-half second after the drogue-gun trip rod is extracted, a small controller drogue chute is drawn from the pack by a metal piston fired from the drogue gun. The controller drogue chute then tows the stabilizer chute out of its container. The stabilizer chute is secured to the seat by a scissor shackle until released by the time-release mechanism.

When the seat is below an altitude of 11,500 to 14,500 feet (barostat setting), the time-release mechanism starts a 2.0-second time delay, after which the time-release mechanism releases the drogue parachute from the scissors shackle, allowing the drogue chute to pull the personnel parachute withdrawal line out of the guillotine and deploy the personnel parachute. At the same time, the harness

Section I
Part 2

NAVAIR 01-F14AAA-1

EJECTION SEAT — MK-GRU-7A

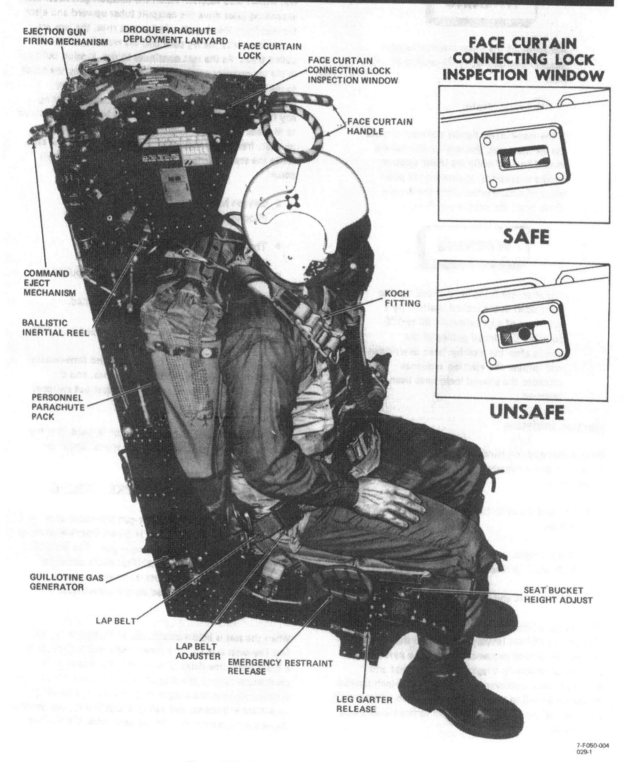

Figure 1-82. Ejection Seat - MK-GRU-7A (Sheet 1 of 2)

NAVAIR 01-F14AAA-1

Section I
Part 2

EJECTION SEAT — MK-GRU-7A

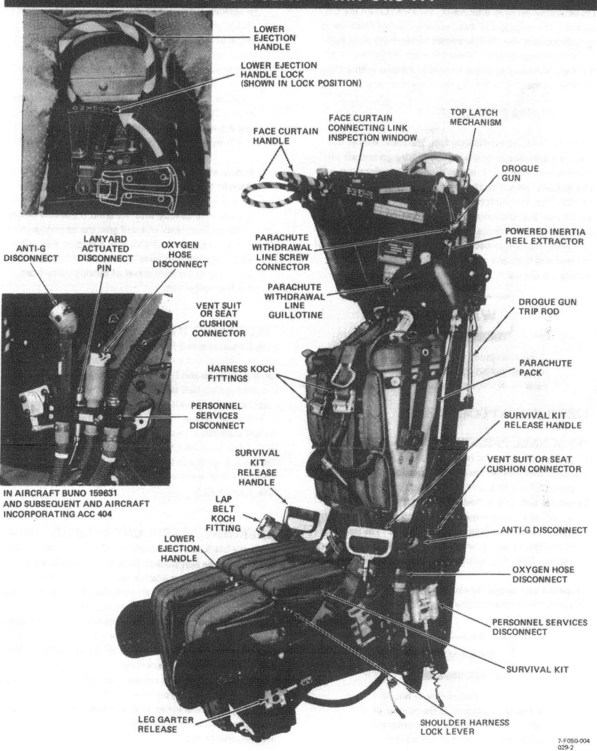

Figure 1-82. Ejection Seat - MK-GRU-7A (Sheet 2 of 2)

1-155

restraints, lap belt, leg restraints, survival kit, upper block of the personnel services disconnect, and face curtain are unlocked. The occupant is then free to be pulled from the seat sticker clips by the line stretch of the main parachute as the seat rotates away. When the occupant is free from the seat, he makes a normal parachute descent with the survival kit connected to him by the integrated torso harness.

MANUAL SEAT SEPARATION

If the time-release mechanism fails, the crewmember can manually separate from the seat by pulling up and aft on the emergency restraint release handle (front right side of seat bucket). Rotating the emergency harness release handles, fires the guillotine severing the personnel parachute withdrawal line and releasing the lap belt, shoulder harness, leg-line straps, parachute container, survival kit, and personnel services disconnect. The occupant must kick free of the seat and manually deploy the personnel parachute by pulling the D-ring.

If the emergency restraint release handle has been pulled, the personnel parachute must be deployed manually.

EJECTION SEAT COMPONENTS

PERSONNEL PARACHUTES. The personnel parachute is in a rectangular back pack aft of the occupant on a seat-mounted support shelf. It consists of a 28-foot canopy, suspension lines, and personnel harness connections. The parachute pack is attached to the seat by two lockpin fittings. On ejection, the locks are automatically released by the time-release mechanism at both points. The parachute pack is attached to the occupant by two straps connected to the top Koch fittings of the integrated torso harness.

On parachute assemblies equipped with ACSE 383 (four-line release) capabilities, the following procedures apply during descent.

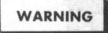

The four-line release should not be activated if damage to the canopy or broken suspension lines are observed after deployment of a full canopy.

1. After visual inspection of parachute and deciding to activate the four-line release, the flightcrewman should grasp the release lanyard loops on the inside of rear risers and break the release ties by a sharp pull (approximately 20 pounds force). This action will free the rear four suspension lines from the connector links, allowing the canopy to form a lobe in the rear center and permit a steady escape of air. This will eliminate oscillation and give minimal directional control of the canopy to the right or left by pulling on the respective release lanyard.

2. In preparation for landing, the flightcrewman should attempt to determine the surface wind direction by any means possible; that is, smoke, dust, etc., and turn the canopy into the wind if possible before landing. Over land, this will give the advantage of missing obstacles and choosing a landing area. Over water this will minimize the chance of line entanglement due to forward travel of canopy away from the flightcrewman.

INTEGRATED TORSO HARNESS. The integrated torso harness is a vest-like garment worn by the crewmember. It eliminates the need for the crewmember to wear the parachute to and from the aircraft and takes the place of a separate lap belt and shoulder harness. The upper torso harness is connected by Koch fittings to the parachute riser-shoulder release fitting, which is attached to the locking reel assembly. Two buckles on the lower part of the torso harness connect to the seat lap belt Koch fittings. The lap belt girth can be adjusted to accommodate each individual flightcrew member by adjusting each belt strap. (See figure 1-82.)

EMERGENCY RESTRAINT RELEASE. This control is the black-and-yellow-striped handle forward on the right side of the seat bucket. In the forward (locked) position, the occupant and his personal equipment are secured to the seat. When the emergency restraint release is lifted and rotated to the full aft (unlocked) position, the occupant and his personal equipment are disconnected from the seat except for the sticker clips. The initial travel of this handle fires the parachute withdrawal-line guillotine, severing the personnel parachute from the drogue parachute. The emergency restraint release simultaneously releases the lap-belt harness locks, personal parachute container retention straps, survival kit attachment lugs, and sticker clips, inertia-reel straps, dual leg-restraint cords, and personnel services disconnect.

WARNING

- In aircraft with ACC 390 incorporated, actuating the emergency restraint release handle no longer locks the face curtain handle.

- When the emergency restraint release is actuated, the lower firing handle is automatically locked and ejection cannot be initiated from the seat using the lower handle. Ejection may still be initiated with the upper handle but will be extremely hazardous and will require manual parachute deployment.

- The emergency restraint release may be manually repositioned to the unactuated position, which will enable the occupant to unlock the face-curtain handle and the lower ejection handle. However, in this situation, the lap-belt harness locks, parachute withdrawal line guillotine, inertia-reel straps, dual leg-restraint cords, and personnel services disconnect will remain released. If ejection is initiated in this condition, it will be extremely hazardous and will require manual parachute deployment from that seat.

This lever is normally in the locked position. Emergencies such as over-the-side bailout, or failure of the time-release mechanism will require operation of this control.

SHOULDER HARNESS LOCK LEVER. This control is on the left side of the seat. In the forward (locked) position, forward movement of the occupant is restricted and any slack created by rearward movement is taken up by the inertial reel. The control is locked in this position by a detent. In the spring-loaded center position, the occupant can move forward freely, unless the reel locks owing to excessive forward velocity. When the forward velocity decreases sufficiently, the inertia straps are released without the necessity of repositioning the manual control. Both straps feed from the same shaft and it is impossible for one to lock without the other. If the reel is locked manually the control must be positioned full aft to the unlocked position to release the straps.

LEG RESTRAINTS. The leg garters and restraint cords keep the occupant's legs firmly against the leg rests during ejection. The garters are placed around the leg above the calf and above the ankle. They should be tight enough so they do not slip down over the calf.

The leg-restraint cords are attached to the aircraft deck and routed through the snubber box seat structure. They are then passed through garter rings and snapped into the leg line locks. The garter rings are snapped into the bayonet fitting when strapping in. Leg-line snubber release is accomplished by pulling the release lever located on the outer side of the leg-line snubber boxes.

VENTILATED CUSHIONS. Each seat is equipped with a ventilated back and seat cushion to provide comfort for the seat occupant. Air from the pressure ventilation suit system flows through the internal ducting in the seat cushion and is discharged through the perforated surface.

SEAT ADJUSTMENT. Seat adjustment is controlled by a three-position, momentary-contact switch on the right side of the right thigh support. Moving the switch forward (FWD) lowers the seat and the AFT position raises the seat. Seat adjustment is limited to 5 inches of vertical movement. Electrical power is supplied from the right main ac bus, through the ACM LT/SEAT ADJ/STEADY POS LT circuit breaker (214).

Note

The seat height-adjustment actuator is an intermittent-duty motor with duty cycle of 1 minute on and 10 minutes off.

SURVIVAL KIT. An RSSK-7 or SKU-2/A survival kit packed within a two-piece fiberglass container and attached to the occupant by strap-harnessing, is in the seat bucket of each crewman's ejection seat. (See figure 1-83.) The upper half of the kit is a contoured lid assembly covered by a ventilated seat pad and contains an emergency oxygen bottle, which can be activated automatically by seat ejection, or manually by pulling the emergency oxygen actuator. The lower half of the kit contains the following:

Lift raft	Space blanket
Dye markers	Desalter kit or canned water
Signal flares	50 feet of nylon cord
Beacon AN/URT-33	Bailing sponge
Morse code and signal card	SRU 31/P flightcrew survival kit

Note

Emergency provisions included in the RSSK-7 or SKU-2/A survival kit are subject to local operation and may be altered at the discretion of the area commander.

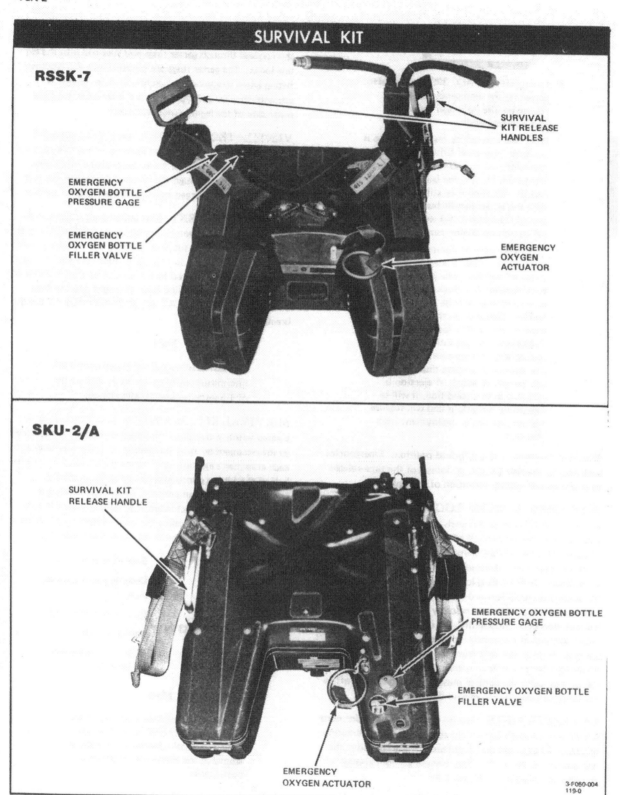

Figure 1-83. Survival Kit

After the seat clears the aircraft during the ejection sequence, a time release mechanism releases the flightcrew member's harness and survival kit from the seat. The opening shock of the personnel parachute deployment snaps the flightcrew member and the survival kit clear of the seat. During descent, when the flightcrew member pulls the survival kit release handle (one on each side of the kit), the lower half of the kit falls away but remains connected to the upper half by a lanyard approximately 23 feet long. As the lower half reaches the end of its free fall, the life raft inflation mechanism is actuated and the raft inflates.

When ejecting over water, the life raft should be inflated before entering the water. If the crewman enters the water before the survival kit release handle is pulled, the life raft can be inflated only by first pulling the release handle and then pulling the life raft lanyard until the raft is inflated.

LIGHTING SYSTEM

Exterior Lights

The exterior lights include position lights, formation lights, anticollision lights, a taxi light, approach lights, and an air refueling probe light. All exterior lighting controls except for the air refueling probe light and approach lights, are located on the MASTER LIGHT panel on the pilot's right console. The exterior lights master switch on the outboard throttle must be on for any exterior light to function. The pilot's light control panel is shown in figure 1-84. A two-channel flasher unit is used for flashing lights. One channel flashes the anticollision and position lights and has circuit protection from the ANTI COLL/SUPP POS/POS LTS circuit breaker (2I1). The second flasher channel flashes the angle-of-attack indexer and approach lights and has circuit protection from the ANGLE OF ATTK IND circuit breaker (3C3).

Note
The anticollision, position and supplementary position, formation, and taxi lights are inoperative when operating on emergency generator.

POSITION LIGHTS

The position lights consist of a red light on the left wing tip, a green light on the right wing tip, and a white position light in the fin cap assembly. In aircraft BUNO 159025 and earlier, the white tail light is located on the boat tail. Supplemental position lights include upper and lower red lights on the left wing glove, and upper and lower green lights on the right wing glove. When the wing sweep angle is forward of 25°, the wing tip position lights are operational; when the wings are swept aft of 25°, the wing tip position lights are disabled and the glove position lights are operational. When operating in the steady mode with the landing gear down and the wings forward of 25°, both the position lights and the glove lights are operational. The position lights are powered from the right main ac bus through the exterior lights master relay.

Note
When the anticollision lights are on, the flasher for the position lights is disabled and the lights revert to steady.

ANTICOLLISION LIGHTS

There are three red, flashing anticollision lights. One anticollision light is installed in the bottom of the IR pod on the lower forward fuselage. Another anticollision light is installed in the top forward part of the left vertical stabilizer. The third anticollision light is on the top aft part of the right vertical stabilizer and directs its anticollision beacon up and down.

In aircraft BUNO 161133 and subsequent and aircraft incorporating AFC 582, the lower fuselage forward anticollision light remains off during takeoff and landing with the nosewheel door open. With the nosewheel door closed, the lower fuselage forward anticollision light will operate with the ANTI COLLISION light switch set to ON. The anticollision lights are powered through the right main bus with circuit protection on the RIO's ac right main circuit breaker panel (2I1).

FORMATION LIGHTS

The formation lights consist of wing tip lights on each wing, aft fuselage lights, and vertical fin tip lights on both sides of the aircraft.

All formation lights are green. Intensity of the lights are controlled by the FORMATION thumbwheel on the MASTER LIGHT panel. Electrical power is supplied through the right main bus with circuit protection on the RIO's ac right main circuit breaker panel (2H2).

TAXI LIGHT

The taxi light installed on the nosewheel is a fixed position light. A limit switch on the nose gear door will turn the light off when the gear is retracted. A two-position, ON and OFF switch is on the MASTER LIGHT panel. Electrical power is supplied through the right main bus with circuit procection on the RIO's circuit breaker panel (2H2).

Interior Lights

The interior lighting of the cockpit consists of red instrument panel and console panel lights, red and white flood

Section I
Part 2
NAVAIR 01-F14AAA-1

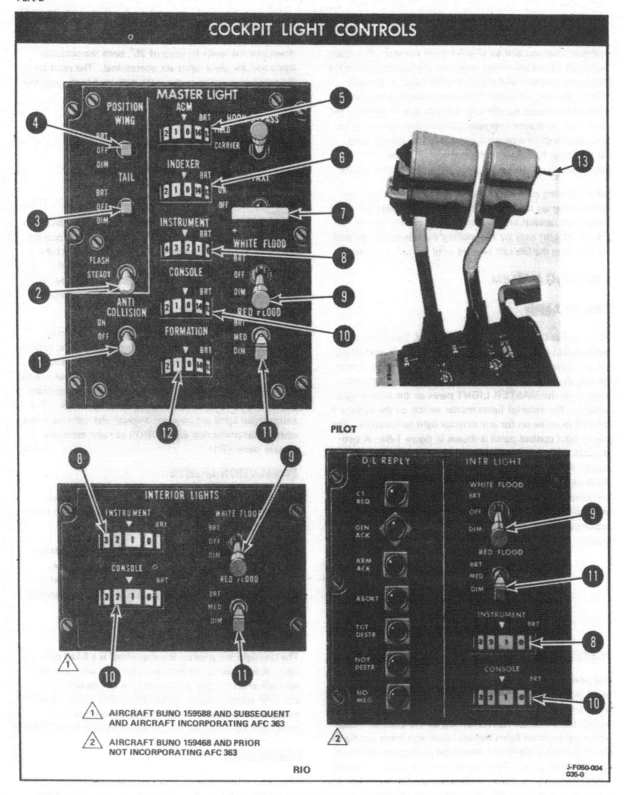

Figure 1-84. Cockpit Light Controls (Sheet 1 of 2)

NOMENCLATURE		FUNCTION
(1) ANTI COLLISION light switch	ON and OFF	— Energizes or deenergizes the anticollision lights. When anticollision lights are on, the flasher unit for the position lights is disabled.
(2) POSITION lights flasher switch	FLASH	— Causes the wing or supplementary tail and position lights to operate in a flashing mode with landing gear up. With gear down supplementary lights operate steady only.
	STEADY	— With the wing and tail (or either) position lights on, lights are on steady.
(3) TAIL POSITION light switch	BRT OFF DIM	— Bright tail lights — Deenergizes tail position lights — Dim tail lights
(4) WING POSITION light switch	BRT OFF DIM	— Bright wing lights — Deenergizes wing lights — Dim wing lights
(5) ACM panel light thumbwheel	0 to 1 1 to 14	— Turns lights on — Night variable intensity
(6) INDEXER thumbwheel	0 to 14	— Variable increase in intensity of indexer lights.
(7) TAXI light switch	ON	— Nose gear must be down and locked and the master exterior light switch must be on.
	OFF	— Turns light off.
(8) INSTRUMENT lights thumbwheel	0 to 1 1 to 14	— Turns instrument panel lights on. — Variable increase of intensity to a maximum brightness at 14.
(9) WHITE FLOOD lights switch		**Note** Switch must be pulled out to be moved to BRT or DIM.
	BRT DIM OFF	— Bright light. — Dim light. — Turns light off.
(10) CONSOLE lights thumbwheel	0 to 1 1 to 14	— Turns console lights and red floodlights on. — Variable increase of console lights intensity to maximum brightness at 14.
(11) RED FLOOD lights switch	BRT MED DIM	— Bright red instrument flood and console floodlights only. — Red console floodlights only. — Dim red console floodlights only.
(12) FORMATION lights thumbwheel	0 to 1 1 to 14	— Turns formation lights on. — Variable increase of light intensity to maximum of 14.
(13) Exterior lights master switch	ON	— Enables all exterior lights. Dims approach lights.
	OFF	— Permits pilot to turn off all exterior lights. Increases intensity of approach lights.

Figure 1-84. Cockpit Light Controls (Sheet 2 of 2)

lights for additional console and instrument panel lighting, and a utility and map light for each flightcrew station. At the pilot's station the interior lights are controlled from the MASTER LIGHT panel on the right outboard console. The RIO can control his interior lighting from the interior light panel on his right outboard console.

INSTRUMENT AND CONSOLE PANEL LIGHTS

All flight instruments in the pilot's and RIO's instrument panel and console panel lights are lighted by aviation red lighting. Individual thumbwheel controls are provided for the pilot and RIO instrument and console lighting. The thumbwheels have 14 variable selections from 0 to 14. Initial rotation from 0 to 1 activates the circuitry and provides a low intensity light. Further rotation up to a maximum intensity (14) increases the brightness. The INSTRUMENT thumbwheel also controls the intensity of the CAUTION ADVISORY panels and the digital data indicator lights, which consists of high and low intensity lighting. The console lights thumbwheel turns power on for both the console lights and the red floodlights. The pilot's and RIO's instrument and console lights are protected by circuit b breakers on the RIO's ac right main circuit breaker panel (216, 213, and 212).

Note

- In aircraft BUNO 161416 and subsequent and earlier aircraft modified by AFC 620, lighting for the pilot's turn-and-slip indicator is controlled by the INSTRUMENT lighting thumbwheel.
- In earlier aircraft not modified by AFC 620, turn-and-slip lighting is controlled by the ACM lighting thumbwheel.

FLOODLIGHTS

The floodlights consist of 4.2-watt red and white and 20-watt white lights that illuminate the instrument and console panels. When navigating around thunderstorms the white floodlights should be turned on bright to assist in preventing temporary blindness from lightning. The WHITE FLOOD toggle switch on the pilot's master light panel and another on the RIO's light panel are safety interlock switches that must be pulled up to be positioned to BRT or DIM. In the DIM position low intensity floodlighting is provided.

Note

When the white floodlights are on (BRT or DIM), the intensity of the CAUTION and ADVISORY panel lights is increased to a day (bright) illumination mode.

Red console and instrument panel flood lights are available in the BRT position. In the MED and DIM position, only console floodlights are available. The red floodlights are protected by a circuit breaker on the RIO's ac essential no. 1 circuit breaker panel. The white floodlights are protected by the WHITE FLOOD LT circuit breaker on the RIO's ac right main circuit breaker panel (2H6).

UTILITY AND MAP LIGHTS

The pilot's utility and map light is on a bracket above the right outboard console. The RIO's utility and map light is in a bracket above and midway along the right console. Each light has a rheostat control including an ON and OFF position on the rear of the lamp. A red filter may be selected by rotating the face of the lamp. Pressing the locking button on top of the lamp permits rotating the face of the lamp to reselect a white light with a flood or spot illumination option. An alligator clip and swivel mounting allow the light to be positioned on a clipboard or other convenient location. A flasher button on the heel of the lamp allows either crewmember to use the light as a signal lamp. The utility and map lights are supplied electrical power from the ac essential no. 2 bus and are protected by the UTILITY LT circuit breaker (4D5).

Warning And Indicator Lights

Warning, caution, and advisory lights (figures 1-85 and 1-86) are provided in both cockpits to alert the pilot and RIO of aircraft equipment malfunctions, unsafe operating conditions, or that a particular system is in operation.

The warning lights illuminate red with black letters to warn of hazardous conditions that require immediate corrective action. All caution lights are on the pilot and RIO CAUTION ADVISORY panel. These lights show yellow letters on an opaque background to indicate an impending dangerous condition. The lower half of the CAUTION ADVISORY panel consists of advisory lights that show green letters on opaque background. Advisory lights indicate degraded operations which may require corrective action.

WARNING

Radiation hazard exists on deck when the RDR ENABLE caution light is illuminated. The light indicates that the RADAR TEST ENABLE switch (maintenance switch) is in the "A" (radiate and scan) position. This condition permits the weight-on-wheels interlock to be bypassed, allowing the transmitter to radiate out the antenna when XMT is selected on the hand control unit. Illumination of the light does not indicate a weight-on-wheels failure.

In addition, the digital data indicator (DDI) on the RIO's right console contains 40 advisory lights associated with data link communication. Repeater data link advisory lights are on the left side of the pilot's VDI. A functional description of each light is included in the applicable system description.

Refer to NAVAIR 01-F14AAA-1B for a complete summary of indicator light meaning and action to be taken when illuminated.

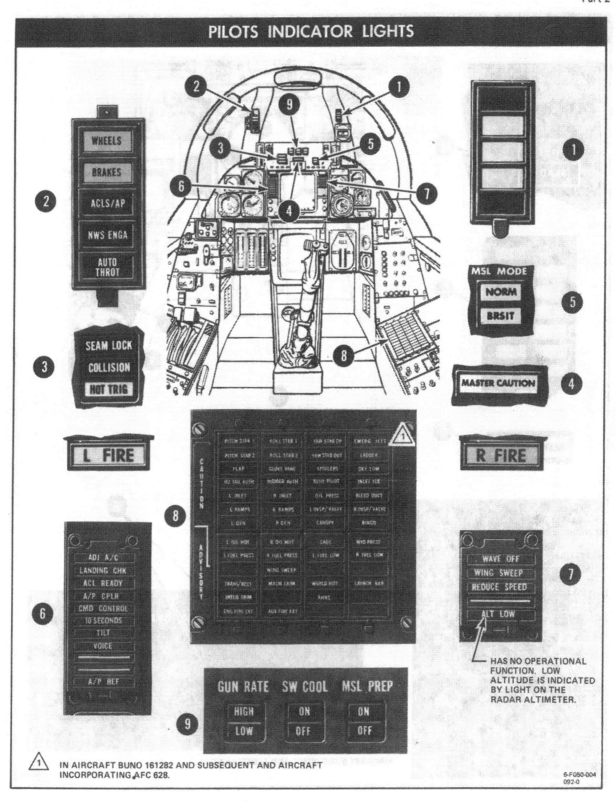

Figure 1-85. Pilot's Indicator Lights

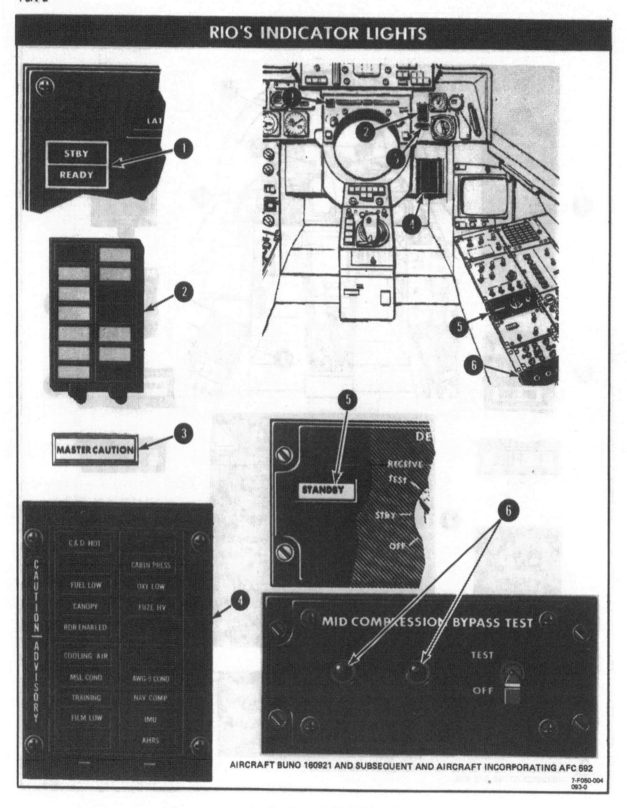

Figure 1-86. RIO's Indicator Lights

MASTER CAUTION LIGHT

The pilot's MASTER CAUTION light is centrally located on the air combat maneuver panel, and in the aft cockpit the RIO's MASTER CAUTION light is on the right instrument panel. When the lights are illuminated, yellow letters show on an opaque background. Individual MASTER CAUTION lights flash whenever a caution light on the respective CAUTION and ADVISORY panel illuminates. A MASTER CAUTION light may be turned off by depressing its lens. This will activate a reset switch that rearms the master circuit for a subsequent caution light. A caution light lit on the CAUTION and ADVISORY panel will not be turned off by resetting the master caution light.

INDICATOR LIGHTS TEST

A check of all indicator lights can be performed while airborne or during ondeck operations. The pilot's caution and advisory lights, the MASTER CAUTION light, and all associated circuitry are tested through the MASTER TEST panel. The test is initiated by selecting the LTS position and pressing the master test knob. Electrical power is routed through the circuitry to provide simulated failure signals to the caution and advisory lights. Illumination of each warning, caution, and advisory light verifies proper continuity of the indicator lights. A malfunction is indicated by failure of a light to illuminate.

Illumination of any caution light causes the MASTER CAUTION light to flash. If the MASTER CAUTION light illuminates steadily during the LTS test it indicates a failure of the MASTER CAUTION light, primary power failure, failure of the flasher module, or that failure has been detected by the BIT circuits.

The following indicator lights are also illuminated by the LTS test through the MASTER TEST panel:

    L FIRE
    R FIRE
    APPROACH INDEXER
    GO/NO GO LIGHT
    SAM
    WHEELS
    EMERG STORES JETT BUTTON
    HOOK LIGHT
    HOT TRIGGER
    REFUEL PROBE LIGHT AUTO THROT
    WING SWEEP
    LDG GEAR LIGHT
    REDUCE SPEED
    ACLS/AP
    A/P REF

All indicator lights located on the sides of the VDI (including the DDI repeater lights) are tested with the caution and advisory lights.

WARNING

Failure of the EMERG STORES JETT button to illuminate during the LTS check could indicate that the jettison button bulb is burned out or that the test circuit is defective. In this event, status of the emergency stores jettison circuit cannot be determined. If the switch is activated, stores will jettison when weight is off wheels. Under some lighting conditions it may be difficult to determine when the light is illuminated. Ensure that the light goes out when LTS MASTER TEST is deselected. Failure of the light to go out indicates emergency jettison is selected and stores will jettison with weight off wheels.

Note

- Upon completion of each test, return the MASTER TEST switch to OFF.
- The DATA LINK power switch must be on to check the DDI lights.

The RIO's caution and advisory lights are tested in the same manner through the IND LT position on the TEST panel on the right console.

Also the RIO's left and right MCB lights on the mid compression bypass panel will be tested on aircraft BUNO 160921 and subsequent and aircraft incorporating AFC 592.

JETTISON SYSTEM

Four jettison modes are provided in the aircraft:

- Emergency
- ACM
- Selective
- Auxiliary

Weapon arming and fuzing and missile motor ignition are safed during all jettison release modes.

Stores shall be jettisoned above the minimum fragmentation clearance altitude, when possible, even though weapon arming and fuzing is safed in all jettison modes.

External fuel tanks, Phoenix, and Sparrow missiles can be released through all jettison modes except auxiliary jettison. Figure 1-87 shows the aircraft loading station locations.

Note

Sidewinder missiles cannot be jettisoned with AAC 688 incorporated.

Emergency Jettison

Emergency jettison is used to separate from the aircraft as fast as possible all external stores except Sidewinder. The emergency jettison circuit is interlocked only through weight-off-wheels, and overrides any previous logic in the armament system. This mode is actuated by the pilot through the EMERG STORES JETT switch on the landing gear control panel (figure 1-88) and does not require MASTER ARM switch to ON. This applies an emergency jettison command to the station select sequencer in the armament control panel. The station select sequencer logic automatically generates the selected station jettison commands for all stations (except those that carry Sidewinders) independent of the STA SEL switches.

The station jettison release sequence is: 1B, 8B, 2, 7, -4D, -5D, -4A, -5A, -4C, -5C, -4B, -5B, -3D, -6D, -3A, -6A, -3C, -6C, -3B, -6B.

Note

- The time interval between stations indicated by (-) is 100 ms.
- Stations 1B, 8B, 2, 7 are jettisoned simultaneously.

ACM Jettison

The ACM jettison mode selectively jettisons all external stores except Sidewinders. ACM jettison is similar to emergency jettison except instead of weight off wheels, it requires landing gear handle UP, and RIO selection of those stations to be separated. As in emergency jettison, MASTER ARM need not be ON. This mode is used to separate from the aircraft, as fast as possible, all those selected external stores not needed for the air combat situation.

After station selection by the RIO, this mode is actuated by the pilot's placing the ACM switch on the ACM panel to ON and depressing the ACM JETT pushbutton under the ACM cover.

Note

- Sidewinder is inhibited from ACM JETT even if selected by the RIO.
- To ensure release of all selected stores, the ACM JETT pushbutton must be depressed and held at least 2 seconds.

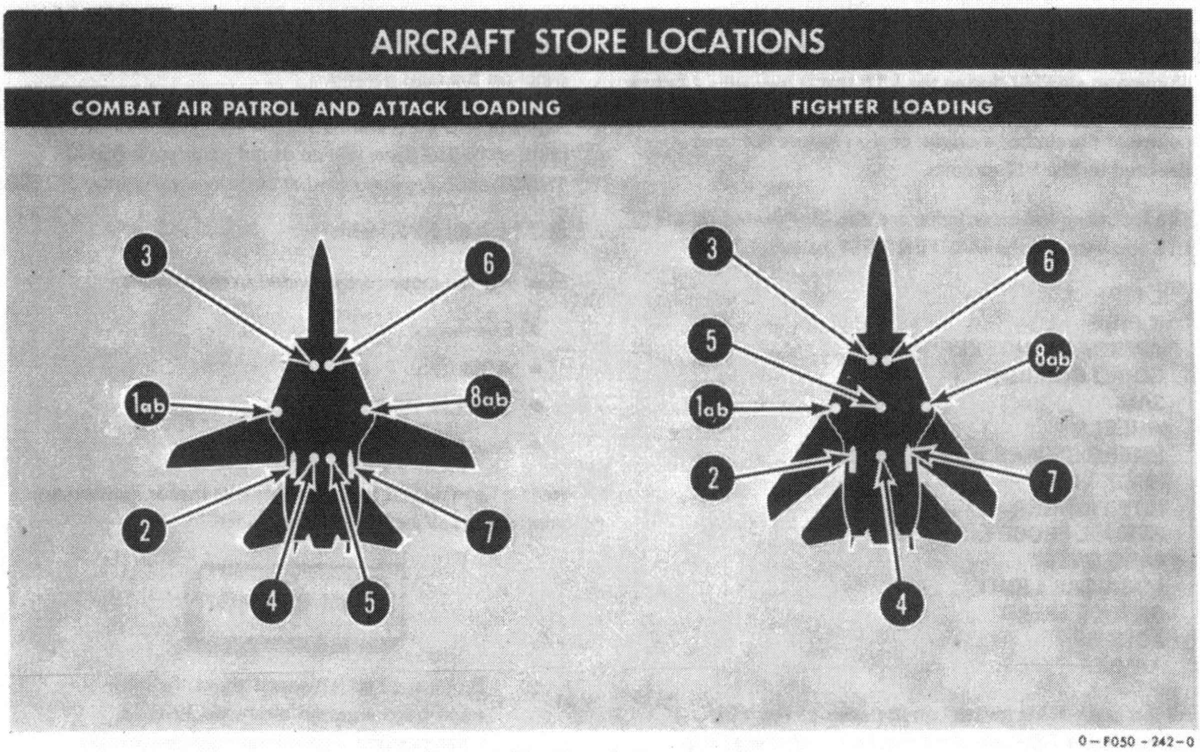

Figure 1-87. Aircraft Store Locations

NAVAIR 01-F14AAA-1

JETTISON CONTROLS AND INDICATORS

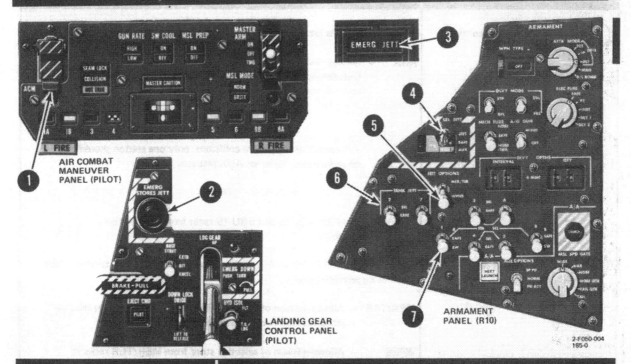

NOMENCLATURE	FUNCTION
① ACM JETT pushbutton	Enables ACM jettison. Pushbutton is under ACM switch cover. In order to activate this jettison mode, cover must be up which places the WCS in the ACM mode. When pressed only those stores selected on the armament control panel are jettisoned. To ensure release of all selected stores the ACM JETT pushbutton must be depressed and held for at least 2 seconds. ACM jettison will not release any Sidewinder missiles even if their stations are selected.
② EMERG STORES JETT pushbutton	Enables emergency jettison. When pressed, the switch is illuminated. This function is checked by MASTER TEST switch in the LTS position. In aircraft BUNO 161282 and subsequent and aircraft incorporating AFC 628, the MASTER CAUTION light and the EMERG JETT caution light illuminate when the EMERG STORES JETT pushbutton is activated.
③ EMERG JETT caution light (In aircraft BUNO 161282 and subsequent and aircraft incorporating AFC 628.)	Indicates EMERG STORES JETT pushbutton is activated.
④ SEL JETT switch	Allows RIO to jettison from selective stations. It is a three-position, lever-locked switch with guarded positions.
	JETT — Releases weapons from all stations selected according to established outboard to inboard, aft to forward sequence, with a 100-ms interval automatically set between releases.

Figure 1-88. Jettison Controls and Indicators (Sheet 1 of 2)

NOMENCLATURE	FUNCTION
④ SEL JETT switch (cont)	SAFE — Inhibits jettison in selective mode. This is normal switch position. AUX — Backup nonejection mode used when selective jettison has failed. Guarded position which provides for freefall release from parent racks of stores from selected stations. **CAUTION** To prevent store-to-store collisions, only one station should be selected at a time for AUX jettison. **Note** Only BRU-24/25 and BRU-19 racks have AUX jettison capability.
⑤ JETT OPTIONS switch	Provides the option of jettisoning just the store or the store and its parent rack when using selective jettison. MER/TER — Allows jettison of MER/TER rack including stores in selective jettison mode from stations 1, 3, 6, and 8. WPNS — Allows jettison of selective store from MER/TER racks in selective jettison mode.
⑥ TANK JETT switch	Selects stations 2 and/or 7 for selective or ACM jettison. SEL — External fuel tanks selected for selective or ACM jettison. SW — Not used.
⑦ STA SEL switches 1B, 8B 3,6 4,5	Allows selective, auxiliary, or ACM jettison of Phoenix or Sparrow missiles. Selects lower station for selective or ACM jettison. SAFE — Inhibits all release and jettison except emergency. SW — Not used SEL — Bombs or flares selected for normal release, selective, auxiliary, or ACM jettison. Selective or ACM jettison for Sparrow or Phoenix missiles.

Figure 1-88. Jettison Controls and Indicators (Sheet 2 of 2)

The station release sequence is:

1B, 8B in pairs	Salvo
2, 7 in pairs	Salvo
4, 5, ripple singles	Sequential
3, 6 ripple singles	Sequential

Stores are released only from the parent rack as the position of the JET OPTIONS switch is ignored in this mode.

Selective Jettison

Selective jettison is used to separate selected external stores from one or more store stations. This mode provides options and capabilities not provided by the other jettison modes. It may be used to clean the aircraft of empty rocket pods, fuel tanks, or other stores selected to be jettisoned. Unlike emergency and ACM jettison, in addition to landing gear handle UP, this mode requires the pilot to set MASTER ARM switch to ON. Selective jettison is actuated by the RIO by setting the SEL JETT switch to JETT, after having first selected those stations to be jettisoned and selecting MER/TER or WPN on the JETT OPTIONS switch. The fixed interval between releases until completion of the separation program is 100 ms. If a TER is carried on a particular store station, selecting MER/TER will route the jettison circuit to the parent bomb rack and jettison the TER with or without stores attached.

CAUTION

Do not attempt jettison of external fuel tanks until wing fuel tanks are depleted. Wing fuel may be lost if the external tank quick-disconnect valve sticks in the open position.

Selecting WPN on the JETT OPTIONS switch will route jettison circuit to the TER and release the stores carried, allowing the aircraft to return with empty TERs. Stores on those selected stations without a TER will be jettisoned with the JETT OPTIONS switch set to MER/TER.

The station release sequence is:

1A/8A in pairs	Salvo
1B/8B in pairs	Salvo
2 and 7 in pairs	Salvo
4 and 5 ripple singles	Sequential
3 and 6 ripple singles	Sequential

Auxiliary Jettison

Auxiliary jettison is a nonejection backup mode to be used when other release or jettison modes have failed. The fuel tanks and the missiles cannot be separated in auxiliary jettison. All air-to-ground stores can be jettisoned through this mode. Unlike selective jettison, stores cannot be jettisoned from the TER but are separated as a total package (including the TER). Like selective jettison, this mode requires MASTER ARM switch set to ON in addition to landing gear handle up, but is actuated by the RIO placing the SEL JETT switch to AUX rather than JETT, after having first selected the station to be jettisoned. Unlike selective jettison, the JETT OPTION switch is overridden when the TER is carried. Stores carried on these multiple racks cannot be separated individually but are separated as a total package, including the TER.

All stations selected are released simultaneously. If stores and assemblies are not ejected, the parent rack hooks merely open.

WARNING

Since auxiliary jettison is a gravity drop rather than ejected separation, the aircraft will be restricted in its flight envelope when jettisoning through this mode. Since this mode is also a true salvo rather than fast ripple, no more than one station should be selected by the RIO for each release.

Note

When more than one store is on a weapon rail, the flight envelope will be further restricted.

MISCELLANEOUS EQUIPMENT

Boarding Ladder

A boarding ladder consisting of three folding sections is housed in the left fuselage between the two cockpits. It is held in the closed position by two mechanical locking pins actuated by the ladder control handle in the face of the boarding ladder. The ladder must be manually released or stowed from the ground level. Unfolding the remaining two sections places the ladder in a fully extended position. The bottom rung of the ladder is approximately 26 inches above the deck when in a fully extended position, with the nose gear unkneeled, and 12 inches above the deck if the nose gear is kneeled. A LADDER caution light on the pilot's CAUTION ADVISORY panel advises the pilot that the boarding ladder is not in a full up-and-lock position.

Section I
Part 2
NAVAIR 01-F14AAA-1

BOARDING STEPS AND HANDHOLD

There are two positive locking board steps, one on either side of the boarding ladder directly below each cockpit. They may be opened or closed from either cockpit or standing on the boarding ladder. A single handhold is directly above the boarding ladder. It is a spring-loaded door that fairs with the fuselage when released.

Nose Radome

The nose radome is attached to the aircraft by a top hinge and bottom mounted latches, permitting it to be rotated up for access and maintenance. A jury strut attached to the lower part of the dome can be fastened to the aircraft bulkhead to hold the dome open. A minimum overhead clearance of 16 feet is required when opening the radome. The radar antenna must be stowed before opening the radome. Antenna stow position is 0° azimuth and 60° tilted down.

Note

After the nose radome is raised and the jury strut fastened in position, release hydraulic pressure to take load off hydraulic system.

Ground Lock Stowage Container

A ground lock stowage container is located on the forward bulkhead in the nosewheel well to provide stowage for the arresting hook, nose, gear, and main gear ground safety locks. The container cover is secured closed by a quick-release latch.

Systems Test And System Power Ground Panel

The SYS TEST and SYS PWR ground check panel (figure 1-89) is on the RIO's left knee panel (accessible from the boarding ladder with the canopy open) for controlling the activation of electrical circuits using ground external power. The panel cover is designed so that when it is closed the switches inside are in the proper position for flight. In addition, when the landing gear handle is in the UP position all switches are deactivated. The panel serves a maintenance and preflight purpose and is not intended for use by the flightcrew.

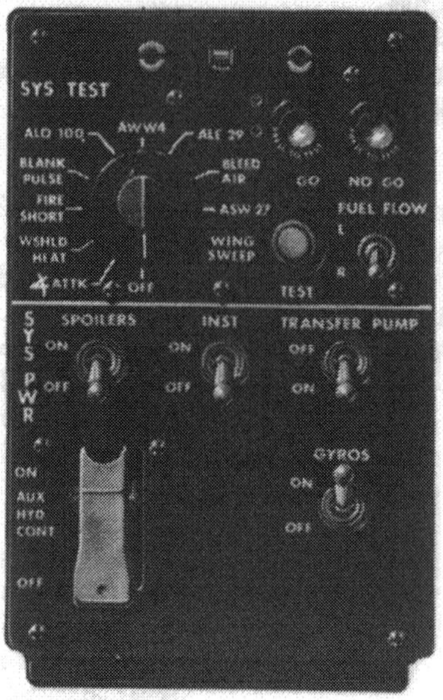

Figure 1-89. Systems Test and System Power Ground Panel

1-170

TACTICAL AIR RECONNAISSANCE POD SYSTEM (TARPS)

The tactical air reconnaissance pod system establishes the F-14A as a multisensor reconnaissance aircraft with the flexibility for a wide range of reconnaissance missions. Specific missions include target generation, prestrike and postrike photography, order-of-battle maintenance, lines of communication coverage, and maritime surveillance.

TARPS sensors and ancillary equipment are in four individual compartments within the pod (figure 1-90). The pod is configured with three major sensors: Serial Frame Camera KS-87B, Low- To Medium-Altitude Panoramic Camera, KA-99, and Infrared Line Scanner AN/AAD-5.

TARPS Pod (LA-610)

The TARPS pod (figure 1-91) is 207.5 inches long with a maximum cross section of 26.5 inches. Total weight is approximately 1,625 pounds including sensor equipment. It is attached to the aircraft by an integral adapter that provides the pod with sensor control signals, data annotation signals, electrical power, and ECS support from the aircraft. Circuit breaker protection is provided through the ac left and right main circuit breaker panels and the dc main circuit breaker panel. The non-jettisonable pod is mounted to the aircraft on the station 5 weapon rail with an integral pylon adapter. Doors on the adapter provide access to electrical and ECS power test systems and sensor system circuit breakers.

The pod is designed for carriage throughout the flight envelope of the F-14A and is easily integrated with the aircraft weapon system. The subsonic drag index of the pod with or without inert Sparrow ballest is 25 counts which is congrous with all NATOPS performance charts and operating limitations.

The F-14/TARPS functional interface is shown in figures 1-92 and 1-93.

TARPS Equipment Circuit Breakers

The main power circuit breakers that control TARPS equipment are in the aft cockpit. FO-10 and FO-11 show their location. The circuit breakers are numbered and labeled as follows:

    1F4 — RECON POD
    2B1 — RECON HTR PWR PH A
    2D1 — RECON HTR PWR PH B

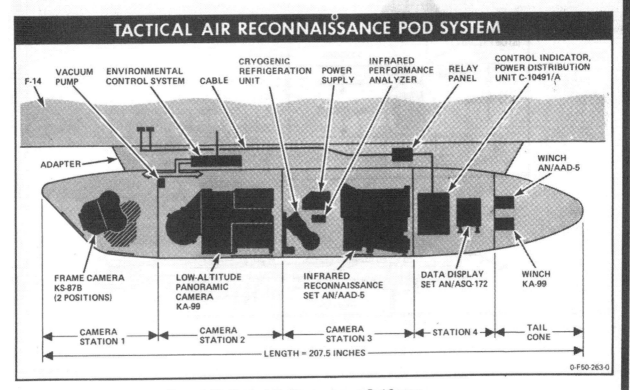

Figure 1-90. Tactical Air Reconnaissance Pod System

Section I
Part 2

NAVAIR 01-F14AAA-1

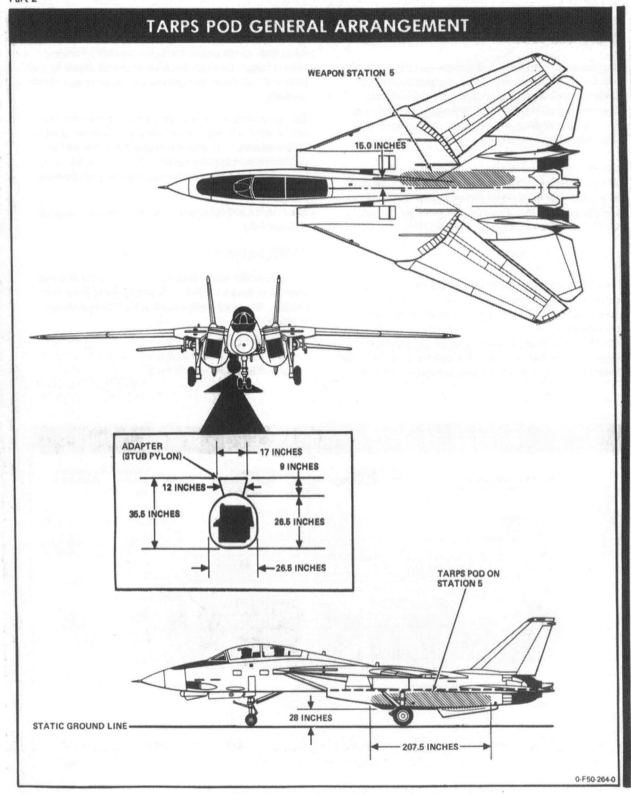

Figure 1-91. TARPS Pod General Arrangement

1-172

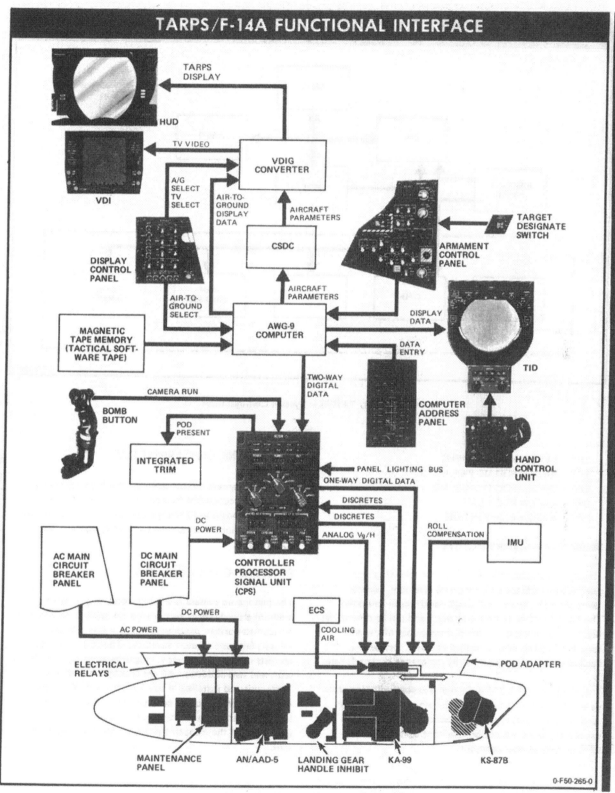

Figure 1-92. TARPS/F-14A Functional Interface

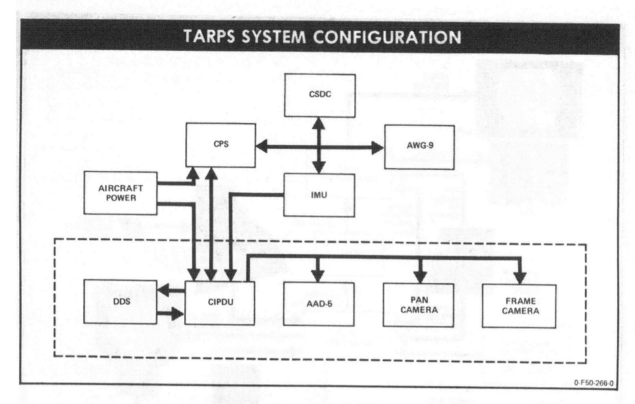

Figure 1-93. TARPS — System Configuration

2F1 — RECON HTR PWR PH C
8E1 — RECON POD DC PWR NO. 2
8E2 — RECON POD DC PWR NO. 1
8E7 — RECON POD CONT
8F7 — RECON ECS CONT DC

Serial Frame Camera (KS-87B)

The serial frame camera consists of the following major assemblies: camera body, magazine assembly, cassette assembly, light sensor, and 6-inch focal length lens cone. The KS-87B frame camera is in bay 1 and can be directed in flight either forward, to obtain photographs of the area seen by the pilot, or to a vertical position, for use as a backup sensor in case of a KA-99 panoramic camera failure.

The 77.6-pound KS-87B serial frame camera will accommodate 1,000 feet of 2.5-mil thick, 5-inch roll film, and can record up to six frames per second for a total of 2,400 exposures. Data annotation is printed using a standard CRT recording head assembly.

CAMERA MOUNT ASSEMBLY

The KS-87 camera mount assembly holds the camera and provides the capability to move the camera in flight from the vertical position to the forward position. Controls for camera positioning are on the controller processor signal (CPS) unit.

Panoramic Camera (KA-99A)

The panoramic camera is a 9-inch focal length low- to medium-altitude aerial reconnaissance camera. It offers full horizon-to-horizon panoramic imagery over a broad velocity/altitude mission envelope. The 250-pound camera consists of the camera body, magazine assembly, electronic unit, and two cassettes. All are packaged together as a single unit and installed in bay 2 of the pod. Interface of the adapter frame to the pod structure is two, in-line pivot bearings, permitting pivoting of the camera system for replacement of the magazine assembly and the electronic unit.

Infrared Line Scanner Set (IRLS) (AN/AAD-5)

The IRLS set provides a high resolution film record of terrain being traversed by the aircraft. Scanning optics receive IR energy from the area under surveillance and are focused onto two detector arrays enclosed in a vacuum sealed dewar. Electrical signals, representing the scanned area, are displayed on a cathode ray tube as video. The displayed video is reflected and focused onto film to provide a continuous strip image of the terrain. Film transport speed is determined by a V/H signal provided by the aircraft inertial navigation system. The AN/AAD-5 system is roll stabilized to $\pm 20°$ using inputs from the aircraft IMU. If IMU failure occurs, IR imagery will be badly degraded.

The system consists of six WRAs installed in bay 3 of the pod. The AN/AAD-5 system is provided with a door on the underside of the pod, which offers protection for the IR scanning optics during takeoff and landing. The door opens automatically when IR system power is applied and the landing gear handle is in the UP position. It closes when IR system power is removed or the gear handle is down.

Data Display System (AN/ASQ-172)

The data display system (DDS) performs three basic TARPS functions: (1) provides annotation on the sensor film recordings for future interpretation of the recorded intelligence data; (2) supplies necessary control signals to the individual sensors; and (3) provides the signal flow interface between the aircraft systems and the pod equipment.

TARPS Environmental Control System

The environmental control system is required to supply conditioned air for pod cooling and heating, and for defogging the camera windows. Cool air is provided by tapping into the aircraft ECS system and running a cooling air line to the adapter. The air provided to the adapter is approximately $40°F$ from sea level to 30,000 feet and $0°$ above 30,000 feet. The air is further conditioned in the adapter by the use of a water separator to remove approximately 75% of the moisture, then is heated as required to maintain pod temperature. Conditioned air is routed into the pod in bay 1 and is dumped to the atmosphere via vents in the aft section of the pod.

Air for defogging is supplied on demand. The defogging air is heated in the same manner as that for the pod compartment. A sensor that determines the window glass temperature and dew point turns on heated airflow to the three pod windows. There is a defogging shutoff valve that shuts off air to the windows when the cameras are operating.

Control Indicator Power Distribution Unit (CIPDU)

The CIDPU consists of a sensor test module, three sensor control modules, an aircraft simulator module, and printed circuit boards. In bay 4, the CIPDU provides manually resettable fail indicators for each of the various pod sensors and major pod equipment to guide maintenance personnel in the identification of faulty WRAs. The CIPDU provides the capability of initiating and monitoring the BIT functions within the pod and a means of operating the sensors, individually or together, as a verification of proper operation following corrective maintenance or for preflight checkout purposes. All pod equipment circuit breakers are on the front panel. An aircraft simulator unit is provided so that the pod equipment can be functionally checked without the aircraft computer operating.

Controller Processor Signal (CPS) Unit

The CPS and cockpit displays provide the controls and information required by the RIO and pilot for operation and checkout of TARPS. The CPS is in the aft cockpit left console (figure 1-94) and contains the primary TARPS controls and indicators. Using the CPS with the tactical information display (TID), the RIO has full control of TARPS except for circuit breakers. The controls and indicators on the CPS are described in figure 1-94. Panel illumination of the CPS is accomplished with the rest of the cockpit lighting using the same controls.

Section I
Part 2

NAVAIR 01-F14AAA-1

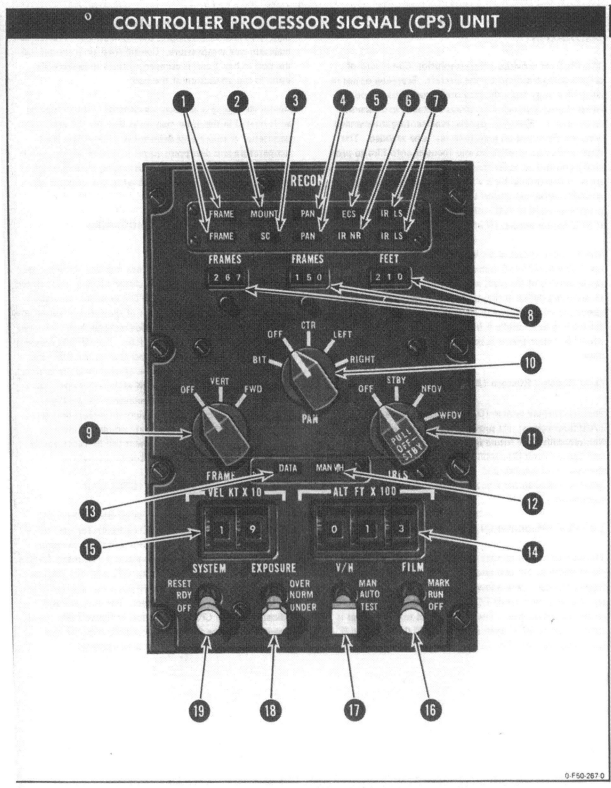

Figure 1-94. Controller Processor Signal (CPS) Unit (Sheet 1 of 4)

NOMENCLATURE	FUNCTION
① FRAME lights 1 Amber 1 Green	Green FRAME light flashes once per camera cycle when KS-87B is activated and no failure exists. Amber FRAME light illuminates if failure exists in KS-87B and green FRAME light goes off.
② MOUNT light Amber	Illuminates indicating mount failure. This occurs when KS-87B fails to achieve directed position. (It may be firmly locked in position opposite to directed one.) CIPDU internal failure can also give mount failure indication.
③ SC (sensor control) light Amber	Illuminates when SC/DDS has failed to furnish film motion compensation (FMC) or cycle commands to sensors. Failure to deliver formatted data on command to sensors will not show SC failure. Consequently, SC go indication can result in good sensor imagery operation but without data annotation.
④ PAN lights 1 Amber 1 Green	Green PAN light flashes once per camera cycle when KA-99A has been activated and no failure exists. Amber PAN light illuminates and green light goes out if failure occurs.
⑤ ECS (environmental control system light) Amber	Illuminates only under failure condition (compartment temperature below 0°C or above 51°C). ECS is automatically activated on takeoff by weight-on-wheels switch.
⑥ IR NR (IR not steady) light Amber	Illuminates when sensor is not sufficiently cooled. Cool down period is a maximum of 17 minutes. After cooldown is completed the IR NR light goes out for 25 seconds, then on for 80 seconds during BIT testing.
⑦ IR LS (IR line scanner) light 1 Green 1 Amber	Green IR LS light illuminates when infrared sensor is activated. Green IR LS light flashes once per each foot of film exposed. Green IR LS indicator goes out and amber IR LS light illuminates if failure occurs in infrared sensor.
⑧ Frame and feet (indicators)	Display number of frames remaining in frame and pan cameras, and number of feet of film remaining in infrared sensor. Indicators are set initially as part of sensor servicing via reset knobs directly under indicators. Each frame or pan camera cycle decreases indication by 1. Each foot of film cycled through IR sensor decreases feet indication by 1.
⑨ Frame camera switch	OFF — Frame camera is shut off. VERT — SYSTEM switch in RDY. Power applied to frame camera. Mount placed in vertical position. FILM switch in RUN; camera is cycling. FWD — SYSTEM switch in RDY; power is applied to frame camera. Mount is placed in forward position (depressed 16). FILM switch in RUN; camera is cycling. **Note** Requires 15 seconds to transition between FWD and VERT positions.

Figure 1-94. Controller Processor Signal (CPS) Unit (Sheet 2 of 4)

NOMENCLATURE	FUNCTION
⑩ Pan camera switch	BIT (momentary position) — SYSTEM switch must be in RDY to get BIT. Applies power to pan camera. Initiates 12-second BIT. With FILM switch in RUN, BIT will not function. OFF — Pan camera is shut off. CTR — SYSTEM switch in RDY. Pan camera enabled. Awaiting operate command. FILM switch in RUN; pan camera cycling. Exposure, average of left and right light sensors. Camera set for 55% overlap at NADIR. LEFT — SYSTEM switch in RDY. Pan camera enabled. Awaiting operate command. FILM switch in RUN; pan camera cycling. Exposure controlled by left light sensor. Camera set for 55% overlap at 30° below left horizon. RIGHT — SYSTEM switch in RDY. Pan camera enabled. Awaiting operate command. FILM switch in RUN; pan camera cycling. Exposure controlled by right light sensor. Camera set for 55% overlap at 30° right horizon. **Note** LEFT and/or RIGHT positions should only be selected for high altitude standoff, or low sun angle photography. Sensor resolution is degraded when LEFT and/or RIGHT is selected.
⑪ IRLS switch	OFF — Sensor is shut off. STBY — SYSTEM switch in RDY. CMM mode is activated. System is in cooldown mode. (Maximum of 17 minutes.) IR door is closed if gear handle is down, or if system cooldown, or BIT check is incomplete. NFOV/WFOV — SYSTEM switch in RDY. Sensor in ready mode. Sensor selects narrow (or wide) field of view in response to switch position. Mirror spin motor energized. Sensor is ready to cycle if cooldown has occurred. IR door will open if gear handle is UP. FILM switch in RUN. If cooldown has not occurred and IR NR is illuminated, sensor responds as above. If cooldown has occurred, IR NR is off, sensor will cycle. Green IR LS light flashes once per foot of film travel.

Figure 1-94. Controller Processor Signal (CPS) Unit (Sheet 3 of 4)

NAVAIR 01-F14AAA-1

Section I
Part 2

NOMENCLATURE	FUNCTION
(12) MAN V/H light	Off — Vg/H from aircraft computer within acceptable limits. Illuminated amber: • V/H switch in TEST. With V/H set to a 1.8 ratio on the thumbwheel, thumbwheel circuitry has failed if light does not go out. • V/H switch in AUTO. Computer fail discrete received. Computer failed. Manual Vg/H being used. Set correct values to Vg/H under thumbwheels. Set V/H switch to MAN. • V/H switch in MAN. Manual Vg/H intentionally selected. Values set under thumbwheels being used. Set correct values under thumbwheels.
(13) DATA light	Off — Data received from aircraft computer. Illuminated amber: Data from aircraft computer failed.
(14) ALT FT X100 thumbwheel	Used to set manual altitude inputs to pod. Counter range is from 000 to 999, read in multiples of 100 feet.
(15) VEL KT X10 thumbwheels	Used to set manual ground speed in knots inputs to pod. Counter range is from 00 to 99, read in multiples of 10 knots.
(16) FILM switch	MARK (momentary position) — Allows RIO to mark special interest frame with X in data block. RUN — Activates selected sensor when SYSTEM switch is set to RDY. OFF — Stops TARPS sensor operation unless activated by pilot's bomb button, if sensors are selected to operate position on CPS.
(17) V/H selector switch	MANUAL — Selects manual thumbwheel inputs. AUTO — Selects aircraft computer inputs. TEST (momentary position) — Tests proper functioning of thumbwheels Vg/H circuitry.
(18) EXPOSURE selector switch	UNDER — One f-stop exposure for doubled S/C setting. NORM — Normal exposure for doubled S/C film setting. OVER — +1 exposure for doubled S/C film setting.
(19) SYSTEM switch	OFF — Aircraft power denied to TARPS. No sensors can be operated. RDY — Aircraft power available at sensor connectors. If respective sensor moved from OFF position, sensor is placed in standby or ready mode. Pilot can cycle any sensor placed in ready mode by pressing bomb button. RESET — Clears TARPS failure signal and removes POD acronym from TID. If failure is other than transient, CPS failure (amber) lights illuminate.

Figure 1-94. Controller Processor Signal (CPS) Unit (Sheet 4 of 4)

PART 3 — AIRCRAFT SERVICING

SERVICING DATA

The following servicing data is for use of the flightcrew members and for maintenance crews who are unfamiliar with servicing the aircraft. (See figure 1-95.) When operating in and out of military airfields, consult the current DOD IFR Supplement for compatible servicing units, fuel, etc. Figure 1-96 contains a tabulation of servicing data and power units required to support the aircraft.

Ground Refueling

Single-point refueling is provided for pressure filling of all aircraft fuel tanks through a standard refueling receptacle on the lower right side of the forward fuselage. Ground refueling is controlled by two precheck selector valves and the vent pressure gage adjacent to the refuel receptacle on the ground refuel and defuel panel. Positioning of these valves can be used for selective ground refueling of either the fuselage or wing and drop tanks. The direct reading vent pressure gage indicates pressure in the system vent lines. When aircraft fuel tanks are full, fueling stops automatically. For hot refueling procedures, refer to Section III. For defueling procedures, refer to NAVAIR 01-F14AAA-2-1.

The maximum refueling rate is approximately 500 gallons per minute at a pressure of 50 psi. Nominal and minimum pressure is approximately 15 psi, maximum pressure is 50 psi.

WARNING

Ensure that both the fueling unit and the aircraft are properly grounded, bonding cable is connected between aircraft and refueling source, and that fire extinguishing equipment is readily available.

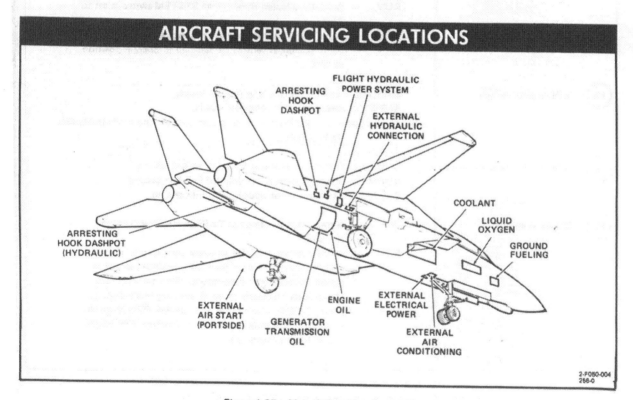

Figure 1-95. Aircraft Servicing Locations

AIRCRAFT SERVICING

ITEM	DESIGNATION SPECIFICATION	NATO CODE	COMMERCIAL EQUIVALENT	DOD IFR SUPPLEMENT CODE	REMARKS
FUEL	MIL-T-5624 (JP-5) MIL-T-5624 (JP-4) MIL-T-83133 (JP-8) (STRAW COLOR)	F-44 F-40 F-34	JET B JET A JET A-1	JP-5 JP-4 JP-8	IF JP-5 OR JP-8 IS USED, SELECTOR (MAIN FUEL CONTROL) ON BOTH ENGINES SHALL BE SET FOR JP-5 FUEL.
ENGINE OIL	MIL-L-23699 MIL-L-7808	O-156 O-148	NONE NONE	O-156 O-148	USE MIL-L-7808 WHEN GROUND TEMPERATURE IS −40°F (−40°C).
GENERATOR-TRANSMISSION OIL	MIL-L-23699 MIL-L-7808	O-156 O-148	NONE NONE	O-156 O-148	USE MIL-L-7808 WHEN GROUND TEMPERATURE IS −40°F (−40°C).
HYDRAULIC FLUID	MIL-H-83282 MIL-H-5606 (RED)	NONE H-515	NONE NONE	NONE NONE	MIL-H-83282 IS PRIMARY. EITHER FLUID MAY BE USED.
OXYGEN (LIQUID)	MIL-O-27210 TYPE II	NONE	NONE	LOX	
OXYGEN (GASEOUS)	MIL-O-27210 TYPE I	NONE	NONE	HPOX LHOX	SURVIVAL KIT SHALL BE REMOVED FROM AIRCRAFT FOR SERVICING EMERGENCY OXYGEN BOTTLE.
NITROGEN	BB-N-411 (TYPE I, GRADE A)	NONE	NONE	NONE	USE CLEAN, OIL-FREE FILTERED DRY AIR, IF NITROGEN IS NOT AVAILABLE.
LIQUID COOLANT	COOLANT 25 (MONSANTO CHEMICAL CO) CHEVRON FLO-COOL 180 (CHEVRON CHEMICAL CO)	NONE NONE	NA NA	NONE NONE	EITHER COOLANT MAY BE MIXED WIHOUT ADVERSE REACTION.
WINDSHIELD RAIN REPELLANT	MIL-R-81261	NONE	NONE	NONE	DO NOT USE ON DRY WINDSHIELD

POWER UNITS

	PNEUMATIC STARTING	ELECTRICAL POWER	AIR CONDITIONING	HYDRAULICS
ACCEPTABLE USN/USAF UNITS	GTC85-72 MA-3MPSU MA-2 A/M 32A-60	NC8A MD-3 MD-3A MD-3M MA-3MPSU A/M32A-60 A/M32A-60A	NR5C/8 MA-1 MA-1A A/M32C-5 A/M32C-6	AHT-63/64 TTU-228/E (AHT-73) MJ-3
GROUND SUPPORT EQUIPMENT REQUIREMENTS	120 lb/min AT 49 ± 3 PSI AND 50°F	115±20 V ac, 400 ± 25 Hz, 60 kVA, 3 PHASE ROTATION	70 lb/min AT 3 psi AND 60°F	50 gal/min MAXIMUM AT 3000 PSI

Figure 1-96. Aircraft Servicing (Sheet 1 of 2)

PNEUMATIC PRESSURE

SYSTEM	PRESSURE
EMERGENCY LANDING GEAR	3000 psi AT 70°F
COMBINED HYDRAULIC	1800 psi AT 70°F
FLIGHT HYDRAULIC	1800 psi AT 70°F
CANOPY NORMAL (1200 psi MINIMUM)	3000 psi AT 70°F
CANOPY AUXILIARY (800 psi MINIMUM)	3000 psi AT 70°F
WHEEL BRAKE ACCUMULATORS (2) (AIRCRAFT BUNO 161167 AND EARLIER WITHOUT AFC 622)	950 psi AT 70°F
WHEEL BRAKE ACCUMULATORS (2) (AIRCRAFT BUNO 161168 AND SUBSEQUENT AND AIRCRAFT INCORPORATING AFC 622)	1900 psi AT 70°F
ARRESTING HOOK DASHPOT	800 ± 10 psi
MAIN GEAR SHOCK STRUTS (2)	980 psi
NOSE GEAR SHOCK STRUT	1300 psi
WINDSHIELD RAIN REPELLANT	75 psi

CAUTION

REPELLANT SHALL NOT BE SPRAYED ON A DRY WINDSHIELD BECAUSE IT MAY DAMAGE THE WINDSHIELD DURING CLEANING.

NOTE

DRY NITROGEN, SPECIFICATION BB-N-41B, TYPE 1, GRADE A, IS PREFERRED FOR TIRE INFLATION AND CHARGING PNEUMATIC SYSTEMS SINCE IT IS INERT, AND THEREFORE WILL NOT SUPPORT COMBUSTION.

TIRES

TYPE	OPERATION	PRESSURE
NOSE (2) 22 × 6.6-10 20 PLY	ASHORE	105 PSI
	AFLOAT	350 PSI
MAIN (2) 37 × 11.50 -16 28 PLY	ASHORE	245 PSI
	AFLOAT	350 PSI

Figure 1-96. Aircraft Servicing (Sheet 2 of 2)

NAVAIR 01-F14AAA-1

Section I
Part 3

CAUTION

During ground refueling operations, the direct-reading vent pressure indicator shall be observed and refueling stopped if pressure indication is in the red band (above 4 psi).

Note

- If the aircraft is being regularly serviced with JP-4 type fuel, the fuel-control fuel-grade (specific gravity adjustment) selector on each engine should be reset to the JP-4 position. If the aircraft is being regularly serviced with JP-8 fuel, the fuel-control fuel-grade (specific gravity adjustment) selector on each engine should be reset to the JP-5 position. Satisfactory engine performance depends upon trimming of the engine fuel controls to ensure rated thrust, to prevent exceeding engine temperature limits, and to ensure airflow compatibility with the air inlet duct opening.

- Removal of JP-8 type fuel from the aircraft is not required before refueling with JP-5. If removal of JP-8 from the aircraft aboard ship is necessary, it shall not be defueled into the storage tanks containing JP-5.

Engine Oil

Engine oil level is checked with a dipstick (figure 1-97, sheet 2) on the outboard side of the accessory gear box of each engine. Servicing is by the gravity method, using the outboard filler on each engine. Normal oil consumption rate is 0.3 gallon per hour.

CAUTION

Proper use of the correct tapered alignment tool (3/8 inch diameter at narrow tip) is needed to open inspection door latches. Do not use a screwdriver, punch, or other implement as a substitute for the alignment tool. Any damage to the doors or one of the latches requires that the aircraft be downed for structural damage.

Note

Engine oil quantity should be checked within 15 minutes after engine shutdown to ensure proper servicing.

Generator-Transmission (CSD) Oil

The generator-transmission (CSD) oil level sight gage is on the accessory gear box of each engine, aft of the engine oil filler, (see figure 1-97, sheet 2.)

To obtain a valid fluid level check, the sight gage must be checked no sooner than 5 minutes and no later than 60 minutes after engine shutdown. CSD oil can be serviced at the gravity fill cap or the pressure fill fitting.

Hydraulic Systems

The main hydraulic systems are serviced at the flight and combined hydraulic system ground servicing panels. A hydraulic pressure filling cart is required to service the systems with fluid, and an air-nitrogen cart to preload the reservoirs. The outboard spoiler backup module is serviced at the servicing panel on the outboard nacelle of the port engine. Additional hydraulic servicing is required at the main landing gear shock strut (figure 1-97, sheet 2), the nosewheel shock strut (figure 1-97, sheet 2), and the arresting hook dashpot. (See figure 1-97, sheet 1.)

The flight reservoir fill and ground power access panel and the flight system filter module (figure 1-97, sheet 1) are on the starboard side. Indication of hydraulic system fluid contamination can be detected by the position of the buttons on the Delta-P type filter units.

The combined hydraulic system reservoir fill and filter module (figure 1-97, sheet 3) are on the port side of the aircraft. A temperature recording gage in the same area indicates the maximum temperature attained by the hydraulic fluid during the last turnup or flight. After a reading has been taken, the temperature gage must be reset prior to the next turnup.

Pneumatic System

The pneumatic power supply systems, which provide for normal operation of the canopy and for emergency extension of the landing gear, are ground charged through a common filler in the nosewheel well. (See figure 1-97, sheet 3.) The auxiliary canopy open pneumatic bottle is in the turtleback behind the cockpit. (See figure 1-97, sheet 1.) Additional

Section I
Part 3

NAVAIR 01-F14AAA-1

AIRCRAFT SERVICING

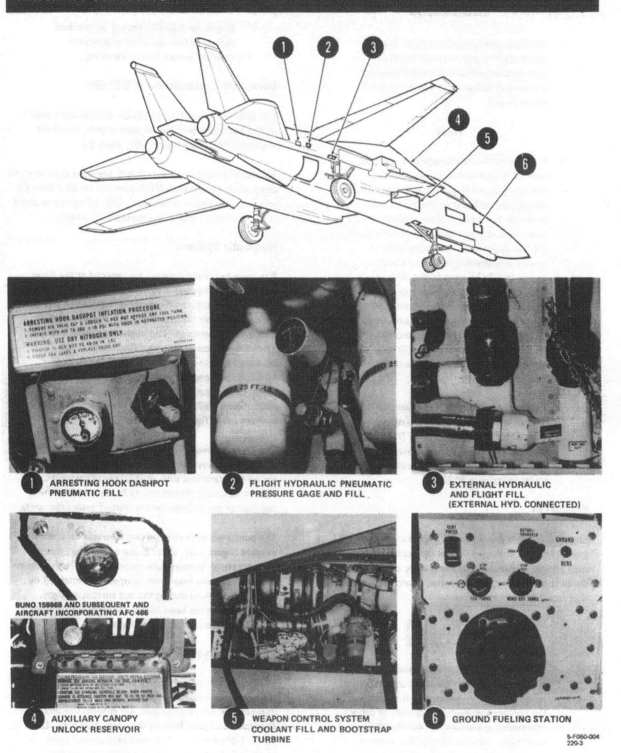

1. ARRESTING HOOK DASHPOT PNEUMATIC FILL
2. FLIGHT HYDRAULIC PNEUMATIC PRESSURE GAGE AND FILL
3. EXTERNAL HYDRAULIC AND FLIGHT FILL (EXTERNAL HYD. CONNECTED)
4. AUXILIARY CANOPY UNLOCK RESERVOIR
5. WEAPON CONTROL SYSTEM COOLANT FILL AND BOOTSTRAP TURBINE
6. GROUND FUELING STATION

Figure 1-97. Aircraft Servicing (Sheet 1 of 3)

NAVAIR 01-F14AAA-1

AIRCRAFT SERVICING CONT'D

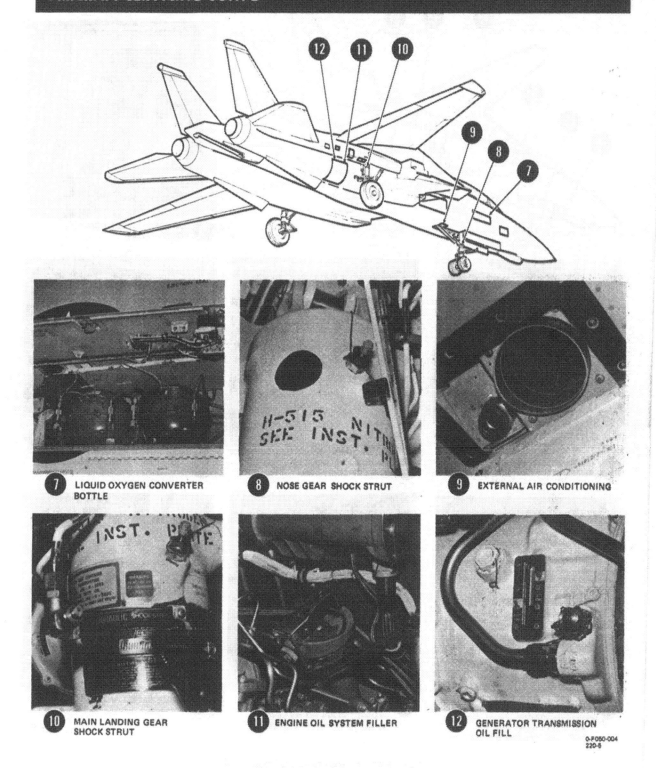

Figure 1-97. Aircraft Servicing (Sheet 2 of 3)

Section I
Part 3

NAVAIR 01-F14AAA-1

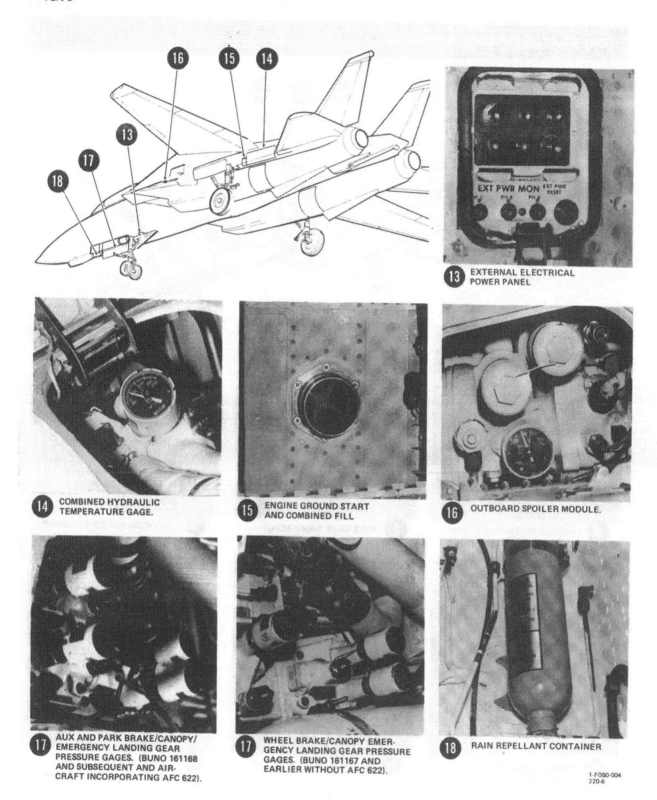

⑬ EXTERNAL ELECTRICAL POWER PANEL

⑭ COMBINED HYDRAULIC TEMPERATURE GAGE.

⑮ ENGINE GROUND START AND COMBINED FILL

⑯ OUTBOARD SPOILER MODULE.

⑰ AUX AND PARK BRAKE/CANOPY/ EMERGENCY LANDING GEAR PRESSURE GAGES. (BUNO 161168 AND SUBSEQUENT AND AIRCRAFT INCORPORATING AFC 622).

⑰ WHEEL BRAKE/CANOPY EMERGENCY LANDING GEAR PRESSURE GAGES. (BUNO 161167 AND EARLIER WITHOUT AFC 622).

⑱ RAIN REPELLANT CONTAINER

Figure 1-97. Aircraft Servicing (Sheet 3 of 3)

pneumatic servicing points are at both hydraulic systems, servicing panels, brake systems, and arresting hook.

In aircraft BUNO 16118 and subsequent and aircraft incorporating AFC 622, individual pneumatic servicing points and pressure gages are provided for the auxiliary and parking brake systems.

Note

Dry nitrogen, specification BB-N-411B, Type 1, Grade A, is preferred for tire inflation and for charging pneumatic systems since it is inert and therefore will not support combustion.

GROUND HANDLING

Danger Areas

Engine exhaust and intake danger areas are shown in figure 1-98. Noise danger areas are shown in figure 1-99. Figure 1-98, sheet 1 shows temperature distribution with afterburners at maximum nozzle opening for idle, military, and maximum power (zone 5). Figure 1-98, sheet 2 shows exhaust jet wake velocity distribution with afterburner at maximum nozzle opening for idle, military, and maximum power (zone 5).

Radiation Hazard Areas

Radiation hazards from high-power radio and radar radiation are included in Section VIII, NAVAIR 01-F14AAA-1A.

Fuel Ignition Hazard

When performing fueling or defueling operations, use minimum safe distances outside of radiation hazard areas. Fuel ignition hazard occurs within 90 feet of the aircraft where rf radiation induced sparks could ignite flammable vapors of fuels. Fuel ignition hazard is based on $5W/cm^2$ peak power density.

Good housekeeping operations are of utmost importance in areas where radar transmission is anticipated. RF radiation may cause steel wool to be set afire or metallic chips to produce sparks, which in turn may ignite spilled fuels or oils around aircraft and buildings. Keep all areas clean and refuse in approved containers.

Transmission Aboard Carrier

Radar transmission aboard carrier shall be limited to over the side operation at the discretion of the commander. The aircraft shall be spotted so the nose radome overhangs the side of the carrier. All necessary safety precautions shall be enforced to prevent injury to personnel and damage to equipment aboard the carrier and on adjacent ships that may accidentally stray into the main beam of the radar.

Towing Turn Radii and Ground Clearances

Forward and rearward towing (figure 1-100) can be accomplished with a standard tow bar (NT-4 aircraft universal tow bar) and the tow tractor. The pilot cockpit shall be manned with qualified personnel during towing operations.

CAUTION

Before and during towing, ensure that the needle(s) in the AUX/PARK Brake pressure gauge(s) remain in the green band to ensure sufficient pressure to lock the wheels.

Tiedown Points

Aircraft tiedown points are illustrated in figure 1-101. When mooring a parked aircraft, do not depend upon chocks alone to hold the aircraft in position. Tiedowns shall be installed in a symmetrical pattern being careful not to chafe against the aircraft structure.

The normal six-point tiedown (figure 1-101, sheet 1) locations permit all maintenance servicing including engine removal, jacking and weapons loading. Standard chain-type tiedowns are used for an 18-point symmetrical tiedown during heavy weather (figure 1-101, sheet 2).

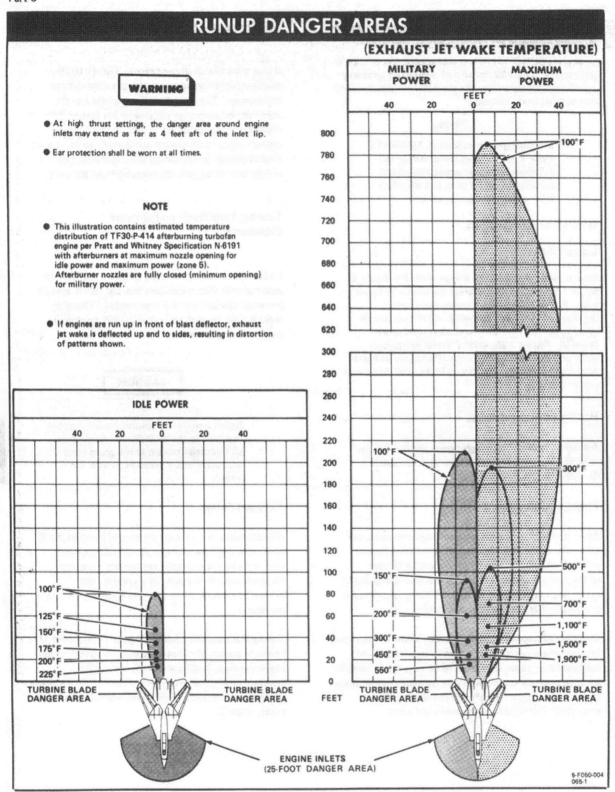

Figure 1-98. Runup Danger Areas (Sheet 1 of 2)

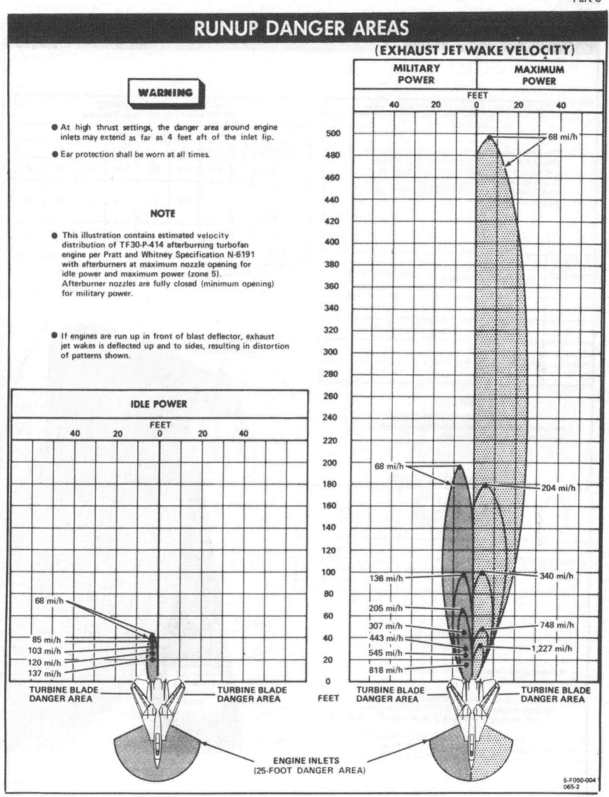

Figure 1-98. Runup Danger Areas (Sheet 2 of 2)

NOISE DANGER AREAS

SOUND LEVEL - IN ADJUSTED DECIBELS (dBa)

This table contains estimated sound levels, in dBa, for TF30-P-414 afterburning turbofan engines. Sound level contour letters A, B, C, and D (shown on this illustration) represents specific dBa value. When dBa value is located on the table, it shall be substituted in place of contour letter.

CONTOUR	MAXIMUM AFTERBURNER POWER	MILITARY POWER
A	134 dBa	123 dBa
B	130 dBa	119 dBa
C	127 dBa	116 dBa
D	123 dBa	112 dBa

NOTE

Sound levels contours shown are for single engine operation. Contours are symmetrical about engine centerlines during engine operation.

EAR PROTECTIVE DEVICES

UNIVERSAL FIT EARPLUG	EARPLUG V-51-R TYPE OR SIMILAR	TYPICAL EARMUFF	FLIGHT DECK SOUND-ATTENUATING HELMET (INCLUDES EARMUFF)

WARNING

- Earplugs and earmuffs shall be worn together when performing engine runup.

- If engines are run up in front of blast deflector, exhaust jet wake and sound shall be deflected up and to sides, resulting in distortion to contour patterns shown.

- At maximum power, afterburners are at maximum nozzle opening (zone 5). At military power, the nozzles are fully closed (minimum opening).

ALLOWABLE NOISE EXPOSURE
SOUND LEVEL IN dBa (Slow Response)

TYPE EAR PROTECTIVE DEVICES	EXPOSURE TIME, HOURS*						
	1/4	1/2	1	2	4	6	8
NO PROTECTION	115	110	105	100	95	92	90
EARPLUGS WITH AVERAGE SEAL	127	122	117	112	107	104	102
EAR PLUGS AND EARMUFFS	135	130	125	120	115	112	110

*Duration of exposure per day
Ref. BUMED Inst 6260.6B, 5 March 1970

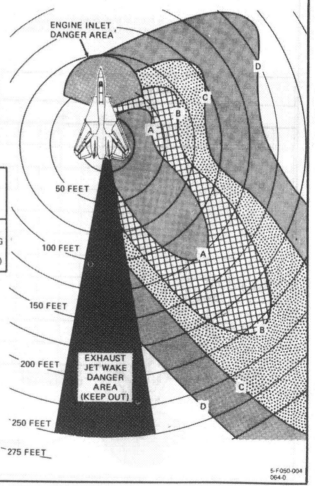

Figure 1-99. Noise Danger Areas

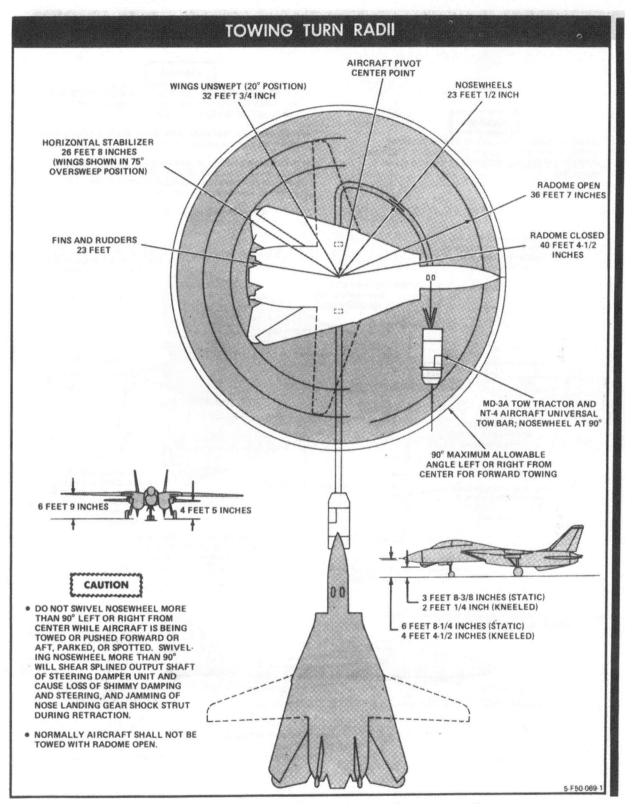

Figure 1-100. Towing Turn Radii and Ground Clearances (Sheet 1 of 2)

Section I
Part 3

NAVAIR 01-F14AAA-1

TOWING

CAUTION
- AUTHORIZED BRAKE RIDER SHALL MAN PILOT STATION AT ALL TIMES DURING TOWING, EXCEPT WHEN TOWING AIRCRAFT WITH INOPERATIVE BRAKES.
- WING AND TAIL WALKERS, AND MAIN LANDING GEAR CHOCK MEN SHALL BE AVAILABLE AT ALL TIMES DURING TOWING.
- IN NOSEWHEEL WELL, BRAKE ACCUMULATORS PRESSURE GAGE SHALL INDICATE 3000 psi. ON PILOT CENTER CONSOLE, BRAKE PRESSURE INDICATOR NEEDLE SHALL BE AT RIGHT END OF AUX GREEN BAND.

CAUTION
DO NOT EXCEED 120° MAXIMUM AVAILABLE NOSEWHEEL SWIVEL ANGLE DURING PARKING OR SPOTTING IN TIGHT SPACES; STEERING DAMPER UNIT WILL BE DAMAGED.

- WINGS SHOWN IN 68° SWEPT POSITION
- NOSE MINIMUM TURN RADIUS 39 FEET 7 INCHES WHEN PIVOTING ABOUT CENTER POINT
- 23 FEET 1/2 INCH AIRCRAFT PIVOT CENTER POINT
- HORIZONTAL STABILIZER MINIMUM TURN RADIUS 26 FEET 8 INCHES WHEN PIVOTING ABOUT CENTER POINT
- 90° MAXIMUM ALLOWABLE ANGLE FOR FORWARD TOWING (AIRCRAFT WILL PIVOT ABOUT CENTER POINT)
- 110° ANGLE REQUIRED FOR PIVOTING ABOUT LEFT MAIN LANDING GEAR WHEEL
- 16 FEET 5 INCHES
- 72° MAXIMUM TOW TRACTOR SWIVEL ANGLE AVAILABLE FOR AFT TOWING. TOW BAR (TILLER BAR) IS REQUIRED AT NOSEWHEELS FOR DIRECTIONAL CONTROL
- 120° MAXIMUM AVAILABLE NOSEWHEEL SWIVEL ANGLE LEFT OR RIGHT FROM CENTER
- EMERGENCY AFT TOWING FROM MAIN LANDING GEAR TIE DOWN RING (2). TOW BAR (TILLER BAR) IS REQUIRED AT NOSEWHEELS FOR DIRECTIONAL CONTROL

NOTE
AIRCRAFT CAN BE TOWED WITH WINGS IN ANY SWEEP ANGLE FROM 20° (UNSWEPT) TO 75° (OVERSWEPT)

- MD-3A TOW TRACTOR, TA-75 TOW TRACTOR, OR ANY TRACTOR OVER 3 FEET 9 INCHES
- AIRCRAFT TOW FITTINGS INCORPORATED IN LOWER STRUCTURE OF EACH NACELLE
- MD-3A TOW TRACTOR
- NT-4 AIRCRAFT UNIVERSAL TOW BAR

Figure 1-100. Towing Turn Radii and Ground Clearances (Sheet 2 of 2)

NAVAIR 01-F14AAA-1

Section I
Part 3

TIEDOWN ARRANGEMENT

PERMANENT (TEN-POINT) TIEDOWN ARRANGEMENT

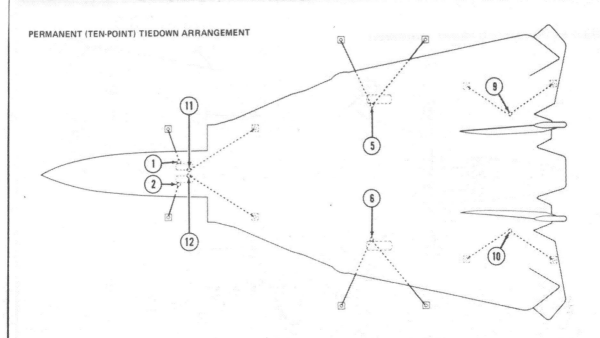

AIRCRAFT TIEDOWN FITTINGS

LOCATION	FITTING NO.	ACCESS NO.	HEIGHT ABOVE DECK	FUSELAGE STATION
Fuselage	1	2221-4	62 INCHES	296.0
	2	1221-6		
Main Gear Shock Strut (Outboard Side)	5	—	43 INCHES	567.0
	6			
Main Gear Shock Strut (Lower Inboard Side)	7	—	10.5 INCHES	569.0
	8			
Sponson (Below Horizontal Stabilizer)	9	6232-3	60 INCHES	747.0
	10	5232-3		
Nosegear Drag Brace	11	—	44 INCHES	297.9
	12			

Figure 1-101. Tiedown Arrangement (Sheet 1 of 2)

TIEDOWN ARRANGEMENT

HEAVY WEATHER (18-POINT) TIEDOWN ARRANGEMENT

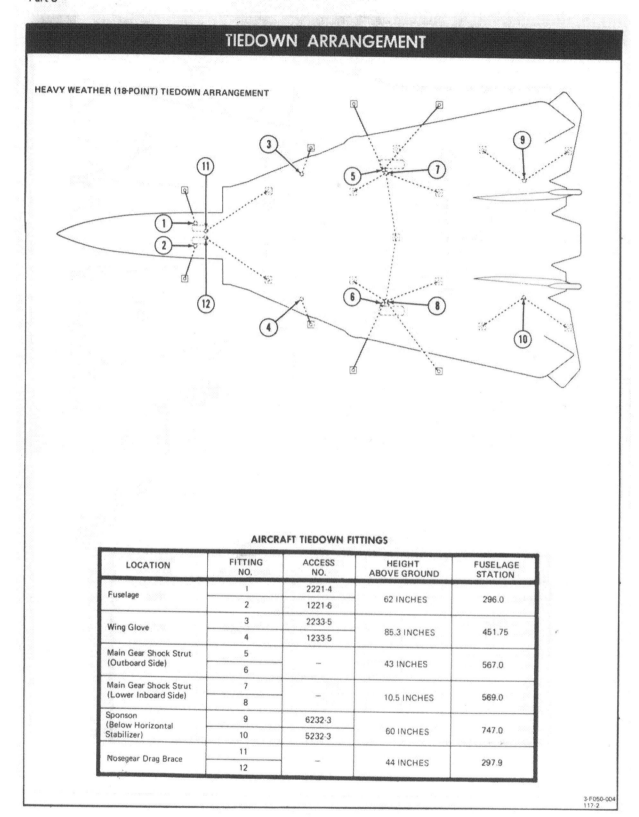

AIRCRAFT TIEDOWN FITTINGS

LOCATION	FITTING NO.	ACCESS NO.	HEIGHT ABOVE GROUND	FUSELAGE STATION
Fuselage	1	2221-4	62 INCHES	296.0
	2	1221-6		
Wing Glove	3	2233-5	85.3 INCHES	451.75
	4	1233-5		
Main Gear Shock Strut (Outboard Side)	5	—	43 INCHES	567.0
	6			
Main Gear Shock Strut (Lower Inboard Side)	7	—	10.5 INCHES	569.0
	8			
Sponson (Below Horizontal Stabilizer)	9	6232-3	60 INCHES	747.0
	10	5232-3		
Nosegear Drag Brace	11	—	44 INCHES	297.9
	12			

Figure 1-101. Tiedown Arrangement (Sheet 2 of 2)

PART 4 — OPERATING LIMITATIONS

LIMITATIONS

This section includes the aircraft and engine limitations that must be observed during normal operations. Engine instrument markings for various operation limitations are shown in figure 1-102. Engine inspection requirements for overtemperature as a result of engine stall are shown in figure 1-103. Engine operating limitations are shown in figure 1-104.

Note

The aerodynamic and structural limitations in this section apply only to aircraft BUNO 158620 and subsequent for the store configurations in figure 1-105. Engine limitations apply to all aircraft with the TF30-P-414 engine.

Operating Limitations Technical Directives

AFC	Title
166	Super Single CADC Processor
255	Nose Landing Gear Drag Brace Redesign
291	Preclude Inadvertent Firing Signals to the GCU
304	Add Counter Weights to Eliminate Static and Dynamic Unbalance of Lateral Control
341	Negative g Flapper Valve
364	Maneuver Slat Installation
365	Integrated Trim System
371	AFCS Interim Autopilot Relief Mode
400	Aileron to Rudder Interconnect
400	Teflon Lines to Alpha Computer

AFC	Title
441	AICS Revision 4 Program
454	Increase Strength of Longeron and Eliminate Fatigue Cracks on Upper Deck
461	Reinforce Upper Nacelle Longeron
467	Gun Bay General Cleanup
488	AFCS With Relief Mode Changes
536	Forward Fixed Cowl Skin Reinforcement

AVC	Title
1539	Modification of Control Stick Trigger Switch
1565	Spoiler Ground Roll Brake Test
1588	Approach Power Control Set Redesign
1709	Modification of Control Stick Trigger Switch

AFB	Title
198	Inspection of Flight Control Systems Angle-of-Attack Filter

Starter Limits

Starter Cranking Limits:

1 minute or less ON, allowed 5 cycles } then 5 minutes OFF
2 minutes ON, allowed 1 cycle

Windmill Airstart Envelope

The engine windmill airstart envelope is shown in section V, part 3 Emergency procedures.

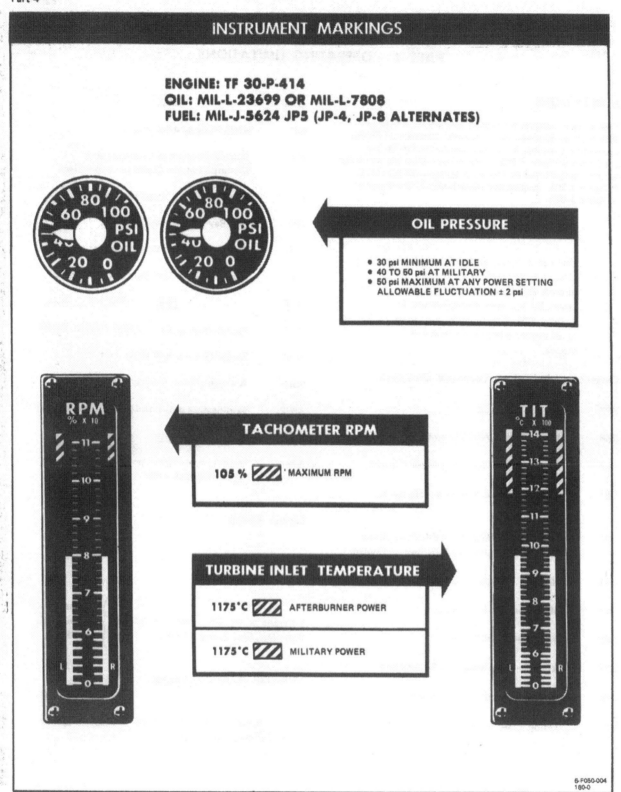

Figure 1-102. Instrument Markings

NAVAIR 01-F14AAA-1

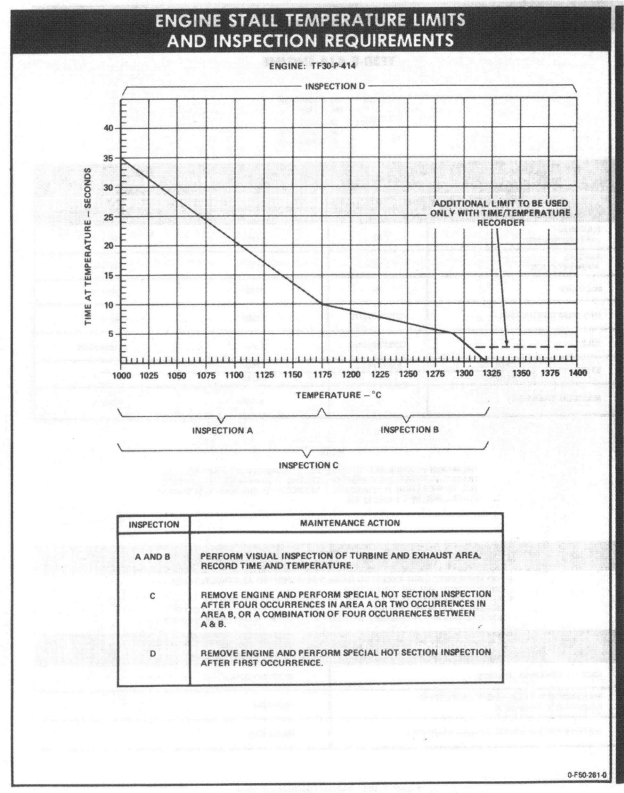

Figure 1-103. Engine Stall Temperature Limits and Inspection Requirements

Section I
Part 4

NAVAIR 01-F14AAA-1

ENGINE OPERATING LIMITS

TF30-P-414 ENGINE

OIL: MIL-L-23699 OR
MIL-L-7808

FUEL: MIL-J-5624
JP-5 (JP-4, JP-8 ALTERNATES)

OPERATING CONDITIONS		OPERATING LIMITS	
THRUST SETTING	TIME LIMIT (MINUTES)	MAXIMUM MEASURED TURBINE-INLET TEMP (C°)	NORMAL OIL PRESSURE (PSIG)
MAXIMUM (AFTERBURNING)	45	1175°	45 ± 5
PARTIAL AUGMENTATION	45	1175°	45 ± 5
MILITARY	45	1175°	45 ± 5
MAXIMUM CONTINUOUS	CONTINUOUS	1020°	45 ± 5
IDLE	CONTINUOUS	—	30 MINIMUM
STARTING — GROUND	5 SECONDS	750°	—
STARTING — FLIGHT	—	870°	—
MAXIMUM TRANSIENT	2	1195°	45 ± 5

NOTE

THE MAXIMUM MEASURED TURBINE INLET TEMPERATURE LIMIT FOR THE TRANSIENT OPERATING CONDITION FOR THE TF30-P-414 ENGINE IS 1195°C. THIS TEMPERATURE IS PERMISSIBLE, IN EXCESS OF THE MAXIMUM STEADY-STATE LIMIT, UP TO 2 MINUTES.

RPM LIMITS

- ANY OVERSPEED LIMIT EXCEEDED SHOULD BE REPORTED AS A DISCREPANCY AND MAXIMUM RPM AND TIME NOTED.
- N_2 RPM EXCURSIONS UP TO 106% FOR TEN SECONDS AFTER THROTTLE MOVEMENT FROM AFTERBURNER TO INTERMEDIATE POWER DURING SUPERSONIC FLIGHT ARE PERMISSIBLE WITHOUT SUBSEQUENT MAINTENANCE ACTION.

OPERATING CONDITIONS	OPERATING LIMITS
IDLE (SEA LEVEL STATIC)	63 TO 74% RPM
MAXIMUM RPM SPEED, ENGINE INSPECTION REQUIRED IF EXCEEDED.	105% RPM
RPM OVERSPEED, ENGINE CHANGE REQUIRED.	106.2% RPM

Figure 1-104. Engine Operating Limits

NAVAIR 01-F14AAA-1

Section I
Part 4

STORE STATION CONFIGURATION

STORE CONFIGURATION	AIRCRAFT STORE STATION						
	1A	1B	2	3 4 5 & 6	7	8B	8A
1A(*)	–	–	–	–	–	–	–
1B1	AIM-9	AIM-9	–	–	–	AIM-9	AIM-9
1B2	AIM-9	–	–	–	–	–	AIM-9
1C	–	–	TANK	–	TANK	–	–
2A(*)	–	–	–	4 AIM-7	–	–	–
2B1	AIM-9	AIM-9	–	4 AIM-7	–	AIM-9	AIM-9
2B2	AIM-9	–	–	4 AIM-7	–	–	AIM-9
2B3	AIM-9	AIM-7	–	4 AIM-7	–	AIM-7	AIM-9
2B4	–	AIM-7	–	4 AIM-7	–	AIM-7	–
2C(*)	–	–	TANK	4 AIM-7	TANK	–	–
2C1	AIM-9	AIM-9	TANK	4 AIM-7	TANK	AIM-9	AIM-9
2C2	AIM-9	–	TANK	4 AIM-7	TANK	–	AIM-9
2C3	AIM-9	AIM-7	TANK	4 AIM-7	TANK	AIM-7	AIM-9
2C4	–	AIM-7	TANK	4 AIM-7	TANK	AIM-7	–
3A(*)	–	–	–	4 AIM-54	–	–	–
3B1	AIM-9	AIM-9	–	4 AIM-54	–	AIM-9	AIM-9
3B2	AIM-9	–	–	4 AIM-54	–	–	AIM-9
3B3	AIM-9	AIM-7	–	4 AIM-54	–	AIM-7	AIM-9
3B4	–	AIM-7	–	4 AIM-54	–	AIM-7	–
3B5	AIM-9	AIM-54	–	4 AIM-54	–	AIM-54	AIM-9
3B6	–	AIM-54	–	4 AIM-54	–	AIM-54	–
3C(*)	–	–	TANK	4 AIM-54	TANK	–	–
3C1	AIM-9	AIM-9	TANK	4 AIM-54	TANK	AIM-9	AIM-9
3C2	AIM-9	–	TANK	4 AIM-54	TANK	–	AIM-9
3C3	AIM-9	AIM-7	TANK	4 AIM-54	TANK	AIM-7	AIM-9
3C4	–	AIM-7	TANK	4 AIM-54	TANK	AIM-7	–
3C5	AIM-9	AIM-54	TANK	4 AIM-54	TANK	AIM-54	AIM-9
3C6	–	AIM-54	TANK	4 AIM-54	TANK	AIM-54	–

- (*) These store configuration limits also apply when multipurpose stub pylons are carried at stations 1 and 8.
- Flight operating limitations applicable to the above configurations are also applicable to down loadings therefrom. In all cases the center-of-gravity position must remain within the center-of-gravity position limits.
- For captive carriage of inert or live AIM-54, installation of ejector cartridges in LAU-93 is mandatory in order to provide jettison capability.

WARNING

With MASTER ARM ON and all conditions satisfied for AIM-54 launch, an ATM-54 (training round) will be ejected if the trigger or launch button is depressed.

Figure 1-105. Store Station Configuration

Section I
Part 4

AIRSPEED LIMITATIONS

The limits and restrictions contained in this part represent the maximum capability of the aircraft comensurate with safe operations. Aerodynamic and structural excesses of these limits shall be logged in the Yellow Sheet, Part A (OPNAV Form 3760-2) for appropriate maintenance action.

Maximum Airspeeds

Maximum speeds are presented in indicated knots and Mach number. These values are derived from the position-error-correction curves of the production pitot-static-operated airspeed and altitude system. Angle of attack is presented utilizing the conventional indicated units AOA while sideslip angle limits are presented in terms of degrees rudder deflection.

Note

Unless otherwise specified, the limits presented herein pertain to flight with stability augmentation system (SAS) on.

CRUISE CONFIGURATION

With the wing sweep in the MANUAL or AUTO mode, the maximum allowable airspeeds are shown in figure 1-106.

In the emergency wing sweep mode the following combination of Mach and wing sweep schedule must be used:

⩽ 0.4 IMN	20°
⩽ 0.7 IMN	25°
⩽ 0.8 IMN	50°
⩽ 0.9 IMN	60°
> 0.9 IMN	68°

APPROACH CONFIGURATION

Landing gear	280 KIAS
Landing flaps and slats	225 KIAS

CAUTION

- In aircraft BUNO 159637 and earlier without AFC 364, after takeoff, move the FLAP handle to the UP position prior to passing 200 KIAS to ensure flap and slat structural limits are not exceeded.

- In aircraft BUNO 159825 and subsequent and aircraft incorporating AFC 364, after takeoff, move the FLAP handle to the UP position prior to passing 180 KIAS to ensure flap and slat structural limits are not exceeded.

IN-FLIGHT REFUELING

Refueling probe	400 KIAS/0.8 IMN
In-flight refueling (cruise configuration)	200 to 300 KIAS/0.8 IMN
In-flight refueling (approach configuration)	170 to 200 KIAS
KC-97 or KC-135 (cruise configuration)	255 KIAS

ACCELERATION LIMITS

Note

- Limits are based on a gross weight of 49,548 pounds. See figure 1-107 for the variation of maximum load factor with gross weights greater than 49,548 pounds.

- Coordinated turns with small rudder and lateral stick inputs are defined as symmetrical flight.

MAXIMUM ALLOWABLE AIRSPEEDS

DATE: 1 FEBRUARY 1978
DATA BASIS: FLIGHT TEST

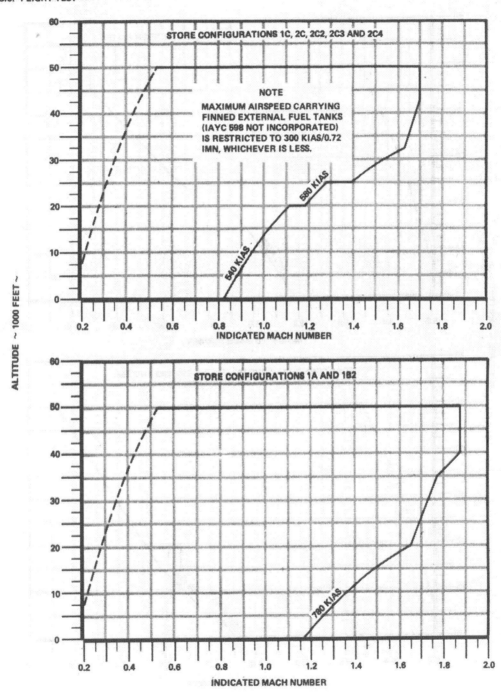

Figure 1-106. Maximum Allowable Airspeeds (Sheet 1 of 4)

Section I
Part 4

NAVAIR 01-F14AAA-1

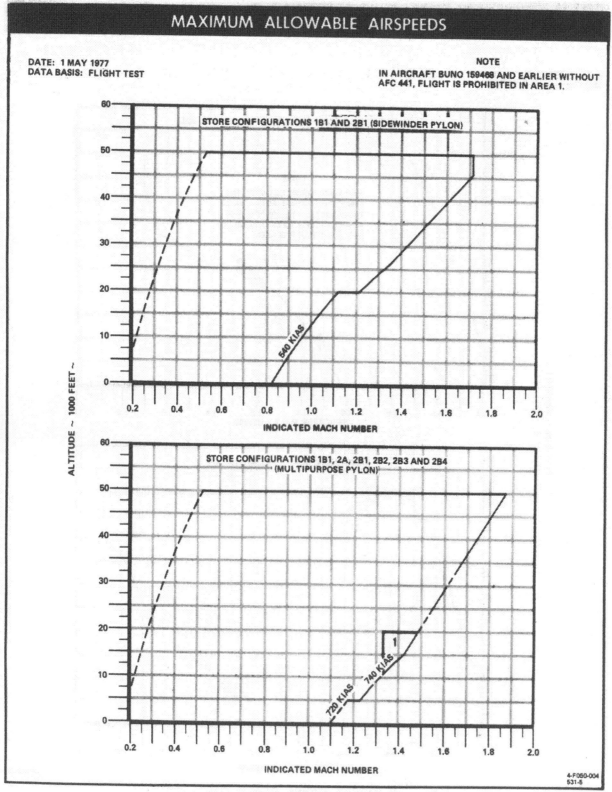

Figure 1-106. Maximum Allowable Airspeeds (Sheet 2 of 4)

NAVAIR 01-F14AAA-1

MAXIMUM ALLOWABLE AIRSPEEDS

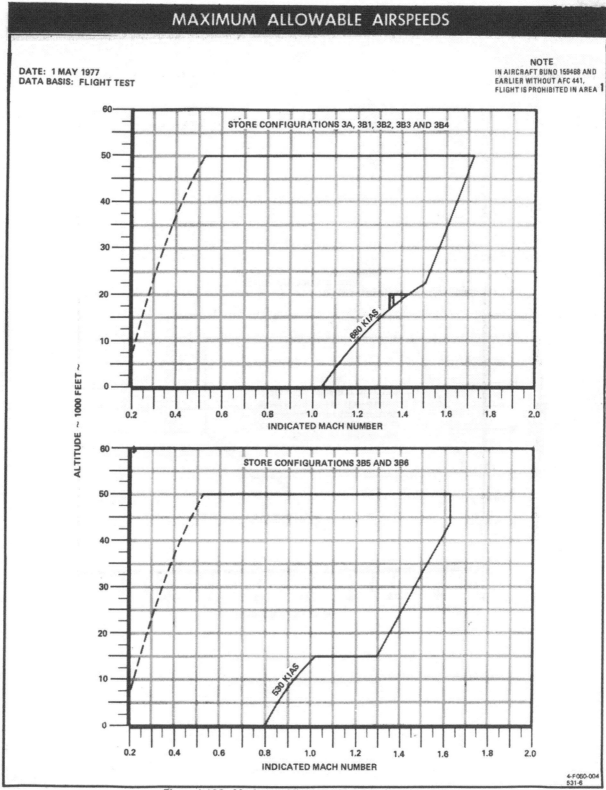

Figure 1-106. Maximum Allowable Airspeeds (Sheet 3 of 4)

Section I
Part 4

NAVAIR 01-F14AAA-1

MAXIMUM ALLOWABLE AIRSPEEDS

DATE: 1 FEBRUARY 1978
DATA BASIS: FLIGHT TEST

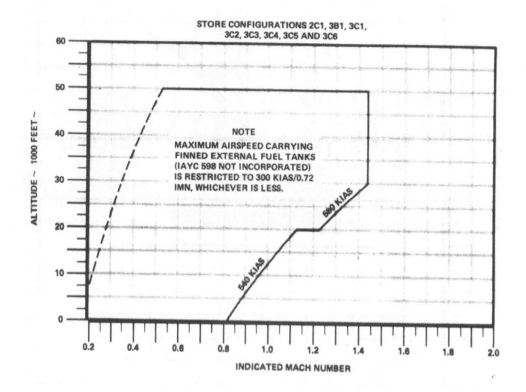

Figure 1-106. Maximum Allowable Airspeeds (Sheet 4 of 4)

NAVAIR 01-F14AAA-1

Section I
Part 4

VARIATION OF MAXIMUM ALLOWABLE NORMAL LOAD FACTOR WITH GROSS WEIGHT

DATE: DECEMBER 1980

DATA BASIS: FLIGHT TEST

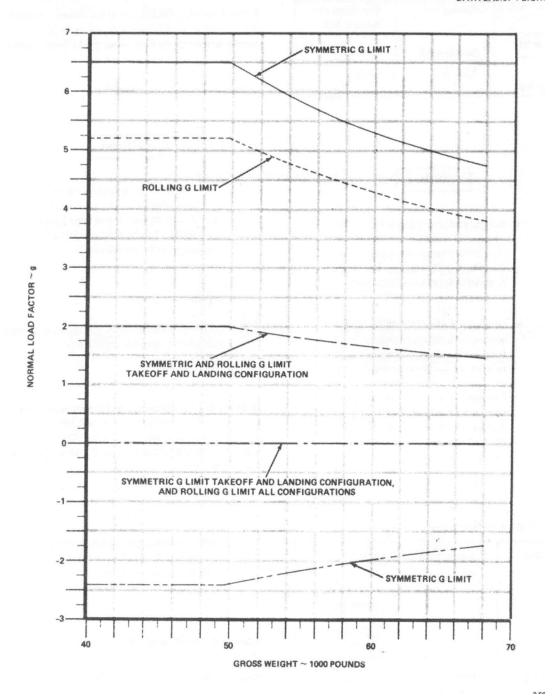

Figure 1-107. Variation of Maximum Allowable Normal Load Factor With Gross Weight

Section I
Part 4

Cruise Configuration

Cruise	Symmetrical	-2.4 to +6.5 g
	Rolling	0 to limits of figure 1-101
Maneuver slats and flaps extended	Symmetrical	LBA
	Rolling	LBA

> **CAUTION**
>
> - In aircraft BUNO 159637 and earlier without AFC 365, an increase in normal load factor will occur when speed brakes are extended (particularly with fuselage stores) at transonic speeds and low altitudes.
>
> - In aircraft BUNO 159468 and earlier without AFC 341, sustained pushovers followed by abrupt pull-ups are prohibited. (A sustained pushover is less than 0 g for greater than 2.0 seconds.) Exceeding these limits will result in adverse fuel cell pressures, failure of fuel cell attachments, and potential fuel cell damage.

Approach Configuration

Landing gear	Symmetrical	0 to 2.0 g
	Rolling	0 to 2.0 g (≤ 225 KIAS) coordinated turns only (225 to 280 KIAS)
Landing flaps	Symmetrical or rolling	0 to 2.0 g

ANGLE-OF-ATTACK LIMITS

Cruise Configuration

Angle of attack is limited by the maximum allowable load factor of figure 1-107, the rolling limits of figure 1-110 and the sideslip limits only under the following conditions:

> **CAUTION**
>
> With roll SAS ON, departure resistance is reduced due to adverse yaw generated by roll SAS inputs. Intentional operation above 17 units AOA should be conducted with roll SAS OFF.

- Wing sweep - AUTO

- Clean or all air-to-air store loadings except those which include wing-mounted AIM-54

Under all other conditions, maneuvering is permitted only to the angle-of-attack limits of figure 1-109 or the maximum allowable load factor of figure 1-107, whichever occurs first, and the rolling limits of figure 1-110 and the sideslip limits.

Note

Refer to High Angle-of-Attack Flight Characteristics and Engine Operating Characteristics in Section IV.

Approach Configuration

Maximum allowable angle of attack with landing gear and flaps extended is shown in figure 1-108.

NAVAIR 01-F14AAA-1

MAXIMUM ALLOWABLE ANGLE-OF-ATTACK RUDDER DEFLECTIONS

LANDING GEAR AND/OR FLAPS EXTENDED

DATE: 1 MAY 1977
DATA BASIS: FLIGHT TEST

NOTE
- NORMAL STALL APPROACH IN 1.0g FLIGHT AT NO GREATER THAN 1.0 KNOT PER SECOND DECELERATION RATE.
- LATERAL CONTROL INPUTS ABOVE 18 UNITS AOA WILL PRODUCE NOSE-UP PITCHING MOMENTS AND APPARENT STICK FORCE LIGHTENING.
- NO INTENTIONAL SIDESLIPS OTHER THAN 1.0g WINGS LEVEL.
- ABRUPT CONTROL REVERSALS PROHIBITED.
- ABRUPT YAWS (FULL ALLOWABLE CONTROL DISPLACEMENT IN LESS THAN 1.0 SECOND) PROHIBITED.
- NORMAL ±30° RUDDER DEFLECTION IS AVAILABLE AND PERMISSIBLE IN ORDER TO MAINTAIN AIRCRAFT CONTROL IN THE EVENT OF AN ENGINE FAILURE.
- DIVERGENT WING ROCK OCCURS ABOVE 25 UNITS AOA.

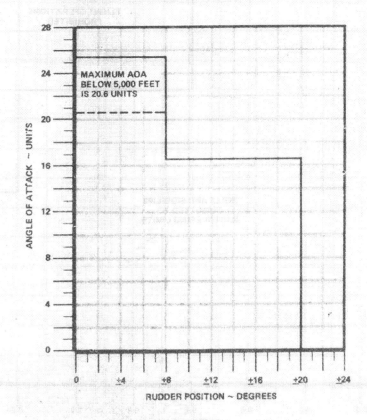

Figure 1-108. Maximum Allowable Angle-of-Attack Rudder Deflections

Section I
Part 4

NAVAIR 01-F14AAA-1

ANGLE OF ATTACK LIMITS

CRUISE CONFIGURATION

AIRCRAFT CONFIGURATION:
- ALL WING MOUNTED AIM-54 LOADINGS
- WING SWEEP NOT IN AUTO

DATE: 1 MAY 1977
DATA BASIS: FLIGHT TEST

NOTE
ABOVE 0.93 IMN, AOA IS LIMITED BY MAXIMUM ALLOWABLE LOAD FACTOR.

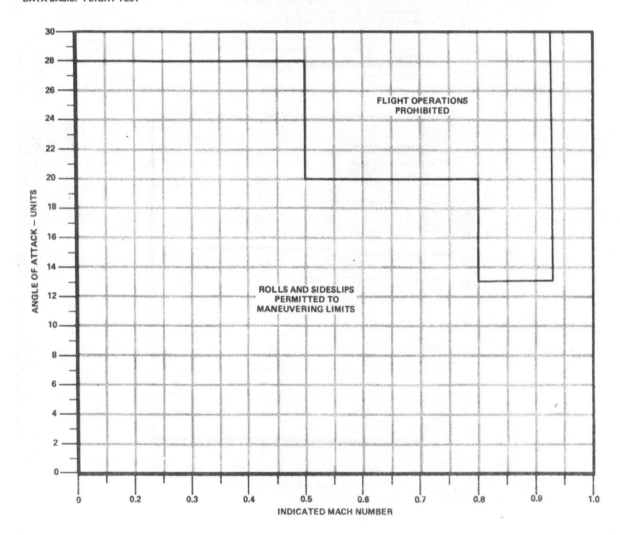

Figure 1-109. Angle-of-Attack Limits

MANEUVERING LIMITS

Cruise Configuration

With maneuver slats/flaps extended, maximum allowable load factor is 6.5 g or the limits of figure 1-107, whichever is less.

Subsonic maneuvering with roll SAS ON shall not be conducted above 15 units AOA with the landing gear retracted.

Cross control inputs shall not be used in the area of the flight envelope indicated in figure 1-109A.

When auto maneuvering flaps/slats are not operating or are not incorporated, uncoordinated lateral control inputs shall not be used in the area of the flight envelope indicated in figure 1-109A.

Rolling Limits

With maneuver slats and flaps extended, maximum allowable load factor is 5.2 g or the limits of figure 1-110, whichever is less. Rolling limits are shown in figure 1-110.

> **CAUTION**
>
> - Do not initiate full lateral stick inputs above 4.5 g if a 5.2-g limit applies or above 3.5 g if a 4.0-g limit applies. Control system dynamics may cause load factor to increase beyond limits.
>
> - If outboard spoilers fail with wing sweep less than 57°, limit lateral stick deflection to 1/2 pilot authority.

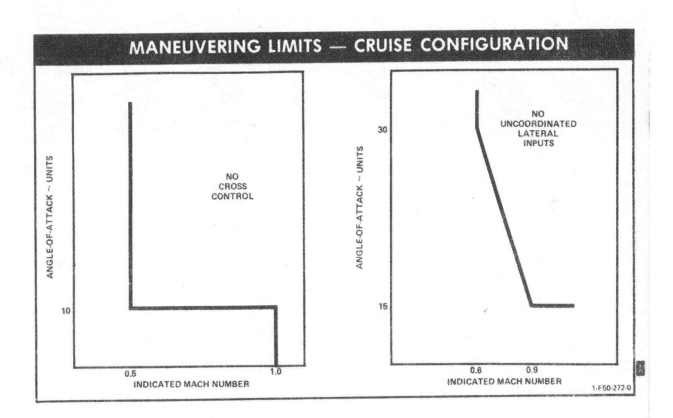

Figure 1-109A. Maneuvering Limits - Cruise Configuration

THIS PAGE INTENTIONALY LEFT BLANK.

Note

Angle-of-attack limitations shown in figure 1-109 apply to designated configurations.

Sideslip Limits

ALL EXTERNAL STORE CONFIGURATIONS

Below 0.7 IMN - rudder inputs as required to maneuver aircraft at high angle of attack.

Above 0.7 IMN - rudder inputs coordinated with lateral stick as required to reduce or eliminate the effects of adverse yaw at high angle of attack.

CAUTION

Rapid rudder inputs cause nose up pitch and may generate snap roll departures.

Above 1.7 IMN - Intentional sideslips prohibited.

WARNING

If a supersonic engine stall and/or failure occurs, maintain wings-level attitude with lateral stick only. Yaw SAS will maintain sideslip angle within acceptable limits.

Note

Use of full available rudder is permitted at all airspeeds if required to counteract adverse yaw encountered in maneuvering flight.

SAS LIMITS

Cruise Configuration

Pitch SAS off . . . . LBA

Yaw SAS off . . . . 0.93 IMN

Roll SAS off . . . . . Aircraft not incorporating AFC 80, 166, 454, 461, 536, and AVC 1565 and/or for external store configuration 3E5, 3B6, 3C5, and 3C6; limit is 0.93 IMN.

Aircraft incorporating AFC 80, 166, 454, 461, 536, and AVC 1565 and all configurations except 3B5, 3B6, 3C5, and 3C6: limit is as shown in figure 1-106 or 700 KIAS/1.33 IMN, whichever is less.

Roll SAS off maneuvers are authorized to 5.2 g rolling, 6.5 g symmetric limits. Full stick 360° level rolls and rolling maneuvers to 5.2 g are authorized.

MANEUVERING LIMITS – ROLLING

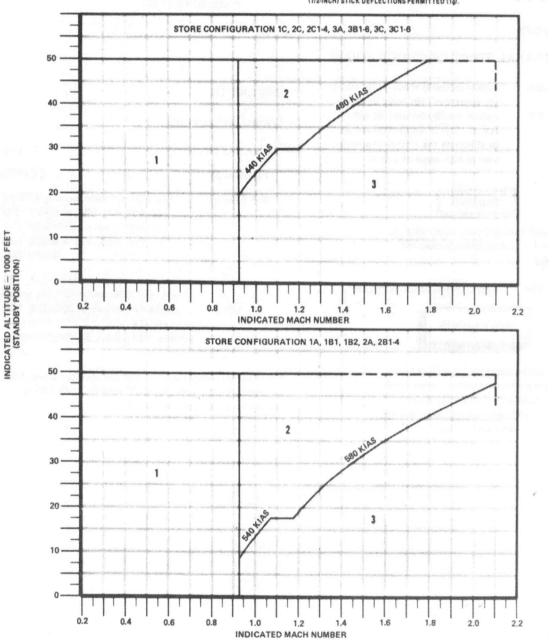

Figure 1-110. Maneuvering Limits-Rolling

NAVAIR 01-F14AAA-1

- Landing approaches to touchdown should not exceed 17 units AOA.
- Only normal flared landings (minimum sink speed) are permitted while carrying AIM-7E/F and/or AIM-9 on the multipurpose pylon, or AIM-7E/F missiles on fuselage stations until the following ACC and AAC are incorporated:

ACC 618 NAM	Modifies multipurpose pylon
AAC 673	Modifies fuselage backup structure
AAC 688	Modifies pylon-mounted swaybraces
AAC 684	Modifies fuselage-mounted AIM-7 swaybraces

BARRICADE ENGAGEMENT LIMITS

Wings at full forward .... 51,800 pounds
sweep angle (20°) (maximum)

- Flaps and slats extended or retracted.
- No external stores, except AIM-7 or AIM-54A on fuselage stations only.
- Empty nacelle tanks permitted only for landing gear malfunction.

Wing sweep angle greater than ...... 46,000 pounds
20° up to 35° (maximum)

- Flaps and slats extended or retracted.
- No external stores, except empty external fuel tanks, for landing gear malfunction.

Wing sweep angle greater .... Not permitted
than 35°

CENTER OF GRAVITY POSITION LIMITS

Unless otherwise stated the following cg limits apply:

Store Configurations	Maximum Forward	Maximum Aft
Clean (no stores) and up to four fuselage mounted AIM-7	6.3% MAC	18.5% MAC
All other configurations	6.3% MAC	17.0% MAC

Throughout these flight operating limitations, all cg positions are quoted at the following reference conditions:

- Zero fuel gross weight (includes weight of stores carried on flight).

- Wing sweep angle = 20°.
- Landing gear and flaps extended.

WARNING

The aft cg limit will be exceeded if all stations are configured for AIM-54 missiles and only stations 4 and 5 are loaded or remain as a result of firing or jettison of stations 1, 3, 6, and 8.

CROSSWIND LIMITS

Crosswind takeoffs and landings are permitted with a crosswind component not to exceed 20 knots at 90°.

GROUND OPERATION LIMITS

Maximum tire speed .... 190 knots

Maximize canopy open speed .... 60 knots

WARNING

Use of antiskid must be in accordance with the following procedures:

- Select antiskid while stopped on the runway in the takeoff position; after landing, clear the runway and turn antiskid off.
- Use only during landing or aborted takeoff.
- Use only during landing or aborted takeoff.
- Do not use antiskid while taxiing.

CAUTION

In aircraft BUNO 158997 and earlier without Material Recall Order No. F-14-050, do not engage nose gear steering above approximately 30 knots.

EJECTION SEAT OPERATION LIMITS

Maximum speed .... 450 KIAS/0.9 IMN
(whichever is greater)

Section I
Part 4
NAVAIR 01-F14AAA-1

PROHIBITED MANEUVERS

The following additional maneuvers are prohibited:

1. Intentional spins.

2. Zero- or negative-g maneuvers in excess of 10 seconds with AB operating or 20 seconds MIL or less.

3. AIM-9 launch or jettisons with landing flaps and slats extended.

4. Fuel dumping with AB operating or with speed brakes extended.

5. Use of autopilot without AFC 371/488 incorporated.

6. Glove vane extension between 0.9 and 1.3 IMN below 15,000 feet with pitch SAS off.

EXTERNAL STORES AND GUN LIMITS

Refer to NWP 55-5-F14, Volume I, Appendix A for external stores and gun operating limits.

267-Gallon External Fuel Tank Limits

- Catapult launch with a partially filled external tank is not authorized due to surge load considerations.

- Arrestment with full or partial fuel in the external tanks is not authorized due to overload of nacelle backup structure.

- Carriage of external tanks not incorporating IAYC 598 is limited to 300 KIAS/0.72 IMN.

CAUTION

Dive angles in excess of 10° nosedown with 900 pounds or more fuel in an external tank will result in fuel venting (dumping).

Launch Limits

Maximum flight conditions for launch of external stores are:

AIM-7

Stations 1, 3, 5, 6, and 8 .... V_{min} to V_{max}, all altitudes, 0 g to limits of basic aircraft

Station 4 (AIM-7E) .... Not authorized

Station 4 (AIM-7F without K-9 autopilot) .... Not authorized

Station 4 (AIM-7F with K-9 autopilot) .... V_{min} to V_{max}, all altitudes, positive g to limits of basic aircraft. (Missiles with K-9 autopilot are identified by segmented black line under missile serial number or letters "POP" after serial number.)

Station 4 (AIM-7M) .... Same as AIM-7F with K-9 autopilot

AIM-9

All stations .... V_{min} to V_{max}, all altitudes, -1.0 g to limits of basic aircraft

WARNING

With landing flaps and slats extended, AIM-9 launch is prohibited.

Note

Engine stall may result from firing of AIM-9 missiles. Engine TIT should be monitored after each firing.

AIM-54A

All stations .... V_{min} to V_{max}, all altitudes, 0 g to limits of basic aircraft

Jettison Limits

Flight conditions for jettison (emergency only) of external stores are:

AIM-7

Stations 1 and 8 .... V_{min} to M_{max}, all altitudes, 0 g to limits of basic aircraft

Stations 3, 4, 5, and 6 .... V_{min} to 400 KIAS all altitudes, +1.0 g

AIM-7 on stations 3 and 6 exhibit pronounced outboard movement when jettisoned.

AIM-54A

All stations .... V_{min} to M_{max}, all altitudes 0 go to limits of basic aircraft

267-Gallon External Fuel Tank (Without IAYC 598 Incorporated) (Landing Gear and Flaps Retracted)

Full or empty tanks .... Less than 300 KIAS/ 0.72 IMN, all altitudes, 0 to +3.0 g

Partially full tanks .... Less than 300 KIAS/ 0.72 IMN, all altitudes, +1 g

Landing gear and/or .... Less than 225 KIAS, flaps and slats extended all altitudes, +1 g (emergency only)

Finless 267-Gallon External Fuel Tank (With IAYC 598 Incorporated) (Landing Gear and Flaps Retracted)

Full or empty tanks .... Less than 350 KIAS/ 0.88 IMN, all altitudes, wings level +1 g flight

Partially full tanks .... Less than 250 KIAS/ 0.72 IMN, all altitudes, wings level +1 g flight

Landing gear and/or .... Less than 225 KIAS/ flaps and slats extended all altitudes, +1 g (emergency only)

BANNER TOWING RESTRICTIONS

Airspeed .... 220 KIAS maximum recommended

Maximum angle of bank .... 30°

Use of speed brakes .... Prohibited in flight

Note

During takeoff, adequate clearance exists to use speed brakes for takeoff abort without contacting the tow cable.

THIS PAGE INTENTIONALY LEFT BLANK.

NAVAIR 01-F14AAA-1

SECTION II – INDOCTRINATION

TABLE OF CONTENTS

Ground Training Syllabus 2-1
Flight Training Syllabus 2-2
Operating Criteria 2-2
Flightcrew Member Flight Equipment Requirements 2-4

GROUND TRAINING SYLLABUS

Minimum Ground Training Syllabus

The ground training syllabus sets forth the minimum ground training that must be satisfactorily completed prior to operating the F-14A. The ground training syllabus for each activity will vary according to local conditions, field facilities, requirements from higher authority, and the immediate Unit Commander's estimate of the squadron's readiness. The minimum ground training syllabus for the pilot and for the RIO is set forth below.

FAMILIARIZATION

- Flight physiological training as appropriate
- F-14A flightcrew academic course
- F-14A OFT/COT/WST (within 5 days)

FLIGHT SUPPORT LECTURES

- F-14A flightcrew academic course

INTERCEPT FLIGHT SUPPORT

- F-14A flightcrew academic course

WEAPONS FIRING FLIGHT SUPPORT LECTURES

- Weapons preflight procedures
- Arming/de-arming procedures
- Firing procedures
- Safety procedures
- Jettison/dump areas

FCLP/CARQUAL FLIGHT SUPPORT LECTURES

- Mirror and Fresnel lens optical landing system
- Day landing pattern and procedures
- Night landing pattern and procedures
- Shipboard procedures and landing patterns
- CCA/ACLS procedures
- In-flight refueling (day/night)

WAIVING OF MINIMUM GROUND TRAINING REQUIREMENTS

All F-14A qualified flightcrew members shall be instructed on the differences from model in which qualified and comply with those items listed below, as directed by the Unit Commanding Officer.

Where recent crewmember experience in similar aircraft models warrant, Unit Commanding Officers may waive the minimum ground training requirements provided the flightcrew member meets the following mandatory qualifications:

- Has obtained a current medical clearance
- He is currently qualified in flight physiology
- Has satisfactorily completed the NATOPS Flight Manual open and closed-book examinations

- Has completed at least one emergency procedure period in the OFT/COT/WST (within 10 days)
- Has received adequate briefing on normal and emergency operating procedures
- Has received adequate instructions on the use and operation of the ejection seat and survival kit

FLIGHT TRAINING SYLLABUS

Flightcrew Flight Training Syllabus

Prior to flight, all flightcrew members will have completed the familiarization and flight support lectures previously prescribed. A qualified instructor pilot will be assigned for the first familiarization flight. The instructor pilot will occupy the rear seat. The geographic location, local command requirements, squadron mission, and other factors will influence the actual flight training syllabus and the sequence in which it is completed. The specific phases of training are:

FAMILIARIZATION

- Military and afterburner power takeoffs
- Buffet boundary investigation
- Approach to stalls
- Slow flight
- Acceleration run to Mach 1.5
- Subsonic and supersonic maneuvering
- Investigate all features of the AFCS/Stab Aug
- Formation flight
- Aerobatics
- Single-engine flight at altitude and airstarts
- Nosewheel steering on landing rollout
- Simulated single-engine
- Landing with full and with no flaps
- Acceleration runs at various altitudes

INSTRUMENTS

- Basic instrument work
- Penetrations and approaches
- Local area round-robin (day and night) flights

WEAPONS SYSTEM EMPLOYMENT

In accordance with existing training and readiness directives

FIELD MIRROR LANDING PRACTICE AND CARRIER QUALIFICATION

- Slow flight
- Field mirror landing practice
- Carrier qualifications flights

OPERATING CRITERIA

Ceiling/Visibility Requirements

Before the pilot becomes instrument-qualified in the aircraft, field ceiling, visibility, and operating area weather must be adequate for the entire flight to be conducted in a clear airmass according to Visual Flight Rules. After the pilot becomes instrument-qualified, the following weather criteria apply:

Time-in-Model (hours)	Ceiling and Visibility (feet) (miles)
Less than 10	VFR
10 to 20	800 and 2; 900 and 1 1/2; 1000 and 1
20 to 45	700 and 1; 600 and 2; 500 and 3
45 and above	Field minimums or 200 and 1/2, whichever is higher

Where adherence to these minimums unduly hampers pilot training, Commanding Officers may waive time-in-model requirements for actual instrument flight, provided pilots meet the following criteria:

- Have a minimum of 10 hours in model
- Completed two simulated instrument sorties

- Completed two satisfactory TACAN penetrations
- Completed five satisfactory GCA approaches

Initial Qualification and Currency Requirements

Initial NATOPS qualifications require a formal ground and flight syllabus at a fleet replacement squadron. The syllabus will include 10 flight hours under instruction, 4 of which may be flown in a CNO-approved flight simulator. FRS Commanding Officers may waive the flight-hour requirement for radar intercept officers. Minimum flight hour requirements to maintain pilot and RIO qualifications after initial qualification in each specific phase will be established by the Unit Commanding Officer. Flightcrew members who have more than 45 hours in model are considered current subject to the following criteria:

- Must have a NATOPS evaluation check with the grade of Conditionally Qualified, or better, within the past 12 months and must have flown 10 hours in model and made 5 takeoffs and landings within the last 90 days.
- Must have satisfactorily completed the ground phase of the NATOPS evaluation check, including OFT/COT/WST emergency procedures check (if available) and be considered qualified by the Commanding Officer of the unit having custody of the aircraft.

Requirements For Various Flight Phases

NIGHT

PILOT   Not less than 10 hours in model as first pilot.

RIO   Not less than 3 hours in model as crewmember.

CROSS-COUNTRY

PILOT

- Have a minimum of 15 hours in model as first pilot
- Have a valid instrument card
- Have completed at least one night familiarization flight
- Have completed maintenance checkout for servicing aircraft.

RIO   Have completed at least one night familiarization flight

AIR-TO-AIR MISSILE FIRING

PILOT

- Have a minimum of 15 hours in model
- Be considered qualified by the Commanding Officer

RIO

- Have a minimum of 25 hours in model as crewmember
- Have satisfactorily completed a minimum of two intercept flights on which simulated firing runs were conducted utilizing the voice procedures and clear-to-fire criteria to be utilized in live firing
- Be considered qualified by the Commanding Officer

CARRIER QUALIFICATIONS

Each crewmember will have a minimum of 50 hours in model, of which 15 hours is night, and meet the requirements set forth in the CVA/CVS NATOPS manual.

Mission Commander

The mission commander shall be a NATOPS qualified pilot or RIO, qualified in all phases of the assigned mission and be designated by the Unit Commanding Officer.

Minimum Flightcrew Requirements

The pilot and the RIO (or two pilots) constitute the normal flightcrew for performing the assigned mission for all flights. Unit commanders may authorize rear seat flights for personnel other than qualified pilots and RIOs provided such personnel have received thorough indoctrination in the use of the ejection seat and oxygen equipment and in the execution of rear seat functions and emergency procedures. Where operational necessity dictates, unit commanders may authorize flights with rear seat unoccupied provided the requirement for such flights clearly overrides the risk involved and justifies the additional burden placed on the pilot. In no case is solo flight authorized for shipboard operations, combat, or combat training missions.

Ferry Squadrons

Training requirements, checkout procedures, evaluation procedures, and weather minimum for ferry squadrons are governed by the provisions contained in OPNAVINST 3710.6.

FLIGHTCREW MEMBER FLIGHT EQUIPMENT REQUIREMENTS

Minimum Requirements

In accordance with OPNAVINST 3710.7, the flying equipment listed below will be worn or carried, as applicable, by flightcrew members on every flight. All survival equipment shall be secured in such a manner that it will be easily accessible and will not be lost during ejection or landing. All equipment shall be the latest available as authorized by Aircrew Personal Protective Equipment Manual, NAVAIR 13-1-6.

- Protective helmet
- Oxygen mask
- Anti-g suit (required on all flights where high-g forces may be encountered)
- Fire retardant flight suit
- Steel-toed flight safety boots
- Life preserver
- Harness assembly
- Shroud cutter
- Sheath knife
- Flashlight (for all night flights)
- Strobe light
- Pistol with tracer ammunition, or approved flare gun
- Fire-retardant flight gloves
- Identification tags
- Anti-exposure suit in accordance with OPNAVINST 3710.7
- Personal survival kit
- Other survival equipment appropriate to the climate of the area
- Full pressure suit and MK 4 life preserver on all flights above 50,000 feet MSL
- Pocket checklist
- Navigation packet

SECTION III — NORMAL PROCEDURES

TABLE OF CONTENTS

PART 1 — FLIGHT PREPARATION
- Preflight Briefing ... 3-2

PART 2 — SHORE-BASED PROCEDURES
- Checklists ... 3-3
- Exterior Inspection ... 3-3
- Ejection Seat Inspection ... 3-10
- Pilot Procedures ... 3-12
- Interior Inspection - Pilot ... 3-12
- Prestart - Pilot ... 3-15
- Engine Start - Pilot ... 3-16
- Poststart - Pilot ... 3-18
- Taxiing ... 3-22
- Taxi - Pilot ... 3-22
- Takeoff ... 3-24
- Takeoff Checklist ... 3-27
- Ascent Checklist ... 3-28
- In-Flight OBC ... 3-28
- Preland and Descent ... 3-28
- Pattern Entry ... 3-29
- Landing ... 3-29
- Landing Checklist ... 3-32
- Postlanding - Pilot ... 3-32
- RIO Procedures ... 3-33
- Interior Inspection - RIO ... 3-33
- Prestart - RIO ... 3-35
- Engine Start - RIO ... 3-36
- Poststart - RIO ... 3-36
- Taxi - RIO ... 3-38
- On-Deck Entry of Reference Points or Targets ... 3-38
- Own-Aircraft Altitude Correction (On Deck) ... 3-38
- TARPS Mode Entry and Display Requirements ... 3-38
- Own-Aircraft Altitude Correction (Airborne) ... 3-39
- In-Flight Reconnaissance System Check - RIO ... 3-39
- Steering Display Requirements ... 3-39
- Airborne Entry of Reference Points/Targets (TARPS) ... 3-40
- Nav System Updates via HUD ... 3-40
- Targets of Opportunity ... 3-40
- Mapping Mode Entry ... 3-40
- Unplanned Air-to-Air Photography ... 3-41
- TARPS Altitude Determination ... 3-41
- TARPS Mode Exit ... 3-41
- TARPS Degraded Mode Procedures ... 3-41
- Postlanding - RIO ... 3-43
- Hot Refueling Procedures ... 3-43
- Deck-Launched Intercept Procedures ... 3-44
- Hot Switch Procedures ... 3-44
- Field Carrier Landing Practice (FCLP) ... 3-45

PART 3 — CARRIER-BASED PROCEDURES
- Launch ... 3-48
- Briefing ... 3-48
- Preflight ... 3-48
- Start and Poststart ... 3-48
- Carrier Alignment ... 3-48
- Taxiing ... 3-49
- Catapult Hookup ... 3-49
- Catapult Trim Requirements ... 3-49
- Catapult Launch ... 3-50
- Catapult Abort Procedures (Day) ... 3-51
- Landing ... 3-51
- Bingo Fuel ... 3-55
- Arrested Landing and Exit from the Landing Area ... 3-55
- Carrier-Controlled Approaches ... 3-55
- Waveoff and Bolter ... 3-57
- Night Flying ... 3-57
- Catapult Hookup (Night) ... 3-58
- Catapult Abort Procedures (Night) ... 3-58
- Arrested Landing and Exit from Landing Area (Night) ... 3-59

PART 4 — SPECIAL PROCEDURES
- In-flight Refueling Procedures ... 3-60
- Formation Flight ... 3-61
- Banner Towing ... 3-62
- In-flight Compass Evaluation ... 3-64

PART 5 — FUNCTIONAL CHECKFLIGHT PROCEDURES
- Functional Checkflights ... 3-66
- Checkflight Procedures ... 3-66
- Pilot's Checklist ... 3-67
- RIO's Checklist ... 3-77

PART 1 — FLIGHT PREPARATION

PREFLIGHT BRIEFING

These briefings are presented immediately before the launching of scheduled flights, and therefore must be carried out in the most expeditious manner. It is imperative that all pilots and RIOs be in flight gear and ready for the briefing at the designated time. The briefing shall include, but shall not be limited to, the following:

Note

Information marked with an asterisk (*) shall be displayed on a status board in the briefing or ready room, and should be copied by pilots before the briefing.

1. SCHEDULING
   a. Event Number*
   b. Takeoff, recovery times, and recovery order*
   c. Aircraft-pilot lineup*
   d. Mission assigned to each aircraft*
   e. Marshal information

2. MISSION
   a. CAP station assignment and control*
   b. Target aircraft rules of engagement
   c. Ground target description and procedures

3. COMMUNICATIONS
   a. Channels and frequencies*
   b. Navigational aids*
   c. Lost communication procedures
   d. Reports required
   e. Authenticators, IFF
   f. EMCOM conditions

4. PARTICIPATING UNITS
   a. Voice calls and side numbers*
   b. Disposition
   c. Utilization
   d. Friendly subs and surface units

5. OPERATIONS
   a. Instructions for coordinating other units

6. ORDNANCE
   a. Ordnance carried*
   b. Restrictions on use

7. WEATHER - BASE, ENROUTE, TARGET, AND DIVERT FIELD
   a. Wind - direction and velocity at surface and at applicable altitudes*
   b. Cloud coverage - present and forecast*
   c. Visibility*
   d. Sea state*
   e. Water and air temperature (cold weather)*
   f. Target weather*
   g. Divert weather*

8. MISCELLANEOUS
   a. Other units in the area
   b. Restricted or danger areas
   c. Current NOTAMS, bulletins, and safety-of-flight information
   d. Flight leader brief on takeoff, rendezvous frequency switch, landing procedures, etc.
   e. SAR - Participating units and procedures
   f. BINGO fuel
   g. Tanker aircraft information

PART 2 — SHORE-BASED PROCEDURES

CHECKLISTS

Aircraft checklists are available in two forms, based on the degree of flightcrew familiarization; since the sequence remains the same, the only difference in the forms is the degree of amplification. As the flightcrewman becomes more proficient in type, a more abbreviated form is available to promote operational efficiency, and safety is not compromised since in all instances the thoroughness of checks remains the same. The placarded takeoff and landing checklists on the cockpit instrument panels are a fundamental element in all instances. In the interest of procedural standardization, the shore-based and carrier-based procedures are maintained the same, except for the response relative to the checks. The expanded procedures presented in this flight manual describe in detail those items that should be checked on each flight. Adherence to these procedures will provide the flightcrew with a detailed status of weapons system performance incident to flight. However, it is incumbent on the flightcrew to expand the checks as necessary to verify the corrective status of previously reported discrepancies. Reference should be made to the functional checkflight procedures (part 5, this section) for more detailed tests that can be performed on the aircraft and weapons systems if deemed necessary. The flightcrew should be thoroughly familiar with the details of the procedures outlined herein so that the abbreviated checklists forms of the procedures may be safely employed. As the first level of simplification, NAVAIR 01-F14AAA-1B contains a reprint of the normal procedures, with less amplifying information.

A T in the outer margin preceding a procedural step identifies items pertaining only to TARPS aircraft.

EXTERIOR INSPECTION

A proper preflight inspection begins with a thorough review of aircraft status and past maintenance history. An understanding of previous discrepancies, corrective action and their impact on the flight can best be gained at this time. The flightcrew should ensure that any and all discrepancies have been properly corrected or deferred prior to accepting the aircraft as ready for flight.

Area Around Aircraft

Enroute to the aircraft, attention should be directed to the maintenance effort going on in the line area. The flightcrew should ensure that no hazardous situations exist. The entire area should also be generally examined for FOD hazards.

At the aircraft, the area around the aircraft that may not be visible from the cockpit should be examined. Particular attention should be paid to the aircraft and support equipment adjacent to the aircraft. It should be determined that the wings and flight controls can be safely moved and that the effect of jet blast during start and taxi will not create a dangerous situation.

FOD Inspection

Engine intakes and adjacent deck area are of prime concern since the TF-30 is highly susceptible to FOD damage and the engines are capable of picking up objects from the deck. AICS ramps, bleed doors, ECS cooling intakes and exhausts, and afterburner ducts are catchalls for loose objects. They should be closely inspected for security and foreign objects.

Ground Safety Devices and Covers

The following items should be installed.

1. Main landing gear ground safety locks (2)
2. Nose landing gear ground safety pin
3. Launch abort mechanism lock (if applicable)
4. Wheel chocks
5. AN/ALE-29 or AN/ALE-39 ground safety pins
6. LAU-7/LAU-92 ground safety pins
7. Sidewinder seeker-head covers (if applicable)

The following items should be removed.

1. Intake, probes, bleed door, and ECS duct covers
2. Water-intrusion tape

Surface Condition

All surfaces should be checked for cracks, distortion, or loose or missing fasteners. All lights and lenses should be checked for cracks and cleanliness.

Security of Panels

All fasteners should be flush and secure on all panels.

Leaks

All surfaces, lines, and actuators should be checked for oil, fuel and hydraulic leaks. Particular attention should be paid to the underside of the fuselage, engine nacelles, and outer wing panels.

Movable Surfaces

All movable surfaces (flight controls and high lift devices) should be inspected for position, clearance, and obvious damage.

The following exterior inspections are divided into 10 areas. (See figure 3-1.) Checks peculiar to only one side are designated (L) or (R) for the left or right side. Both the pilot and RIO should preflight the entire aircraft individually.

(A) Forward Fuselage

1. Access panel fasteners forward of engine inlets .... NO LOOSE OR MISSING FASTENERS

2. Gun (L) .. SAFETY PIN INSTALLED IN CLEARING SECTOR HOLDBACK ASSEMBLY

3. Probes .... SECURE, OPENINGS CLEAR, AOA PROBE FREE FOR ROTATION

4. Nosewheel well:

   a. Electrical leads ...... CONNECTED, NO EVIDENCE OF OVERHEATING

   b. Hydraulic lines .. NO CHAFING OR LEAKS

   c. Doors and linkages .... COTTER PINS INSTALLED, NO DISTORTION

   d. Brake accumulator gages (2) .... 900 PSI MINIMUM

   e. Canopy air bottle gage .... 1200 PSI MINIMUM

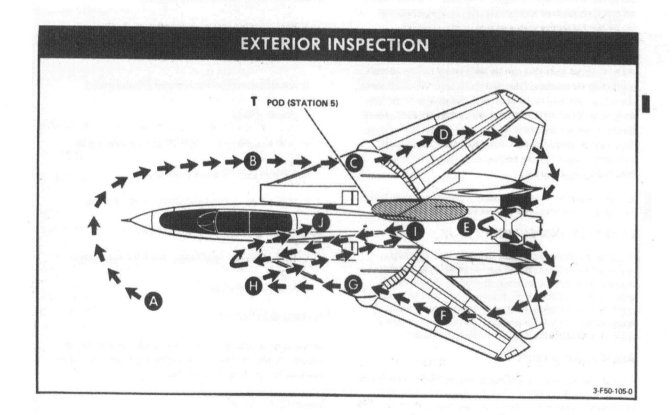

Figure 3-1. Exterior Inspection

NAVAIR 01-F14AAA-1

f. Emergency landing gear
nitrogen bottle gage .. 3000 PSI MINIMUM

g. Retract actuator .... PISTON CLEAN, NO LEAKS

h. Anti-skid control box
BIT flags .... NOT TRIPPED

i. Cabin pressure port screens ...... CLEAN

j. Master arm override ..... COVER CLOSED

5. Nose strut .... PISTON CLEAN, FREE OF CRACKS AND SCORING, AND UPLOCK ROLLER FREE

6. Steering actuator .... SECURE, NO LEAKS

7. Launch bar and holdback fitting:

   a. Abort .... FULL UP (SAFETY LOCK INSTALLED, IF APPLICABLE)

   b. Roller .... FREE ROTATION

   c. Uplatch and holdback .... FREE MOVEMENT

8. Nosewheels and tires .... INFLATION, NO CUTS, BULGES, UNEVEN WEAR, OR IMBEDDED OBJECTS

9. Drag brace .... NO LEAKS, DOOR SECURE

10. Approach lights .... LENSES CLEAN, NO CRACKS, SECURE

11. IR/TV pod .... APN-154 ANTENNA AND ANTICOLLISION LIGHT SECURE

12. Radome .... LOCK HANDLE FASTENED, ROSEMONT PROBE STRAIGHT

13. LOX vent outlet (R) .... NO OBSTRUCTIONS

Ⓑ **Right Inlet**

1. Ramps and seals .... INTACT, FREE OF DIRT AND GRIT

2. Engine compressor face .... BLADES AND STATORS FREE OF NICKS AND CRACKS
Plane captain to verify that all visible damage has been blended.

Section III
Part 2

3. ECS heat exchanger inlet and fan:

   a. Fan .... FREE ROTATION

   b. Overspeed pin .... RECESSED

   c. ECS inlet .... FREE OF FOD, CABLES CONNECTED (2)

4. Inlet .... FREE OF STANDING WATER, DRAINS CLEAR

Ⓒ **Right Nacelle and Sponson**

1. Station 7 and 8 Stores:

   a. Stores .... ALIGNED

   b. Access panels .... SECURE

   c. Sidewinder coolant doors ...... LATCHED

   d. Stores safety pins........... INSTALLED

2. Main wheel well:

   a. Doors and linkages .... SECURE

   b. Uplock microrollers .... FREE

   c. Uplock hooks .... SECURE

   d. Hydraulic lines .. NO CHAFING OR LEAKS

3. Drag brace .... SECURE

4. Side brace......... SEATED IN LATCH, DOWN LOCK SAFETY PIN FORWARD

5. Main struts .... PISTONS CLEAN, FREE OF CRACKS OR SCORING

6. Brakes .... PUCKS; SAFETY-WIRED; WEAR INDICATORS VISIBLE (PINS AT LEAST FLUSH). LOWER TORQUE ARM SWIVEL; KEY AND KEY RETAINER PROPERLY INSTALLED AND SAFETY WIRED.

7. Hubcap .... SECURE, SAFETY-WIRED

8. Main wheels and tires:

   a. Wheels and tires...... INFLATION, CUTS, BULGES, UNEVEN WEAR, IMBEDDED OBJECTS (LOOK BEHIND CHOCKS)

9. Gear down microrollers ...... CONTACT MADE

3-5

10. Engine compartment (if applicable):

   a. Engine oil filler cap .... OUTBOARD NOT SAFETY-WIRED, SECURELY FASTENED, LATCH AT 12 O'CLOCK POSITION

   b. Generator-transmission (CSD) fluid .... FLUID VISIBLE, FILTER PINS FLUSH

   c. Bilges .... NO FOD, EVIDENCE OF OVERHEATING, OR LEAKAGE

   d. Fuel, oil, and hydraulic lines .... FREE OF CHAFING OR LEAKS

   e. Bleed air lines .... NO HEAT DISCOLORATION OR DAMAGE

11. Flight hydraulic reservoir .... 1800 PSI MINIMUM, FILTER PINS FLUSH

12. Flight hydraulic system tape gage .... 7 ON TAPE (AT 70°F)

13. Hook dashpot pressure gage .... 800±10 PSI

14. Ventrals .... NO DAMAGE, OIL COOLER INTAKE CLEAR

(D) Right Glove and Wing

1. Glove vane .... FREE OF CRACKS OR DISTORTION

2. Slats, flaps, and cove doors .... SURFACES AND HINGES SECURE

3. Wing cavity seal .... FREE OF CUTS AND CHAFING

4. Formation and position lights .... INTACT, LENSES CLEAN

(E) Aft and Under Fuselage

1. Horizontal tails .... LEADING EDGES FREE OF DAMAGE

2. Exhaust nozzles and fairings:

   a. Nozzles and fairings .... NO CRACKED OR MISSING LEAVES OR SEALS

   b. Bottom surface .... NO SCRAPES OR CRACKS

   c. Spray rings .... INTACT

   d. Turbine blades .... NO EVIDENCE OF OVERHEATING

3. Fuel vent .... NO LEAKAGE OR FOD

4. Tail hook:

   a. Hook point................ SMOOTH

   b. Nut and cotter pin .... INSTALLED

   c. Safety pin .... REMOVE IF HOOK IS SECURELY LATCHED UP

5. Backup flight control module .... NO LEAKS (FEEL AFT OF INSPECTION DOORS), CLOSE BOTH ACCESS DOORS

6. Fuel dump .... NO LEAKAGE FROM MAST, FREE OF FOD

7. Stations 3 through 6 stores:

   a. Stores .... ALIGNED

   b. Access panels .... SECURE

   c. Stores safety pins .... INSTALLED

8. Fuel cavity drains .... NO LEAKAGE

T 9. Pod................ CHECK FOR SECURITY

T 10. Protective window covers .... REMOVED

T 11. Camera windows................ CLEAN

T 12. Camera S/C .... AS BRIEFED

T 13. Light meter .... FACING OUTBOARD

T 14. Lens filter .... AS BRIEFED

(F) Left Glove and Wing

1. Glove vane .... FREE OF CRACKS OR DISTORTION

2. Slats, flaps, and cove doors .... SURFACES AND HINGES SECURE

3. Wing cavity seal . . . . FREE OF CUTS AND
CHAFING

4. Formation and position
   lights . . . . INTACT, LENSES CLEAN

G Left Nacelle and Sponson

1. Station 1 and 2 racks and stores:
   a. Racks and stores . . . . ALIGNED
   b. Access panels . . . . SECURE
   c. Sidewinder coolant doors. . . . . . . LATCHED
   d. Stores safety pins . . . . INSTALLED

2. Main wheel well:
   a. Doors and linkages . . . . SECURE
   b. Uplock microrollers . . . . FREE
   c. Uplock hooks . . . . SECURE
   d. Hydraulic lines . . . . NO CHAFING

3. Drag brace . . . . SECURE

4. Side brace . . . . SEATED IN LATCH, DOWN
   LOCK SAFETY PIN FORWARD

5. Main struts . . . . PISTONS CLEAN, FREE
   OF CRACKS OR SCORING

6. Brakes . . . . PUCKS; SAFETY-WIRED; WEAR
   INDICATORS VISIBLE (PINS AT
   LEAST FLUSH). LOWER TORQUE
   ARM SWIVEL; KEY AND KEY
   RETAINER PROPERLY INSTALLED
   AND SAFETY-WIRED.

7. Hubcap . . . . SECURE, SAFETY-WIRED

8. Main wheels and tires:
   a. Wheels and tires . . . . . . INFLATION, CUTS,
      BULGES, UNEVEN WEAR, IMBEDDED
      OBJECTS (LOOK BEHIND CHOCKS)
   b. Uplock hooks . . . . SECURE

9. Gear up microrollers . . . . . . CONTACT NOT MADE

10. Engine compartment (if applicable):
    a. Engine oil filler cap . . . . . . OUTBOARD NOT
       SAFETY-WIRED, SECURELY
       FASTENED, LATCH AT
       12 O'CLOCK POSITION
    b. Generator-transmission
       (CSD) fluid . . . . FLUID VISIBLE,
       FILTER PINS FLUSH
    c. Bilges . . . . NO FOD, EVIDENCE OF
       OVERHEATING, OR LEAKAGE

   d. Fuel, oil, and hydraulic
      lines . . . . NO CHAFING OR LEAKS
   e. Bleed air lines . . . . NO HEAT DISCOLORA-
      TION OR DAMAGE

11. Combined hydraulic
    reservoir . . . . 1800 PSI MINIMUM,
    FILTER PINS FLUSH

12. Combined hydraulic system
    tape gage . . . . 7 ON TAPE (AT 70°F)

13. Airstart door . . . . . GROUND HYDRAULIC AND
    ELECTRIC COVERS TIGHT

14. Ventrals . . . . NO DAMAGE, OIL COOLER
    INTAKE CLEAR

H Left Inlet

1. Ramps and seals . . . . INTACT, FREE OF
   DIRT AND GRIT

2. Engine compressor
   face . . . . BLADES AND STATORS
   FREE OF NICKS AND CRACKS
   Plane captain to verify that all visible damage
   has been blended.

3. Ice detector (L) . . . . SECURE

4. ECS heat exchanger inlet and fan:
   a. Fan . . . . FREE ROTATION
   b. Overspeed pin . . . . RECESSED
   c. Inlet . . . . FREE OF FOD, CABLES
      CONNECTED (2)

5. Outboard spoiler module temperature
   indicator and servicing . . . . NO LEAKS,
   FLUID INDICATOR ROD PROTRUDING

6. Inlet . . . . FREE OF STANDING WATER,
   DRAINS CLEAR

I Fuselage Top Deck and Wings

1. Bleed exit doors . . . . FREE OF FOD,
   HARDWARE INTACT

Section III
Part 2
NAVAIR 01-F14AAA-1

2. ECS heat exchanger
exhausts .... FREE OF FOD AND CRACKS

3. Overwing fairings .... NO CRACKED OR BENT FINGERS

4. Eyebrow doors .... INTACT

5. Speed brake .... NO DISTORTION OR LEAKS

6. Vertical tails and
rudders .... NO DISTORTION, LIGHTS INTACT

(J) Canopy

1. Canopy lanyard ...... CONNECTED, YELLOW FLAG ATTACHED AT BOTH ENDS

2. Auxiliary canopy bottle .... RESET CABLE TAUT

3. Canopy hooks and seal .... SECURE, SEAL INTACT

4. Ejection seat safe-
and-arm device
safety pins
(See figures 3-2 and 3-3) .... PULLED

5. Auxiliary canopy
bottle gage .... 800 PSI MINIMUM

6. Blade antennas .... INTACT

7. Canopy .... CLEAN, FREE OF CRACKS AND DEEP SCRATCHES

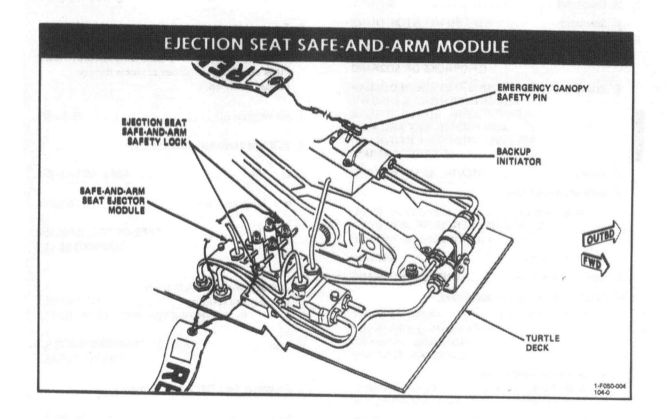

Figure 3-2. Ejection Seat Safe-and-Arm Module

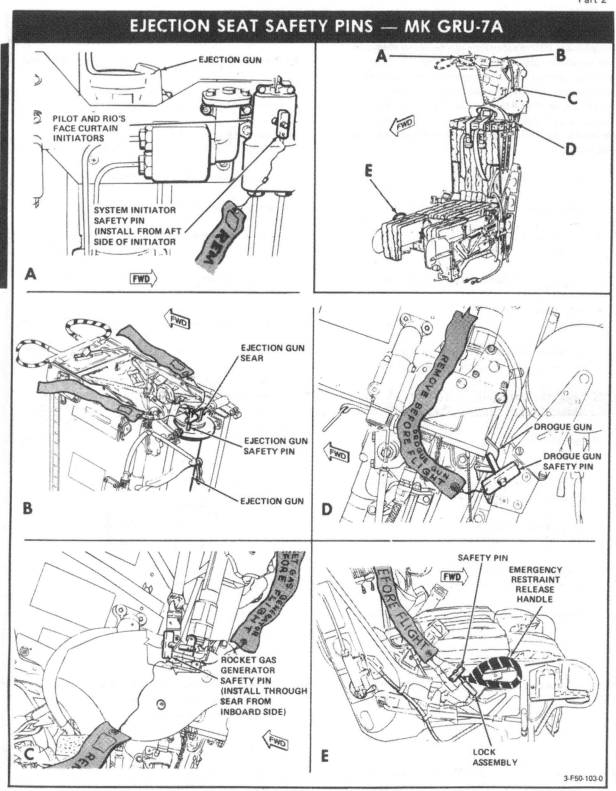

Figure 3-3. Ejection Seat Safety Pins — MK-GRU-7A

EJECTION SEAT INSPECTION

The pilot and RIO shall perform the following checks on their respective ejection seats prior to flight. Ground safety pin removal should be confirmed on all actuation devices where a pin can be inserted to safety the mechanism for ground maintenance (see figure 3-3). Abbreviated preflight checklists for the ejection seat are provided in the pocket checklist and on the ejection seat headbox.

1. Face curtain locking tab . . . . UP (LOCKED)

2. Lower ejection handle. . . . GUARD UP (LOCKED)

3. Face curtain connecting link (R) . . . . SAFE INDICATION

   Tabs adjacent as viewed through inspection window.

4. Time release mechanism trip rod . . . . CONNECTED

   No red indication showing on inner rod and rod attached to ejection gun crossbeam.

5. Face curtain automatic release linkage . . . . DISCONNECTED

6. System initiator. . . . . CONNECTED AND PIN REMOVED

   Check that initiator is connected to withdrawal arms and inner and outer torque shaft.

7. Time release mechanism rod . . . . SCISSOR RELEASE PIN PROTRUDING

   Check that pin is protruding above housing so as to inhibit plunger movement and scissor release.

8. Drogue chute link lines . . . . CLIPS ENGAGED

9. Scissor shackle. . . . . STOWED, DROGUE LANYARD, CONNECTED

10. Ejection gun sear . . . . CONNECTED AND PIN REMOVED

11. Ejection gun sear withdrawal links . . . . CONNECTED

    Check that withdrawal links are connected to sear.

12. Drogue chute housing flaps. TACKED DOWN

13. Face curtain . . . . SECURE

    Check that face curtain cross bar is securely restrained in the headbox.

14. Face curtain connecting link (L) . . . . SAFE INDICATION

    Tabs adjacent as viewed through inspection window.

15. Drogue chute lines (L) . . . . CLIPS ENGAGED

16. Drogue withdrawal line . . . . CONNECTED TO DROGUE SLUG

    Check that shielded withdrawal line is free for the slug to deploy the stabilizing drogue chute.

17. Drogue slug shear pin . . . . HEAD IMPRESSED

18. Top latch mechanism . . . . FLUSH

    Check that top latch dowel is flush with or slightly recessed from the end of the housing.

19. Parachute withdrawal line screw connector. . . . . FINGERTIGHT

    Check that the screw connector is torqued with essentially no gap.

WARNING

If the personnel parachute withdrawal line is not securely connected, automatic deployment of the parachute will not occur after ejection.

20. Guillotine cutter . . . . WITHDRAWAL LINE ROUTED THROUGH GUILLOTINE

    Check that the sheathed parachute withdrawal line is routed through the guillotine cutter under the yellow protective guard.

21. Power inertia reel sear . . . . CONNECTED

    Check that actuating link is connected to the power inertia reel gas generator sear.

22. Fork ends of harness release sear rods . . . . CONNECTED

23. Rocket motor initiator sear . . . . CONNECTED AND PIN REMOVED

24. Drogue gun trip rod . . . . . CONNECTED AND PIN REMOVED

   No red indication showing on inner rod and rod attached to ejection gun cross beam.

25. Rocket motor initiator sear extraction lanyard . . . . ATTACHED

   Check that the lanyard is attached to the sear actuating link and the drogue gun trip rod.

26. Parachute premature deployment lanyard . . . . SECURED

   Check that the lanyard is anchored to the inertia reel housing.

27. Shoulder harness . . . . SECURED

   Check that inertia reel straps are anchored to inertia reel housing.

28. Ripcord . . . . . ROUTED THROUGH PARACHUTE PACK RING

29. Manual ripcord eyelet . . . . CONNECTED TO LEAP RIPCORD PIN

30. Ripcord pins . . . . . . STRAIGHT/NO CORROSION

31. Ventilation hose . . . . CONNECTED TO SEAT OR DISCONNECTED

   Connect as appropriate depending on flight suit configuration..

32. Personnel services blocks . . . . VERIFY CONNECTION TO SEAT BUCKET DISCONNECT LANYARD(S) (2) SECURED TO DECK

33. Survival kit release handles . . . . NOTE TYPE

34. Lap belt restraint . . . . SECURED

   Tug lap restraint fittings to assure that end fittings are secured to seat bucket.

35. Emergency oxygen trip cable . . . . SECURED

   Check that the seat kit cable is connected to the lower block in order to enable automatic oxygen actuation upon ejection.

   Note

   In aircraft BUNO 159631 and subsequent modified by ACC 309 and those aircraft incorporating ACC 404, the emergency oxygen trip cable is attached to the personnel services disconnect.

36. Survival kit sticker clips and lugs . . . . SECURED AND ENGAGED

   Check clips secured to seat bucket fitting and release linkage attached.

37. Survival kit front lock release lever . . . . FULL FORWARD

   Check that survival kit front lock release lever peg is full forward against leg line release linkage.

   WARNING

   Failure of survival kit front lock release lever to return to its full forward position will prevent ejection by means of the secondary firing handle.

   Note

   It may be necessary to raise seat to see lever.

38. Emergency oxygen supply bottle . . . . FULL INDICATION, 1800 PSI (IN BLACK)

39. Emergency oxygen actuator . . . . CONNECTED/STOWED

40. Leg restraint cords . . . . SECURED TO DECK, ROUTED THROUGH SNUBBER, ANCHORED TO SEAT RELEASE FITTING

41. Survival kit forward restraint
    pin .... ENGAGED

    Pull upward on forward edge of survival kit to
    ensure that restraint pin is engaged.

42. URT-33 lanyard .... CONNECTED

43. Emergency restraint release
    handle .... DOWN AND SECURED,
                                   PIN REMOVED

    Check that handle is full down and trigger
    protrudes from grip.

44. Guillotine initiator sear .... CONNECTED

    Check that emergency release handle actuating
    linkage is connected to sear and leg restraint
    release linkage.

45. Ejection and canopy pins .... STOWED

PILOT PROCEDURES

The interior inspection provides a systematic coverage of all cockpit controls to ensure proper setup prior to the application of external power, assuming no external air-conditioning source will be used prior to engine start. These checks correspond to the condition which the plane captain should set up in the cockpit as part of his preflight. Each cockpit setup consists of a sequential sweep of controls on the left console, instrument panel, and right console.

INTERIOR INSPECTION — PILOT

1. Harnessing .... FASTEN

   a. Leg restraint lines and
      garters .... CONNECT

      Connect leg line bayonet fitting to leg garter
      quick-release buckle on the respective side.
      Ensure that leg lines are not twisted.

   b. Personal services (vent and
      anti-g hoses) .... CONNECT

      Insert personal vent and anti-g hoses into seat
      block fittings. In the absense of a suit vent
      requirement the seat block vent hose should
      be inserted into the seat back fitting to
      provide ventilation air through the seat and
      back cushions.

   c. Lap belt .... ATTACH

      Attach lap belt fittings and pull bucket straps
      snug so as to provide secure lap restraint for
      flight and seat kit suspension in the event of
      emergency egress or ejection.

   d. Shoulder harness .... ATTACH

      Attach shoulder harness release fittings to torso
      fittings. Check for proper positioning of
      parachute D-ring on left riser.

   e. Inertia reel .... CHECK

      Position the shoulder harness lock lever forward
      to the lock position and check that both shoulder
      straps lock evenly and securely when leaning
      back. Position lock lever full aft to unlock har-
      ness; release lever to the neutral position.

   f. Oxygen-audio connection .... ATTACH

      Attach composite fitting without causing
      unnecessary twisting of hard hose.

2. Oxygen .... CHECK

   Turn OXYGEN switch ON, purge with mask held
   away from face. Place mask to face and check
   for normal breathing and regulator and mask
   operation. Turn OXYGEN switch OFF, check
   no breathing.

3. VENT AIRFLOW thumbwheel .... SET

   Set thumbwheel as desired to control vent airflow anywhere between no flow (0) and full
   flow (15).

4. TONE VOLUME .... SET

5. TACAN switch .... OFF

   a. Channel .... SET

   b. Function selector .... OFF

   c. VOL knob .... COUNTERCLOCKWISE

6. ICS panel . . . . SET
   a. VOL knob . . . . CLOCKWISE
   b. Amplifier . . . . NORM
   c. Function selector . . . . COLD MIC
7. STAB AUG switches . . . . OFF
8. UHF function selector . . . . OFF

WARNING

With electrical and/or hydraulic power on the aircraft, wings can move if wing control systems fail.

9. Wing sweep mode switch . . . . MAN
10. Left and right throttles . . . . OFF
11. Speed brake switch . . . . RET
12. Exterior lights master switch . . . . SET

   Position switch in accordance with standard procedures for day or night and field or carrier operations.

13. FLAP handle . . . . CORRESPONDING
14. Throttle friction lever . . . . OFF (AFT)
15. THROTTLE TEMP switch . . . . SET

   Ambient air temperature:
   - >80°F - HOT
   - 40° to 80°F - NORM
   - <40°F - COLD

16. THROTTLE MODE switch . . . . BOOST
17. L and R INLET RAMPS switches . . . . AUTO
18. ANTI SKID SPOILER BK switch . . . . OFF

19. FUEL panel . . . . SET
    a. WING/EXT TRANS switch . . . . AUTO
    b. REFUEL PROBE switch . . . . RET
    c. DUMP switch . . . . OFF
    d. FEED switch . . . . . . NORM (GUARD DOWN)
20. LDG GEAR handle . . . . DN
21. NOSE STRUT switch . . . . OFF
22. Parking brake . . . . PULL
23. Radar altimeter . . . . OFF
24. Altimeter . . . . SET

    Set field or carrier elevation as applicable

25. Left and right FUEL SHUT OFF handles . . . . . IN
26. ACM panel . . . . SET
    a. ACM switch . . . . OFF (GUARD DOWN)
    b. MASTER ARM switch . . . . OFF (GUARD DOWN)
27. Weapon select . . . . OFF
28. VDIG and HSD retaining locks . . . . . ENGAGED

Note

Visually check that the retaining locks for the pilot's displays are engaged to ensure safe back-up mounting for catapulting and flight.

29. HUD and VDI filters . . . . AS REQUIRED

    Filter control handle on right side of VDI should be pushed in for day flights and pulled out (filter in place) for night operations.

30. Standby attitude gyro . . . . UNLOCKED
31. G-meter . . . . RESET
32. Clock . . . . WIND AND SET

33. Fuel BINGO .... SET

Set total fuel remaining value for initial activation of fuel BINGO caution reminder consistent with mission profile to be flown.

34. Circuit breakers .... CHECKED

35. Brake accumulator pressure .... CHECK IN GREEN

36. HYD HAND PUMP .... CHECK

Extend handpump handle and stroke to check firmness of pumping action and an indication of pressure buildup on the brake pressure gage. Stow handpump handle in a convenient position for ready access.

37. HOOK handle .... CORRESPONDING

38. DISPLAYS panel .... SET
   a. MODE pushbutton .... T.O.
   b. HUD DECLUTTER switch .... OFF
   c. HUD AWL switch .... ACL
   d. VDI MODE switch. .... NORM
   e. VDI AWL switch. .... ACL
   f. HSD MODE switch .... NAV
   g. HSD/ECM switch .... OFF
   h. STEER CMD pushbutton .... DEST
   i. DISPLAYS POWER switches .... OFF

39. ELEV LEAD knob .... SET

40. INBD and OUTBD spoiler switches .... NORM (GUARD DOWN)

41. L and R generator switches .... NORM

42. EMERG generator switch .... NORM (GUARD DOWN)

43. ARA-63 panel .... OFF
   a. CHANNEL selector .... SET
   b. POWER switch .... OFF

44. COMPASS panel .... SET
   a. Mode selector knob .... SLAVED
   b. Hemisphere N-S switch .... SET
   c. LAT knob .... SET

45. Air condition controls .... SET
   a. TEMP mode selector switch .... AUTO
   b. TEMP thumbwheel control .... AS DESIRED (5 TO 7 MIDRANGE)
   c. CABIN PRESS switch .... NORM
   d. AIR SOURCE pushbutton .... BOTH ENG

46. WINDSHIELD switch .... OFF

47. ENG/PROBE ANTI-ICE switch .... AUTO

48. MASTER LIGHT panel controls .... AS REQUIRED

Set external and interior lighting controls consistent with day or night and field or carrier operating conditions.

49. MASTER TEST switch .... OFF

50. EMERG FLT HYD switch .... AUTO (GUARD DOWN)

51. HYD TRANSFER PUMP switch ..... SHUTOFF (GUARD UP)

52. CANOPY DEFOG-CABIN AIR lever .... CABIN AIR

53. Storage case .... INSPECT

Check adequacy of flight planning documents and storage of loose gear.

PRESTART—PILOT

> **CAUTION**
>
> Wings will move to emergency handle position with circuit breakers pulled.

1. If wings are not in OV SW:

    a. WING SWEEP DRIVE NO. 1 and WG SWP DR NO 2/MANUV FLAP cb . . . . PULL (LE1, LE2)

    b. Emergency WING SWEEP handle . . . . EXTEND AND MATCH WING POSITION

    Captains bars match position tape on indicator.

2. External electrical power . . . . ON

3. MASTER TEST switch . . . . MACH LVR

4. ICS . . . . CHECK

5. Landing gear indicator and transition light . . . . CHECK

    Check gear position indication down and transition light off.

6. MASTER TEST switch . . . . CHECK

    Coordinate with RIO.

    a. LTS

    Check that all warning, caution, and advisory lights illuminate. The brightness of the ACM panel and the indexer lights should be set during the test. The DATA LINK switch must be ON to check the DDI lights.

> **WARNING**
>
> Failure of the EMERG STORES JETT pushbutton to illuminate during LTS check could indicate that the pushbutton light is burned out or that the test circuit is defective. If the switch is actuated, stores will jettison when weight is off wheels. If this occurs, status of the emergency stores jettison circuit cannot be determined. Under some lighting conditions it may be difficult to determine when the light is illuminated. Ensure that the light goes off when LTS position on MASTER TEST switch is deselected. Failure of the light to go off indicates that emergency jettison is selected; stores will jettison when weight is off wheels.

   b. FIRE DET/EXT

   L and R FIRE lights illuminate to verify continuity of respective system. The GO light will illuminate verifying continuity through the four squib lines, that 28 volts dc is available at the left and right fire switches, and that the fire extinguisher containers are pressurized.

   c. INST

   Check for the following responses:

   (1) RPM . . . . .80% ± 2%

   (2) TIT . . . . 1300° ± 20°C

   (initiates engine overtemperature alarm tone at 1215° ± 15°C.)

   (3) FF . . . . 4300 ± 100 PPH

   (4) AOA (UNITS) . . . . 18 ± .5

   Reference and indication

(5) Wing sweep . . . . 45° ± 2.5°

    Program, command, and position.

(6) FUEL QTY . . . . 2000 ± 200 POUNDS
(BOTH COCKPITS)

(7) Oxygen quantity . . . . 2 LITERS

(8) L OVSP/VALVE and R OVSP/VALVE lights . . . . ILLUMINATED

(9) L and R FUEL LOW lights . . . . ILLUMINATED
(BOTH COCKPITS)

(10) OXY LOW light . . . . ILLUMINATED
(BOTH COCKPITS)

  d. MASTER TEST switch . . . . OFF

7. Ejection seats . . . . ARMED

Verify seat armed with RIO.

8. CANOPY handle . . . . CLOSE

WARNING

Flightcrews shall ensure that hands and foreign objects are clear of front cockpit handholds and top of ejection seats and canopy sills to prevent personal injury and/or structural damage during canopy opening or closing sequence. Only minimum clearance is afforded when canopy is transitioning fore and aft.

Note

If the CLOSE position does not close the canopy, depress the grip latch, release and push handle outboard and forward into the BOOST position. If it is necessary to use the BOOST position, the handle shall be returned to the CLOSE position to avoid bleed off of pneumatic pressure.

9. ACM panel . . . . SET

  a. Gun rate . . . . SET AND CHECK ROUNDS REMAINING

  b. SW COOL . . . . OFF

  c. MSL PREP . . . . OFF

  d. MSL MODE . . . . NORM

  e. Station loading status windows . . . . . . CHECK

    Verify proper indication consistent with external store loading condition.

10. EMERG STORES JETT pushbutton light . . . . OFF

Note

In aircraft BUNO 161282 and subsequent and aircraft incorporating AFC 628, the MASTER CAUTION light and the EMERG JETT caution light illuminates when the EMERG STORES JETT pushbutton is activated.

11. LADDER light . . . . OFF

Plane captain shall stow boarding ladder and steps.

12. Inform RIO . . . . READY TO START

13. Starter air . . . . ON

14. Windshield rain removal air . . . . CHECK

ENGINE START — PILOT

Prior to engine start, the pilot and plane captain should ascertain that the turnup area is clear of FOD hazards, adequate fire-suppression equipment is readily available, and engine intakes and exhausts are clear. Although the engines cannot be started simultaneously, either can be started first. The following procedure establishes starting the right engine first in which case all ground support equipment should be positioned on the left side of the aircraft insofar as practical. Whenever possible the aircraft should be positioned so as to avoid tailwinds, which can increase the probability of hot starts.

> **WARNING**
>
> Coordinate movement of any external surfaces and equipment with the plane captain or director.

> **CAUTION**
>
> If engine chugs and/or RPM hangup is encountered with one engine turning during normal ground start, monitor TIT for possible hot start. AIR SOURCE pushbutton should be set for the operating engine until RPM stabilizes at idle; then set to BOTH ENG.

1. ENG CRANK switch . . . . L (LEFT ENGINE)

   Observe auxiliary brake pressure rise. Observe combined hydraulic system pressure rise. Cycle rudders full left and right, verify rudder control surface deflection as viewed on surface deflection indicator. Check for any non-synchronous movement of rudders. With combined system pressure only, rudder actuator with crossed flex line condition present will remain neutral. Opposite rudder will move as commanded with excessive amount of force required to actuate.

2. ENG CRANK switch . . . . OFF

3. EMERG FLT HYD switch . . . . CYCLE

   a. EMERG FLT HYD switch . . . . LOW

   Check that ON flag is displayed in EMER FLT LOW hydraulic pressure window. Verify control over horizontal tail and rudder control surfaces as viewed on surface position indicator.

   b. EMERG FLT HYD switch . . . . HIGH

   Check that ON flag is displayed in EMER FLT HI hydraulic pressure window. Verify control over empennage flight control surfaces and higher surface deflection rate.

   c. EMERG FLT HYD switch . . . . . . AUTO (LOW)

   Check that OFF flags are displayed in both EMER FLT HI and LOW hydraulic pressure windows.

> **CAUTION**
>
> Combined and brake accumulators should be charged prior to backup module checks. Checks should be made slowly enough to ensure continuous on indication in the hydraulic pressure indicator.

4. ENG CRANK switch . . . . R (RIGHT ENGINE)

   Place the crank switch to the R position where the switch is solenoid-held until automatically released to the neutral (OFF) position at the starter cutout speed of 45% RPM. Manual deselect of the switch to the OFF position will interrupt the crank mode at any point in the start cycle. Oil pressure and flight hydraulic pressure rise will become evident at 10% RPM.

   Note

   When using wells system air for engine start, manual deselection of starter crank switch may be required.

5. Right throttle . . . . IDLE AT 22% RPM

> **CAUTION**
>
> Attempting a ground start at lower engine rotor speeds will aggravate hot start tendencies. Exceeding 750°C TIT constitutes a hot start. Advance the R throttle around the horn from OFF to IDLE when the rotor speed exceed 22% RPM; this action automatically actuates the ignition system. An immediate indication of fuel flow (≈500 to 700 PPH) will be exhibited and light off (TIT rise) should be achieved within 5 seconds, but not more than 20 seconds. The rapid rise in TIT should be carefully monitored for over-temperature tendencies. Peak starting temperatures will be achieved in the 30 to 40% RPM range, when, after a slight hesitation, a reduction will return the TIT to the nominal 600°C level. During the initial starting phase, the nozzle should expand to a full-open position indication of 5.

Note

Loss of electrical power may result in smoke entering the cockpit via the ECS.

6. R GEN light . . . . OFF

The right generator should automatically pick up the load on the right main ac bus as indicated by the R GEN light going out at approximately 50% RPM.

7. R FUEL PRESS light . . . . OFF

The fuel-pressure lights should go off by the time the engine achieves idle RPM.

Note

During ground operation, the fuel-pressure lights may flicker when advancing the throttle(s) out of IDLE. This phenomenon is related to low power setting nozzle closure and has no adverse effect on engine fuel feed.

8. Idle engine instrument readings . . . . CHECK

   a. RPM . . . . 63 to 74%

   b. TIT . . . . 600°C (nominal)

   c. FF . . . . 1100 PPH (nominal)

   d. NOZ position . . . . .5 (open)

   e. OIL . . . . 40 to 50 PSI

                               (30 PSI minimum)

   f. FLT HYD PRESS . . . . 3000 PSI

9. External power . . . . DISCONNECT

Removal of ground electrical power causes the right generator to supply power to the right and left main electrical buses. Check for non-synchronous movement of rudders. With flight system pressure only, rudder actuator with crossed-flex-line condition present will fail hardover left and remain locked. Both rudders will indicate full left; however rudder actuator that does not have crossed-line condition will move as commanded, but with excessive amount of force.

10. ENG CRANK switch . . . . L (LEFT, ENGINE)

When combined hydraulic pressure reaches 3000 psi, return switch to neutral (center) position.

11. HYD TRANSFER PUMP switch . . . . . . NORMAL

Hydraulic transfer pump will operate from flight side to maintain the combined side between 2400 to 2600 psi.

CAUTION

If the transfer pump does not pressurize the combined system within 10 seconds, immediately set HYD TRANSFER PUMP switch to SHUTOFF.

12. HYD TRANSFER PUMP switch . . . . SHUTOFF

13. Repeat steps 4 through 8 for left engine.

14. Starter air . . . . DISCONNECT

15. HYD TRANSFER PUMP switch . . . . NORMAL

16. Ground safety pins . . . . REMOVE AND STOW

Plane captain should remove landing gear pins and stow them in nosewheel well.

POSTSTART—PILOT

1. STAB AUG switches . . . . ALL ON

2. MASTER TEST switch . . . . EMERG GEN

Check that NO GO light illuminates for about 1 second until emergency generator power is connected to essential buses and GO light illuminates. When disconnecting this test, the resultant power interruption causes the AHRS light to illuminate momentarily.

Advise RIO that test is complete.

3. VMCU operation . . . . CHECK

Following disengagement of the emergency generator, a proper check of the VMCU is indicated by illumination of the PITCH STAB 1 and 2; ROLL STAB 1 and 2; YAW STAB OP and OUT;

SPOILERS; HZ TAIL AUTH; RUDDER AUTH; AUTO PILOT; MACH TRIM; and R INLET lights during the power transient (approximately 1.25 seconds). Once normal power is restored, all lights will go out except the RUDDER AUTH light (resettable with the MASTER RESET pushbutton) and the PITCH and ROLL STAB AUG switches will be OFF.

4. Advise RIO that test and check is complete.

5. STAB AUG switches................ ALL ON

6. Emergency WING sweep handle .... OV SW

If wings are not in oversweep, move the wings to 68° using wing sweep emergency handle in raised position. Then raise handle to full extension and hold until HZ TAIL AUTH caution light goes out and OVER flag appears on wing sweep indicator. Move handle to full aft OV SW position and stow.

7. Wing sweep mode switch .... AUTO

8. WING SWEEP DRIVE NO. 1 and WG SW DR NO. 2/ MANUV FLAP cb .... IN (LE1, LE2)

9. WING/EXT TRANS .... OFF

10. MASTER RESET pushbutton .... DEPRESS

11. MASTER TEST switch .... WG SWP

Wing sweep program index moves from 20° to 44° and back to 20°. The following lights illuminate at start of test and are out at test completion (approximately 25 seconds): WING SWEEP, FLAP, GLOVE VANE, and REDUCE SPEED.

Note

- During the wing sweep preflight test, both altimeters may fluctuate momentarily.

- In aircraft BUNO 159637 and subsequent, the WING SWEEP advisory light illuminates 3 seconds after the test starts, then goes out and illuminates again 8 seconds into the test.

- In aircraft BUNO 159637 and subsequent, the WING SWEEP advisory light goes out at the end of the test. The RUDDER AUTH, HZ TAIL AUTH, and MACH TRIM lights illuminate for the entire test and remain illuminated at the end of the test.

12. COMM/NAV/GEAR/DISPLAYS .... ON

    a. UHF function selector ...... TR+G OR BOTH

    b. Data link power .... ON

    c. TACAN function selector .... T/R

    d. ARA-63 POWER switch .... ON

    e. DISPLAYS control switches .... ON

    f. RADAR ALTITUDE.................. ON

13. Trim .... SET 000

14. Standby gyro .... ERECT

15. MASTER RESET pushbutton .... DEPRESS

16. MASTER TEST switch .... OBC

Coordinate with RIO.

17. Autopilot .... ENGAGE

The following systems are automatically exercised during the 1-1/2 minutes required to complete the OBC tests. Failures are displayed on the TID display.

- The AICS self-test turns on hydraulic power and exercises the ramps through full cycle - STOW-EXTND-STOW. During the test, the respective RAMP light will be illuminated until the ramps return to the fully stowed position and the hydraulics are shut off. A failure is indicated by an INLET light and/or OBC readout.

- PITCH, ROLL, YAW STAB AUG

  AUTHORITY STOPS

  MACH TRIM COMPENSATOR

  AUTOPILOT, SPOILER

} AFCS COMPUTERS

During the course of the test, the STAB AUG lights will remain illuminated until the test is satisfactorily completed. All lights should be off at the termination of the test.

- AUTO THROTTLE

This test is a computer self-test with output commands inhibited to prevent throttle movement.

18. Speed brake switch . . . . EXT, THEN RET

    Cycle speed brake switch to EXT, release and check for partial extension. Select EXT again, checking indicator for transition to full extension. Select RET and check indicator for an indication of full retraction. In aircraft BUNO 159825 and subsequent and aircraft incorporating AFC 365, check for stabilizer position fluctuation during speed brake extension and retraction to verify integrated trim operation.

19. REFUEL PROBE switch . . . . FUS EXT, THEN RET

    Cycle the probe to the extend position, noting illumination of the probe transition light with switch-probe position disparity. Check probe nozzle head for condition. Retract probe and again check that transition light goes out when fully retracted and doors closed.

20. WINDSHIELD switch . . . . CYCLE

21. OBC . . . . OFF

    If engaged, verify that autopilot disengages automatically.

22. WING/EXT TRANS switch . . . . OFF

23. Trim . . . . CHECKED AND SET 000

Note

For CV operations omit steps 24 through 42.

CAUTION

Ensure adequate clearance before moving wings.

24. Emergency WING SWEEP handle . . . . 20°

    Move the emergency WING SWEEP handle to 20° (full forward) and engage the spider detent. Stow handle and guard. HZ TAIL AUTH light illuminates coming out of OVSW. Light goes off when OVSW stops removed.

25. MASTER RESET pushbutton . . . . DEPRESS

    The WING SWEEP warning and advisory lights go out and the AUTO and MAN modes are enabled.

26. External lights . . . . CHECK (Prior to night/IMC flight)

WARNING

During night operations, aircraft with inoperable tail and aft anticollision lights will not be visible from the rear quadrant even under optimum meteorological conditions, thus increasing mid-air potential.

27. Flaps and slats . . . . DN

    Check for full deflection of the flaps and slats to the down position and automatic activation of the outboard spoiler module.

    In aircraft 159825 and subsequent and aircraft incorporating AFC 365 and 400, check for 3° TEU stabilizer position.

28. Flight controls . . . . CYCLE

    Complete full cycle sweep of longitudinal, lateral, directional, and combined longitudinal-lateral controls while checking for full authority on surface position indicator. Check that all spoilers extend at the same rate with slow lateral stick deflections and extend to full up position.

Observe the following:

   a. Pitch control . . . . HORIZONTAL TAIL
                                                33° TEU TO 12° TED

   b. Lateral control . . . . 24° TOTAL
                                                DIFFERENTIAL TAIL

   c. Directional control . . . . 30° RUDDER

   d. Longitudinal-lateral
      combined . . . . 35° UP, 15° DN
                                                HORIZONTAL TAIL

29. DLC . . . . CHECK

Verify horizontal tail shift with DLC input.

30. ANTI SKID SPOILER BK switch . . . . SPOILER BK

31. MASTER TEST switch . . . . STICK SW

SPOILER light should illuminate and all spoilers should fail down. Verify that the GO light illuminates with 1 inch lateral stick in each direction.

32. Spoilers and throttles . . . . CHECK

33. ANTI SKID SPOILER BK switch . . . . OFF

34. Flaps and slats . . . . UP

35. Maneuver flaps . . . . DN

36. WING SWEEP mode switch . . . . MAN 50°

                        CAUTION

If wing sweep commanded position indicator (captain's bars) does not stop at 50°, immediately select AUTO with wing sweep mode switch.

37. Manuever flaps . . . . CRACK UP

38. Wing sweep mode switch . . . . BOMB

Check maneuver flap and glove vane retraction.

39. Emergency WING SWEEP handle . . . . 68°

40. Emergency WING SWEEP handle . . . . OV SW

41. Wing sweep mode switch . . . . AUTO

42. MASTER RESET pushbutton . . . . DEPRESS

Note

CV checklist resumes.

43. ANTI SKID SPOILER BK switch . . . . BOTH

44. ANTI SKID . . . . BIT

Ensure coarse alignment is complete before releasing parking brake.

45. ANTI SKID SPOILER BK switch . . . . OFF

46. Radar altimeter . . . . BIT

Depress SET knob; check that radar altitude displays 100 feet and indicator green light is illuminated. Release knob and pointer should display 0 feet; warning tone signal (both cockpits) and ALT LOW light illuminated momentarily.

47. Displays . . . . CHECK

48. TACAN . . . . BIT

49. ARA-63 . . . . BIT

50. Altimeter . . . . SET/RESET mode

Barometric setting and error determined. Check in RESET mode.

51. Compass . . . . CHECK

Validate IMU heading on alignment on HUD, VDI, HSD and BDHI. Crosscheck with known references and standby compass.

52. Flight instruments . . . . CHECK

Final Checker (Ashore)

1. NOSE STRUT switch . . . . KNEEL-CHECK
                                             LAUNCH BAR DN

> **CAUTION**
>
> Ensure all tiedowns have been removed before selecting kneel.

2. Hook . . . . DN, THEN UP
3. LAUNCH BAR switch . . . . CYCLE
4. NOSE STRUT switch . . . . EXTD

Final Checker Aboard CV

1. Hook . . . . CYCLE ON DIRECTOR'S SIGNAL

TAXIING

Prior to releasing the parking brake the IMU should be aligned to at least a diamond symbol, otherwise the alignment process will automatically be interrupted until the parking brake is reapplied.

To set the aircraft in motion starting from a static position requires application of about 80% RPM. While departing the line area, both flightcrewmen should clear the extremities of the aircraft and the wings should remain at 68° in OV SW to minimize the span clearance. Once in motion, IDLE thrust is normally sufficient to sustain taxi speeds and full nosewheel steering authority may be realized.

Taxispeed should be maintained at a reasonable rate consistent with traffic, lighting, and surface conditions.

> **CAUTION**
>
> - Before taxiing aircraft with wings in oversweep and full wing fuel tanks, trim stabilizer to zero to prevent wing tip and stabilizer interference.
> - When taxiing across obstacles ensure nosewheel is centered to preclude launch bar from impacting nosewheel well doors.
> - To prevent overheating, do not ride the wheel brakes.

The taxi interval should be sufficient to avoid taxiing in another aircraft's jet wash which presents additional FOD potential. Although the antiskid system is armed at speeds less than 15 knots, the antiskid system is not operative. The nosewheel steering can remain engaged throughout the taxi phase. Application of wheel brakes in conjunction with nosewheel steering should be performed symmetrically to minimize nosetire side loads. In minimum radius turns (figure 3-4) using nosewheel steering, the inboard wheel rolls backwards as the axis of rotation is between the main gear. Because of the distance from the cockpit to the main landing gear, the pilot should make allowance for such in-turns to prevent turning too soon and cutting corners short.

Crew comfort during taxi operations is affected by the nosestrut air curve characteristics which maintains the strut in the fully extended (stiff strut) position except during deceleration. Because of the wide stance of the main gear, differential application of wheel brakes is effective for turning the aircraft without the use of nosewheel steering. Subsequent to flight, while returning to the line at light gross weights, one engine may be shut down to prevent excessive taxi speeds at IDLE thrust.

> **CAUTION**
>
> - On deck engine operations for extended periods can result in an unacceptable buildup in fluid (hydraulic, engine oil, and IDG oil) temperatures by taxing heat exchanger capacities. Since the left IDG supplies the majority of the electrical power it is more susceptible to overheating than the right. Tailwinds or large power demand, or both, at high ambient air temperatures increase the chances of fluid overtemperature.
> - Since the outboard spoiler module is automatically energized with the flap handle down and weight-on-wheels, it is necessary to restrict the amount of flaps down operation on the deck to prevent module fluid overheating.

Note

When shutting down one engine during taxiing, the right engine is normally shut down so that normal braking is maintained.

TAXI – PILOT

1. Parking brake . . . . RELEASE
2. Nosewheel steering . . . . CHECK

   NWS ENGA light illuminates upon engagement. Check control and polarity in static position before commencing to taxi.

   Note

   If nosewheel steering is inoperative, the emergency gear extension air release valve may be tripped, which will prevent gear retraction.

3. Brakes . . . . CHECK

   Check for proper operation by applying left or right brake individually and observing brake pressure recovery to the fully-charged condition.

4. Turn-and-slip indicator . . . . CHECK

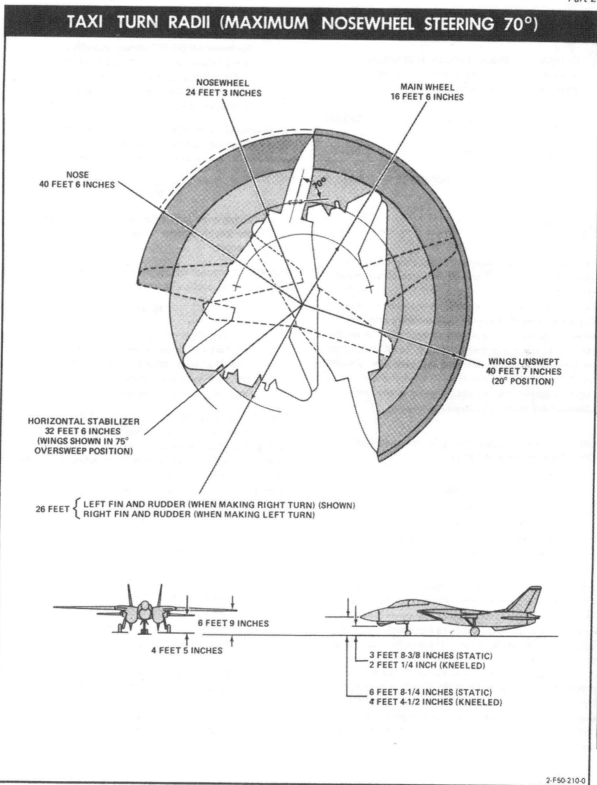

Figure 3-4. Taxi Turn Radii (Maximum Nosewheel Steering 70°)

5. Ordnance.....................SAFE

Perform the following functions at prescribed location prior to takeoff in accordance with base operating procedures:

a. Missile seeker and tuning............CHECK

b. Gun and external stores...........GROUND SAFETY PINS REMOVED AND ARMED

TAKEOFF

The placarded aircraft takeoff checklist should be completed prior to calling for takeoff clearance, and all annunciator lights should be off, except NWS ENGA. Full flaps and slats are optional for all takeoffs regardless of thrust or gross weight conditions. Both flightcrewmen should be operating in HOT MIC during this phase of flight to enhance communications in event of emergency. Upon tower clearance and after visually clearing the approach zone, the pilot should taxi onto the runway (take downwind side if another aircraft to follow) and roll straight ahead to align the nosewheel and to check compass alignment.

Hold in position for takeoff using the tow pedal brakes with nosewheel steering engaged.

Note

Do not use the parking brake to restrain the aircraft under high-power conditions since tire skid might result.

Antiskid and spoiler brakes should have been previously armed, so check spoilers in the drooped position upon throttle advancement for takeoff. The wheel brake is enough to restrain the aircraft in a static condition with both engines in minimum zone afterburning (nozzle position 2); however, the braking effectiveness is dependent on surface condition and on aircraft gross weight. Upon receiving tower clearance for takeoff and after clearing the area to the rear, advance both throttles to MIL and check engine instruments. Cross-check engine instruments (RPM, TIT, FF, and oil pressure) to ascertain that all are within limits.

Note

For heavy weight MIL power takeoffs where single-engine flyaway performance at MIL is marginal, the ability to achieve an afterburner light on both engines should be verified just before brake release.

For an afterburner takeoff, both afterburners should be lit (nozzle position 2) before brake release. Engine instruments should be reverified for afterburner operating conditions.

Note

A minimum nozzle position of 2 is required for takeoff to ensure positive afterburner liftoff.

Ground engine-power checks should be performed as expeditiously as possible to conserve fuel.

Note

The nose strut will compress to the kneel position at high power settings when statically restrained but will recover to the static extended position upon throttle retard or brake release.

After takeoff power checks are completed and at a safe interval behind the preceding aircraft, release the toe pedal brakes. Nosewheel steering may be used for initial takeoff roll until runway heading is established and rudder effectiveness is gained. To minimize stress on the nosewheel steering unit, it should be deselected as soon as it is not required for directional control.

Note

- Takeoffs performed with standing water on the runway may result in unstable engine operation due to water ingestion. Engine stall margin is reduced during engine transients, particulary afterburner staging.

- The nose strut should return to the fully extended position (+1.5° pitch attitude) upon brake release; failure to do so will increase the takeoff ground roll. Use of differential braking to control directional alignment should be avoided due to its attendant effect on ground roll distance.

Minimum ground roll takeoff procedures do not differ from the normal procedures, except for the use of MAX AB thrust. Maintain the control stick at the trimmed condition during the prerotation ground roll phase to minimize aircraft drag. After the pre-computed rotation speed (refer to section XI), pull the control stick straight aft so that 25° symmetrical deflection of the horizontal tails is achieved at the predicted rotation speed. Hold the control stick aft so that 25° horizontal tails is achieved until liftoff, which will occur at approximately a 6° noseup pitch attitude. With the wings swept forward, the aircraft seems to balloon from the runway in a near-level nose attitude with a more docile transition to flight than characteristic of swept-wing aircraft.

Note

- The use of excessive back stick on takeoff may cause the tail surfaces to stall, delaying aircraft rotation and extending takeoff distance.

- Although on-deck pitch attitude rotation in excess of 10° provides marginal tail-ground clearance, the aircraft is airborne well before such a phenomenon becomes a limiting factor.

After lift-off, relax the aft stick force as the aircraft accelerates towards an in-trim condition. Raise the landing gear control handle after ensuring that the aircraft is definitely airborne. Pitching moments associated with gear retraction are negligible and a gear UP indication should be achieved about 15 seconds after initiation.

CAUTION

Illumination of indexer lights is not a positive indication that the main landing gear are clear of the runway. Raising the gear before a positive rate of climb is established will result in blown main tires.

At approximately 180 KIAS (depending on longitudinal acceleration) the FLAP handle can be placed in the UP position. A moderate noseup pitching moment occurs during the flap and slat retraction phase, which takes approximately 8 seconds. Do not attempt to correct immediately after lift-off to counter a lateral drift due to a crosswind condition. The use of large lateral control deflection should be avoided to keep from breaking out the wing spoilers, which have a negative effect on lift and drag. Differential tail authority within the spoiler deadband (1/2-inch lateral stick deflection) is adequate for maintaining wings-level flight or effecting gradual turns with symmetric thrust. Before reaching the flap (225 KIAS for 10° flaps) and gear (280 KIAS) limit speeds, the pilot should ascertain that all devices are properly configured for higher speed flight. The combined hydraulic system non-flight essential components (gear, brakes, and nosewheel steering) may be isolated by selecting FLT position on the HYD ISOL switch. A gradual climbout pitch attitude should be maintained until intercepting the optimum climb speed. A recheck of engine instruments and configuration status should be performed after cleanup during the climbout phase.

WARNING

After takeoff, leave gear retraction HYD ISOL switch in T.O./LDG position. Selection FLT position allows hydraulic pressure to decay in the main landing gear uplock actuators. A zero- to negative-g maneuver can cause the main landing gear to be released from the up-locks and subsequently rest on the landing gear doors. The application of positive-g in this configuration could cause inadvertent extension of one or both main landing gear.

Flaps Up Takeoff

Before the takeoff roll, the procedures for flaps-up takeoff are identical to flaps down, except that the flaps remain retracted and only inboard spoiler brakes are available. During the prerotation ground roll phase, maintain the control stick at the trimmed condition to minimize aircraft drag. At the precomputed rotation speed, smoothly position the stick aft (about 2 inches) to maintain a 5° to 10° pitch attitude until airborne.

Do not exceed 10° of pitch attitude until well clear of the runway, as excessive nose-up attitudes will cause the vertical fins and tailpipes to contact the runway surface.

CAUTION

Due to increased longitudinal control effectiveness with the flaps retracted, overcontrol of pitch attitude during takeoff is possible. Large or abrupt longitudinal control inputs should be avoided until well clear of the runway.

Transition to flight will occur smoothly as compared to the ballooning effect in flaps-down takeoffs. After main gear liftoff, relax the aft stick force as the aircraft accelerates.

Due to the smooth, flat transition to flight, care should be taken to avoid premature landing gear retraction and resulting blown tires. Raise the landing gear control handle only after ensuring that the aircraft is airborne.

Note

- During flaps-up takeoffs, all flap/wing electromechanical interlocks are removed from the CADC and wing-sweep control box, allowing possible inadvertent wingsweep in the event of a CADC failure.

- Outboard spoilers are inoperative with weight on wheels.

Formation Takeoff

Formation takeoffs are authorized in the flaps-up configuration for sections only. However, they shall not be permitted at night, with crosswind component in excess of 10 knots, with standing water on the runway, on runways less than 8000 feet long and 200 feet wide, or with dissimilar aircraft. All aspects of the takeoff must be briefed by the flight leader. Briefing should include flap setting (flaps-up only), power settings, use of nosewheel steering, abort procedures, and signals for power and configuration changes.

MILITARY

With the completion of the takeoff checks (ensure alpha computer circuit breaker pulled), the lead aircraft will take position on the downwind side of the runway with the wingman on a normal parade bearing with no wing overlap. Upon signal from the leader, the engines will be advanced to military power. When ready for flight, the pilots shall exchange a thumbs-up signal. On signal from the leader, brakes are released and the leader reduces power 2%. Directional control is then maintained with nosewheel steering until rudder becomes effective. During takeoff roll, the leader should make only one power correction to enhance his wingman's position. If optimum position cannot be obtained, relative position should be maintained until the flight is safely airborne. At the precomputed rotation speed, the leader should rotate his aircraft 5° to 10° noseup on the HUD or VDI, and maintain this attitude until the flight is airborne.

AFTERBURNER (Preferred method)

With the completion of the takeoff checks (ensure alpha computer circuit breaker pulled), the leader will take position on the downwind side of the runway with the wingman on a normal parade bearing with no wing overlap. The leader will signal engine run up with a two finger turn. After engine runups are complete the wingman will pass a thumbs up to the leader. The leader will signal zone two

NAVAIR 01-F14AAA-1

afterburner for both aircraft by raising his right hand. After checking for proper nozzle indications, and afterburner light is off, the wingman will give a head nod to indicate he is ready for flight. The leader will indicate brake release by dropping his hand. As the aircraft begin their takeoff roll, directional control should be maintained with nosewheel steering until rudders become effective. If optimum position cannot be obtained relative position should be maintained, until the flight is safely airborne. At the precomputed rotation speed, the leader should rotate his aircraft 5° to 10° noseup on the VDI or HUD, and maintain this attitude until the flight is airborne.

Note

- No aux flap electromechanical interlocks exist to prevent inadvertent wing sweep in the event of a CADC malfunction.

- Outboard spoilers are inoperative with weight-on-wheels.

The wingman should strive to match the leader's attitude as well as maintain parade bearing with wingtip separation. When both aircraft are safely airborne, the gear is retracted on signal from the leader. The leader will then signal for deselection of afterburner. Turns into the wingman will not be made at altitudes less than 500 feet above ground level.

CAUTION

- In the event of an aborted takeoff, the aborting aircraft must immediately notify the other aircraft and the tower. The aircraft not aborting should ensure positive wingtip separation is maintained and select maximum afterburner to accelerate ahead of the aborting aircraft. This will allow the aborting aircraft to move to the center of the runway and engage the available arresting gear, if required.

- It is imperative that the wingman be alert for the overrunning situation and take timely action to preclude this occurrence. Should an overrunning situation develop after becoming airborne, the wingman should immediately increase lateral separation

from the leader to maintain wing position. Safe flight of both aircraft must not be jeopardized in an attempt to maintain position.

Takeoff Aborted

See Section V

TAKEOFF CHECKLIST

Prior to takeoff, the checklist will be completed by the challenge (RIO) and reply (pilot) method with the ICS on HOT MIC as a double-check of the aircraft configuration status.

For CV operations, steps 1 through 9 may be completed while tied down. For field operations, steps 1 through 14 should be completed in the warmup area.

RIO CHALLENGE	PILOT REPLY
1. "BRAKES"	"CHECK OK, ACCUMULATOR PRESSURE UP"
2. "FUEL TOTAL _____ lbs."	"NORMAL FEED, AUTO TRANSFER, DUMP OFF, TRANSFER CHECKED, TOTAL _____ WINGS _____ AFT AND LEFT _____ FORWARD AND RIGHT _____ FEED TANKS FULL BINGO SET _____"
3. "CANOPY CLOSED, LOCKS ENGAGED, LIGHT OUT, STRIPES ALIGNED, HANDLE IN CLOSE POSITION"	"CLOSED, LOCKS ENGAGED, LIGHT OUT, SEAL INFLATED, HANDLE IN CLOSE POSITION"
4. "SEAT...ARMED TOP AND BOTTOM COMMAND EJECT _____"(as briefed)	"ARMED TOP AND BOTTOM, PILOT/MCO IN WINDOW" (as indicated)
5. "STAB AUG"	"ALL ON"
6. "ALL CIRCUIT BREAKERS SET"	"ALL IN"

Change 1 3-27

RIO CHALLENGE	PILOT REPLY
7. "MASTER TEST SWITCH"	"OFF"
8. "BI-DIRECTIONAL"	"NORMAL"
9. "COMPASS AND STANDBY GYRO"	"COMPASS SYNCHRONIZED, STANDBY GYRO ERECT"

CV — Approaching

CAT On Director's Signal

10. "WINGS" (visually checked)	"20°, AUTO, BOTH LIGHTS OUT"
11. "FLAPS AND SLATS" (visually checked)	AS REQUIRED
12. "SPOILERS AND ANTI-SKID"	"SPOILER MODULE ON, SPOILER BRAKES SELECTED" (field)
	"SPOILER MODULE ON, SPOILER BRAKES OFF" (CV)
13. "TRIM"	"0, 0, 0," (field) AS REQUIRED (CV)
14. "HARNESS... LOCKED"	"LOCKED"
15. "CONTROLS" (RIO visually check for full spoiler deflection)	"FREE, 33° AFT STICK, FULL SPOILER DEFLECTION, LEFT AND RIGHT, HYDRAULICS 3000 PSI"
16. "ALL WARNINGS AND CAUTION LIGHTS OUT"	"ALL WARNING AND CAUTION LIGHTS OUT"

ASHORE — In Takeoff Position

17. "ANTI-SKID/ SPOILER BRAKES"	"BOTH" (If operable)

ASCENT CHECKLIST

At level-off or 15,000 feet (whichever occurs first):

1. Cabin pressurization .... CHECK
2. Fuel transfer .... CHECK
3. Inflight OBC .... RUN

IN-FLIGHT OBC

The RIO should perform an OBC check as follows:

1. CM failures on lower left corner of TID .... CHECK

Note

Before correcting a malfunction, ensure available flycatcher/CM data is recorded.

2. Verify ACM, NEXT LAUNCH, A/G ...... Not selected
3. OBC disabled on pilot's MASTER TEST PANEL .... CHECK
4. Verify MASTER ARM .... OFF
5. Computer address panel:
   a. CATEGORY switch .... SPL
   b. OBC BIT pushbutton .... DEPRESS

This will initiate the in-flight BIT. After approximately 90 seconds, the TID will display OBC BIT followed by the acronyms of the failed aircraft systems and TEST COMPLETE.

   c. CATEGORY switch .... NAV

PRELAND AND DESCENT

1. STAB AUG switches..................... ON
2. Autopilot .... AS DESIRED
3. Wing sweep mode switch .... AS DESIRED
4. ANTI SKID SPOILER BK switch .... BOTH
   (If operable, CV-OFF)
5. Altimeter .... SET
6. Radar altimeter .... ON/BIT CHECK
7. Fuel quantity and distribution .... CHECK

NAVAIR 01-F14AAA-1

Section III
Part 2

8. Armament .... SAFE

9. CANOPY DEFOG/CABIN
   AIR lever .... DEFOG

10. ENG/PROBE ANTI-ICE switch .... AUTO

11. PDCP .... SET

12. ARA-63A/ACLS .... ON/BIT CHECK

13. ECM switch .... OFF

14. DECM switch .... OFF

15. CHAFF/FLARE dispenser switch .... OFF

WARNING

If LAU-7 mounted stores are loaded and carrier or field arrestment is anticipated, perform steps 16 and 17 with the aircraft in a safe area and headed in a direction where inadvertant gun firing would not cause damage.

16. Weapon Select .... GUN

If trigger is illuminated with MASTER ARM off or CIA CI/PSU acronym was displayed during inflight OBC, perform step 17.

17. STA 1/8 AIM-9 REL
    PWR circuit breaker
    (6D4, 6D3) .... PULL

18. WCS switch .... STBY or XMT PULSE

WARNING

The RIO should place WCS switch to STBY or XMT (Pulse) on final approach to prevent unnecessary exposure of flight deck personnel to RF radiation hazard.

T 19. Resolution run .... COMPLETE

Note

Before reconnaissance system shutdown, run film leader to protect target imagery from inadvertent exposure during film download.

T 20. FRAME switch .... OFF

T 21. PAN switch .... OFF

T 22. IRLS switch .... OFF

T 23. FILM switch .... OFF

CAUTION

Before selecting system switch to OFF, delay 15 seconds for sensor shutdown, IR door to close, and mount to drive to vertical.

T 24. CPS system switch .... OFF

PATTERN ENTRY

Entry to the field traffic pattern will be at the speed and altitude prescribed by local course rules. When approaching the initial for the break, wings may be positioned manually full aft to facilitate multiplane entry and break deceleration. Break procedures shall comply with squadron, field, and/or CV standard operating guidelines.

LANDING

Approach

At the abeam position for landing, the aircraft should be at the prescribed altitude, trimmed up to 15 units angle of attack with the landing checklist completed.

Indicated airspeed should be crosschecked with gross weight in wings-level flight to verify angle of attack accuracy. Direct lift control (DLC) and the approach power compensator (APC) should be engaged as desired and checked for proper operation. The turn off from the 180° position should be made based on surface wind conditions and interval traffic (type, pattern, touch-and-go or final landing, etc.) so as to allow sufficient straightaway on final prior to touchdown.

The quality of the approach and touchdown is enhanced by starting from on-speed and altitude. The low thrust required in the landing approach leaves little margin for corrections from a high, fast position. Therefore, the pilot must control these parameters precisely from the onset of the approach to touchdown. Inertia and tail movement in conjunction with engine thrust response characteristics dictate the use of small, precise corrections on the glide slope for the most effective control technique. Lateral overcontrol produces large yaw excursions which complicate line-up analysis. All turns should be coordinated with rudder inputs.

The landing should be planned for the downwind side of the runway with traffic behind, or opposite the nearest traffic on landing rollout, or on the turnoff side of the runway. When surface wind is not a factor, pilots should practice flying on the field optical landing aid system whenever possible. Fly the aircraft down to the deck without flaring so as to accurately establish a touchdown point and achieve initial compression of main gear struts to arm the spoiler brakes.

Note

Landing with DLC engaged will reduce the amount of aft stick deflection available. DLC should be deselected when established on landing rollout.

Touchdown

To avoid tail-ground clearance problems, pitch attitude should not exceed 15 units AOA. At touchdown, immediately retard throttles to IDLE and confirm spoiler brake deployment. Expeditiously lower the nose gear to the deck and, without allowing the nose to come up, position the stick full aft.

Roll Out

The braking technique to be utilized with or without antiskid selected is essentially the same; a single, smooth application of brakes with constantly increasing pedal pressure. Do not pump the brakes. Directional control during rollout may require some differential braking.

Nosewheel steering may be used during rollout but it must be engaged with the rudder pedals centered to avoid a directional swerve upon engagement. Restrict the use of nosewheel steering during rollout until, or unless required for directional control. Under conditions of normal braking (antiskid selected), the antiskid system is passive and has no effect on wheel brake operation. However, if maximum deceleration is desired, commence braking as the nose is lowered and smoothly apply sufficient pressure to activate the antiskid system. When an impending skid is sensed, antiskid operation will result in a series of short wheel brake releases and a surging deceleration. Constant pedal pressure should be maintained. Approaching taxi speed (about 15 knots), ease brake pressure and deselect antiskid.

CAUTION

- If brakes are lost, release brake pedals and secure antiskid.
- If antiskid is not deselected before 15 knots, continued hard braking could result in blown tires.

Note

If maximum-effort braking or antiskid is not required, or antiskid is not selected, delaying brake application until the aircraft aerodynamically decelerates below 80 knots greatly reduces the possibility of blown tires and overheated brakes.

Follow the postlanding checklist for proper configuration cleanup procedures. Clear the area behind before turning off across the runway. The right engine may be shut down to reduce residual thrust during low gross weight taxiing.

Touch and Go

For touch-and-go landings, MIL thrust is applied after touchdown, and automatic retraction of the speed brakes occurs to configure the aircraft for a go-around. Control for rotation is greater than experienced on takeoff, although the aircraft has the same basic liftoff characteristics. Fuel required per pass is normally 300 pounds, contingent on traffic pattern.

Crosswind Landings

If landing is required at crosswind conditions above the limits or on a wet runway with crosswinds approaching the limits, an arrested landing should be made.

Crosswind landing tests with a 20-knot component with all spoilers extended have been completed. With this limit, crosswind landings present no unusual directional control problems with up to a 20-knot component with all spoilers extended. Touchdown should be on-speed, firm, and within the first 500 feet. Lateral drift before touchdown must be eliminated by the wings-level crab or wing-down technique (this is especially important on a wet runway). Upon touchdown, lower the nose to the runway to improve the straight tracking characteristics. With a 20-knot component, the upwind wing will raise slightly. However, aerodynamic controls (rudders) will effectively maintain a straight track until approximately 80 KIAS where differential braking may be required. If directional control is marginal, nosewheel steering should be used to maintain control. Nosewheel steering will probably be required during crosswind conditions with a wet runway.

If a landing must be made in crosswind conditions in excess of the limits, the techniques must be changed. At some crosswind component the upwind wing will be raised excessively and as a result directional control will be marginal. It is estimated that this will occur with a greater than 25-knot crosswind component. If after touchdown the wing is raised excessively, the spoiler brakes should be turned off and lateral stick applied to maintain a wings-level attitude. If the crosswind component is greater than 25 knots, do not arm the spoiler brakes for landing and again maintain a wings-level attitude with lateral stick. It must be realized that antiskid will not be available.

Landing On Wet Runways

Use of antiskid when landing on wet runways is highly recommended to minimize the probability of blown tires. Standing water greatly decreases braking effectiveness, and may cause total hydroplaning in certain conditions. (Refer to Section V.) Intermittent puddles may cause wheels to lock while braking with antiskid not engaged. As the locked wheel leaves the puddle and encounters a good braking surface, it will skid and blow unless brake pressure is released. The following procedures are recommended when landing on a wet runway:

1. Determine field condition before approach (braking action, crosswind component, arresting gear status).

2. If adverse wind and runway conditions exist, make a short-field arrested landing. In the event of a hook skip-bolter, execute a waveoff.

3. Plan the pattern to be well-established on final in a wings-level attitude (crab, if required) on SPEED. Land on runway centerline, using normal FCLP landing techniques.

4. If a rollout landing is desired, touch down on centerline within the first 500 feet of runway. Landing rollout procedures are the same as in a normal landing. When directional control is clearly

established, utilize normal braking. During the high-speed portion of the landing roll, little or no deceleration may be felt. Do not allow the aircraft to deviate from a straight track down the runway. If a skid develops, release the brakes, continue aerodynamic braking and use rudders or nosewheel steering for directional control. Reapply the brakes cautiously. If the skid continues and adequate runway remains, select power as required and fly away. If conditions do not permit flyaway, use the long field overrun gear if required. If the aircraft is leaving the runway to an unprepared surface, secure both engines.

Note

A blown tire on landing rollout may result in directional control difficulties, particularly at high speeds. Refer to Section V for blown tire emergency procedures.

LANDING CHECKLIST

The placarded landing checklist should be completed in sequence prior to arriving at the 180° position abeam the touchdown point. All checklist items are essential elements to be checked prior to each landing. With the ICS on HOT MIC, the pilot should call out the accomplishment of each step so that the RIO can double-check that all items have been performed.

1. Wing sweep mode switch .... 20° AUTO

   Check wings in AUTO sweep control mode and verify at 20°.

2. Wheels .... THREE DN

   Check for wheels-down indication on all three gear and that gear transition light is out. Check that brake accumulator pressure is fully charged.

3. Flaps .... FULL DN

   Check for flap and slat full down indication and no FLAP light.

4. DLC .... CHECKED

5. Hook .... AS DESIRED

   Transition light should be out.

6. Harness .... LOCKED

7. Speed brakes .... EXT (OUT)

   Check indicator for full speed brake extension.

8. Brakes .... CHECK

9. Fuel .... CHECK

POSTLANDING—PILOT

1. Speed brake switch .... RET

2. ANTI SKID SPOILER BK .... OFF

3. Flaps and slats .... UP

   Move FLAP handle UP and check for complete retraction of main flaps and slats and auxiliary flaps (flaps indicator - 0° and no FLAP caution light). Check auto deactivation of the outboard spoiler module. As soon as the auxiliary flaps are retracted (8 seconds) the wings will sweep aft if commanded.

4. Wing sweep mode switch .... BOMB

CAUTION

Ensure that emergency WING SWEEP handle and wings move to 55° position.

5. Emergency WING SWEEP handle ...... OV SW

   Raise handle and move aft to 68° position. Raise handle to full up extension and hold. When HZ TAIL AUTH caution light goes out and the OVER flag appears, move emergency wing sweep handle full aft (75° sweep position) and stow. Rotate handle guard to stowed position.

6. Avionics .... OFF

   Turn off all avionics (data link, radar altimeter, displays, TACAN, ARA-63), except UHF radio.

7. Right throttle .... OFF

Scavenge at 75% for 10 seconds and OFF.

Note

Care should be taken when shutting down the right throttle (with the left throttle at IDLE) to prevent inadvertent contact with the left throttle, moving it aft to the cutoff position.

8. HYD TRANSFER PUMP switch .. CHECK AND SHUTOFF

Check hydraulic transfer pump operation in the combined-flight direction with the HYD PRESS, OIL PRESS, R GEN, and R FUEL PRESS caution lights illuminated.

9. Ejection seats .... SAFE

Raise guards on face curtain and secondary firing handles to lock seat actuation devices.

10. Ordnance .... DEARM (FIELD)

Dearm and safety ordnance in accordance with local operating procedures.

11. Wheels .... CHOCKED

12. Parking brake .... PULL

CAUTION

Do not pull parking brake subsequent to a field landing if the brakes have been used extensively.

13. UHF function selector .... OFF

14. OXYGEN switch .... OFF

After removing mask, turn oxygen off.

15. EMERG generator switch .... OFF

16. Left throttle (alert RIO) .... OFF

Scavenge at 75% for 15 seconds and OFF.

Alert RIO and upon signal from plane captain, secure left engine. Check emergency generator auto operation upon shutdown.

17. Lights .... OFF

Turn off internal and external light switches.

18. CANOPY handle .... OPEN

Alert RIO.

RIO PROCEDURES

INTERIOR INSPECTION—RIO

1. Left and right circuit breakers .... IN

2. Left and right foot pedals .... ADJUST

3. Harnessing .... FASTEN

   a. Leg restraint lines and garters ... CONNECT

   Connect leg line bayonet fitting to leg garter quick-release buckle on the respective side. Assure that leg lines are not twisted.

   b. Personal services (vent and anti-g hoses) .... CONNECT

   Insert personal vent and anti-g hoses into seat block fittings. In the absence of a suit vent requirement, the seat block vent hose should be inserted into the seat back fitting to provide ventilation air through the seat and back cushions.

   c. Lap belt .... ATTACH

   Attach lap belt fittings and pull bucket straps snug in order to provide secure lap restraint for flight and seat kit suspension in the event of emergency egress or ejection.

   d. Shoulder harness .... ATTACH

   Attach shoulder harness release fittings to torso fittings. Check for proper positioning of parachute D-ring on left riser.

e. Inertia reel .... CHECK

Position the shoulder harness manual control knob forward to the lock position and check that both shoulder straps lock evenly and securely when leaning back. Position control knob full aft to unlock harness and release knob to the neutral position.

f. Oxygen-audio connection .... ATTACH

Attach composite fitting without causing unnecessary twisting of hard nose.

4. Oxygen .... CHECK

Turn OXYGEN switch ON, purge with mask held away from face. Place mask to face and check for normal breathing and regulator and mask operation. Turn OXYGEN switch OFF, check no breathing.

5. VENT AIRFLOW thumbwheel .... OFF
T 6. CPS switches .... OFF
7. KY-28 .... P/OFF
8. ICS panel .... SET
   a. VOL knob .... SET
   b. Amplifier .... NORM
   c. Function selector .... COLD MIC
9. TACAN function selector .... OFF
10. UHF function selector .... OFF
11. LIQ COOLING switch .... OFF
12. EJECT CMD lever .... SET

Determined by squadron policy.

13. ARMAMENT control panel .... SET
   a. WPN type thumbwheel .... OFF
   b. ATTK MODE knob .... MAN
   c. DLVY MODE switch .... STP
   d. DLVY MODE switch .... SGL
   e. ELECT FUZE knob .... SAFE
   f. A/G GUN switch .... OFF
   g. MECH FUZE switch .... SAFE
   h. SEL JETT switch .... SAFE
   i. JETT OPTIONS switch ..... MER TER
   j. INTERVAL .... SET
   k. QTY .... SET
   l. Station select switches 1 to 8 ..... SAFE
   m. MSL OPTIONS switch .... NORM
   n. MSL SPD GATE KNOB .... NOSE QTR

14. SYS TEST-SYS PWR ground check panel .... CLOSED
15. Standby attitude gyro .... UNCAGED
16. NAV MODE knob .... OFF
17. Clock .... SET AND WIND
18. WCS switch .... OFF
19. ALR 45/50 PWR switch .... OFF
20. DECM selector knob .... OFF
21. AN/ALE-39 PWR/MODE switch ...... OFF
22. DATA LINK ON-OFF-AUX ON switch .... OFF
23. APX-76 .... OFF
24. INTERIOR LIGHTS panel .... CHECK
25. IFF MASTER knob .... OFF
26. MODE 4 switch .... OUT
27. IFF ANT and TEST panel .... SET
   a. IFF ANT switch .... OFF (center)
   b. IND LT-DDI BIT switch .... OFF (center)
   c. GND CLG switch .... OFF
28. RADAR BEACON switch .... OFF
29. POWER SYS TEST switch .... OFF
30. MCB test switch .... OFF

PRESTART—RIO

The following checks are performed by the RIO after starting air and electrical power are applied prior to starting engines.

CAUTION

- Starting air, which provides full ECS capability, must be connected to the aircraft with electrical power to cool temperature-critical avionics.

- If starting air is not available, a forced air ground cooling unit and servoair must be connected before turning on avionic equipment.

- If electrical power is not connected with spare starting air, the ECS will drive to full hot.

1. Seat, ICS, and UHF foot switches .... ADJUST

    Adjust seat height such that the helmet can be placed against the headbox without face curtain handle interference. After seat height adjustment, adjust ICS and UHF foot pedal fore-aft position for sitting comfort.

2. External power and air .... ON

3. ICS .... CHECK

    Verify two-way communications between flightcrew members and adjust volume to a comfortable level.

4. DL, TACAN, and UHF .... SET

    Set command control in accordance with mission and flightcrew operating procedures.

5. Fuel quantity .... CHECK

6. Lights .... CHECK

    Check for illumination of console and instrument lighting.

7. LTS test .... CHECK

    Check that all caution and advisory lights, ECM lights, and DDI lights illuminate.

Note

During pilot's INST test, the RIO should observe fuel counter decrease to 2000 pounds and MASTER CAUTION, FUEL LOW, and OXY LOW lights illuminate.

8. Ejection seats .... ARMED

    Arm ejection seat by depressing locking tab on headbox and rotating alternate ejection handle safety guard downward. Visually check that pilot's locking tab is depressed.

9. CANOPY handle .... CLOSE

    RIO will normally close canopy. Ensure verbal clearance from pilot. Check that CANOPY light goes out with full forward transition of canopy into the sill locks.

WARNING

Flightcrews shall ensure that hands and foreign objects are clear of front cockpit handholds, top of ejection seats, and canopy sills to prevent personal injury and/or structural damage during canopy opening or closing sequence. Only minimum clearance is afforded when canopy is transiting fore and aft.

Note

If the CLOSE position does not close the canopy, depress the grip latch, release and push handle outboard and forward into the BOOST position. If it is necessary to use the BOOST position, the handle shall be returned to the CLOSE position to avoid bleed off of pneumatic pressure.

10. Acknowledge .... READY TO START

ENGINE START—RIO

The RIO must monitor pilot procedures and plane captain signals to ensure maximum safety during the engine start sequence.

POSTSTART—RIO

Note

The RIO will ensure that the EMERG GEN check is complete before commencing postart procedures.

1. WCS switch . . . . STBY
   Verify that AWG-9 COND light illuminates.

2. LIQ COOLING switch . . . . AWG-9 OR AWG-9/AIM-54
   Verify that AWG-9 COND light goes off and that AUTO BIT 2 is running.

3. CATEGORY knob . . . . NAV

4. NAV MODE switch . . . . ALIGN
   Select appropriate alignment mode before end of AUTO BIT 2.

5. Communications . . . . ON and SET

    a. UHF function selector . . . . BOTH

    b. DATA LINK switch . . . . ON

T 6. CPS system switch . . . . RDY
   Observe DATA/MAN Vg/H light illuminated.

T 7. IRLS switch . . . . STBY
   Observe IR NR light illuminated for cooldown period (maximum of 17 minutes).

8. TACAN function selector . . . . T/R

9. ECM PWR switch . . . . ON

10. DECM knob . . . . STBY

    a. When STBY light goes off, select HOLD 3 SEC, then ACT for OBC.

11. IFF MASTER knob . . . . STBY

    a. Set CODE knob . . . . AS REQUIRED

    b. IFF panel . . . . TEST

      (1) MC switch . . . . OUT

      (2) M1, M2, and M3 . . . . TEST

         Select NORM and observe that TEST light illuminates.

      (3) MC . . . . TEST

         Observe that TEST light illuminates.

    c. IFF ANT switch . . . . AUTO

12. AUTO BIT 2 . . . . VERIFY COMPLETE

13. CAP . . . . Enter own aircraft latitude, longitude, and field or ship elevation

14. Verify pilot has OBC selected.

    a. Observe SAT display on TID.

    b. Observe failed acronyms on TID in CM

    c. Observe test complete on TID

    d. Altimeter . . . . RESET

15. Computer address panel . . . . ENTER desired data, WP, FP, etc.

T 16. V/H check:

    a. Manual V/H thumbwheels set . . . . 360 knots/200 feet

    b. V/H switch . . . . TEST
       Observe MAN V/H light is out.

    c. V/H switch . . . . MANUAL

T 17. Vertical frame check:

    a. Manual V/H thumbwheel set . . . . 350 knots/1800 feet

    b. FRAME switch . . . . VERT

    c. FILM switch . . . . RUN
       Observe exposure interval of 1.0 second and frame camera green light illuminated; check camera frame counter for proper operation.

    d. FILM switch . . . . OFF

    f. FRAME switch . . . . OFF

T 18. PAN BIT check:

    a. PAN switch . . . . BIT/RELEASE
Observe exposure interval of 1.0 second, green PAN light illumination, and counter decrease as 5 frames are exposed.

T 19. PAN autocycle check:

    a. PAN switch . . . . CTR

    b. FILM switch. . . . . RUN
Observe exposure interval of 1.0 second, green PAN light illumination, and check camera frame counter for proper operation.

    c. PAN switch . . . . LEFT or RIGHT
Observe exposure interval of 2.0 seconds, PAN go light illuminated, and frame counter for proper operation.

    d. FILM switch. . . . . OFF

T 20. PAN pulse mode check:

    a. Manual V/H thumbwheel set . . . . 350 knots/ 13,500 feet

    b. PAN switch . . . . CTR

    c. FILM switch. . . . . RUN
Observe exposure interval of 5.0 seconds, green PAN light illumination, and check camera frame counter for proper operation.

    d. FILM switch. . . . . OFF

    e. PAN switch . . . . OFF

T 21. IR Sensor check:

Note

Before IRLS system check, observe IR NR light out following cooldown and BIT. Observe film counter movement by 1 foot.

    a. IRLS switch. . . . . WFOV

    b. Manual V/H thumbwheel set . . . . 350 knots/ 600 feet

    c. FILM switch. . . . . RUN
Observe green IRLS light flashing at 5 second interval and check proper film counter operation.

    d. FILM switch. . . . . OFF

    e. IRLS switch. . . . . STBY

22. Computer address panel . . . . SET

23. DDD . . . . SET

24. TID controls. . . . . SET

    a. CONTRAST . . . . SET

    b. BRIGHT control . . . . SET

    c. CLSN . . . . OFF

    d. RID/DSBL. OFF

    e. ALT NUM . . . . ON

    f. SYM ELEM . . . . ON

    g. DATA LINK . . . . AS REQUIRED

    h. JAM strobe . . . . AS REQUIRED

    i. NON ATTK . . . . AS REQUIRED

    j. LAUNCH ZONE . . . . AS REQUIRED

    k. VEL VECTOR . . . . AS REQUIRED

    l. RANGE scale . . . . AS REQUIRED

25. ECM panel . . . . SET

    a. TEST button . . . . DEPRESS AND CHECK

    b. BRIGHTNESS . . . . SET

26. ECM DISPLAY panel . . . . SET

    a. CORR . . . . ML

    b. ORIDE . . . . ML/AI

    c. MODE . . . . NAV

    d. DATA/ADF . . . . BOTH

27. Hand control panel . . . . SET

    a. Light test (all AWG-9 lights illuminate) . . . . DEPRESS AND CHECK

    b. El Vernier (category radar position) . . . . SET 0° EL

28. AN/ALE-39 (as required) . . . . SET

    a. BURST switch 3

Section III
Part 2
NAVAIR 01-F14AAA-1

    b. BURST INTERVAL 0.1

    c. SALVO 2

    d. SALVO INTERNAL 0.4

29. CANOPY DEFOG—
CABIN AIR lever . . . . CABIN AIR

30. D/L reply . . . . AS REQUIRED

31. AAI control panel . . . . SET

    a. TEST/CHAL CC switch . . . . TEST
       Check DDD display.

32. Indicator lights . . . . TEST

33. DDI BIT . . . . TEST

34. After alignment completed:

    a. NAV mode . . . . INS

    b. Program restart . . . . DEPRESS

    c. STBY/READY lights . . . . OFF

    d. TID NAV mode . . . . INS

35. DEST data . . . . VERIFY

36. BRG/DIST to destination . . . . CHECK

37. OWN A/C groundspeed . . . . CHECK

38. MAG VAR . . . . CHECK

39. KY-28 . . . . AS REQUIRED

40. Notify pilot . . . . READY TO TAXI

TAXI — RIO

The RIO's primary responsibility during taxiing is to act as co-pilot-safety observer. BIT checks may be performed while taxiing, provided that RIO's attention is not diverted from his co-pilot-safety observer duties.

1. Perform BIT confidence checks 2, 4, 3, 1 if not previously checked and record results on BER form.

Note

Perform BIT sequence 3 as a minimum, to set ACM threshold.

    a. CATEGORY switch . . . . BIT

    b. BIT sequences . . . . SELECT
       Observe failures.

2. OWN A/C ground speed . . . . CHECK
                                         ON ECMD

Check own-aircraft ground speed when stopped. It should be equal to or less than 3 knots for satisfactory alignment.

T 3. OWN A/C altitude . . . . CHECK

ON-DECK ENTRY OF REFERENCE POINTS OR TARGETS.

1. LAT/LONG . . . . ENTER

2. Altitude . . . . ENTER ALTITUDE OF
                                            TARGET AREA

3. NAV MODE switch . . . . AHRS/AM
(after alignment complete)

4. Manual MAG VAR of target area . . . . ENTER

T 5. Heading (CGTL) . . . . ENTER

T 6. Target length of run (1/10 mile) . . . . ENTER
                                   VIA CAP KEYBOARD
                                        SPD PUSHTILE

7. NAV MODE switch . . . . INS

8. Maximum number of reference points is 8. Three waypoints and one each of fixpoint, surface target, home base, hostile area, and defended point.

Note

IP may not be used as a TARPS reference point. The hostile area altitude is used as the lowest priority source for AGL calculations.

OWN-AIRCRAFT ALTITUDE CORRECTION (ON DECK)

1. CAP . . . . NAV CATEGORY

2. OWN A/C . . . . HOOK

3. Altitude pushtile . . . . SELECT

4. Field/ship elevation . . . . ENTER

T TARPS MODE ENTRY AND DISPLAY REQUIREMENTS
PILOT

T 1. PDCP display mode . . . . A/G

T 2. PDCP STEER CMD . . . . DEST OR MAN

T 3. HUD . . . . CHECK ON

3-38

RIO

T 1. CAP CATEGORY . . . . NAV

T 2. CAP function pushbutton 5. . . . . SELECT

T 3. Desired reference points . . . . CHECK

T 4. MAG VAR. . . . . CHECK

Note

Momentary display of M or B indicates AWG-9 tape read to TARPS mode; flashing TARPS acronyn indicates pilot has not selected A/G on the PDCP.

OWN-AIRCRAFT ALTITUDE CORRECTION (AIRBORNE)

1. CAP . . . . NAV CATEGORY

2. Own aircraft. . . . . HOOKED

3. Altitude pushtile . . . . SELECT

4. Altitude. . . . . ENTER
   Enter TID AGL altitude if over water.

IN-FLIGHT RECONNAISSANCE SYSTEM CHECK – RIO

Enroute to target area:

T 1. FRAME switch . . . . VERT

T 2. PAN switch . . . . CTR

T 3. IRLS switch . . . . WFOV

T 4. FILM switch . . . . run

Run only long enough to check operation and observe FRAM, PAN, and IRLS green lights illuminated, and check frame and foot counters.

T 5. FILM switch . . . . OFF

T 6. IRLS switch . . . . STBY

T 7. PAN switch . . . . LEFT OR RIGHT

T 8. FRAME switch . . . . FWD

Note

Prior to selecting FILM switch to RUN, delay 15 seconds for camera positioning.

T 9. FILM switch. . . . . RUN

Run only long enough to check operation and observe FRAM and PAN green lights illuminated and check for proper film counter operation.

T 10. FILM switch. . . . . OFF

T 11. FRAME switch . . . . OFF

T 12. PAN switch . . . . OFF

Note

Keep manual V/H thumbwheels matched with actual altitude and airspeed to avert possible degraded imagery if an automatic shift to the manual mode occurs.

STEERING DISPLAY REQUIREMENTS

Destination Steering (TARPS)

PILOT

T 1. PDCP display mode . . . . A/G

T 2. PDCP STEER CMD . . . . DEST

RIO

T 1. TID DEST switch . . . . ANY POSITION
                                      EXCEPT MAN or IP

T 2. Desired reference point. . . . . HOOK

T 3. Desired CGTL. . . . . CHECK

Note

RIO selection of DEST MAN position ensures that previously selected pilot HUD steering remains, regardless of what reference point or target the RIO may hook. To select a different point for pilot HUD steering, the RIO must cycle out of the MAN position with a new point hooked.

Section III
Part 2

NAVAIR 01-F14AAA-1

Manual Steering (TARPS)

PILOT

T 1. PDCP display mode . . . . A/G

T 2. PDCP STEER CMD . . . . MAN

Note

Pilot HUD steering is direct to hooked point or target, without regard for the CGTL.

RIO

T 1. Desired waypoint . . . . HOOK

AIRBORNE ENTRY OF REFERENCE POINTS/TARGETS (TARPS)

T 1. Reference point which steering to. . . . . . HOOK

T 2. TID DEST switch . . . . MANUAL

T 3. Reference point to be changed. . . . . HOOK

T 4. Reference point data . . . . ENTER

T 5. Reference point to which steering. . . . . . HOOK
      desired

T 6. TID DEST switch . . . . ANY POSITION
                                            EXCEPT MAN

NAV SYSTEM UPDATES VIA HUD

T 1. Desired waypoint . . . . HOOK

T 2. Target . . . . VISUALLY IDENTIFIED

T 3. CAP . . . . NAV CATEGORY

T 4. CAP function button No. 3 . . . . DEPRESS

T 5. Target designate switch. . . . . . . FORWARD TO
                                       UNDESIGNATE TARGET

T 6. Aircraft heading . . . . ALIGN CGTL
                                                WITH TARGET

T 7. Target designate switch. . . . . MOVE UP/DOWN
                                  TO SLEW DIAMOND OVER TARGET

T 8. Target designate switch. . . . . . . FORWARD TO
                                       REDESIGNATE TARGET

T 9. Delta LAT/LONG . . . . EVALUATE

T 10. FIX ENABLE (if update desired) . . . . . . PRESS

Note

The accuracy of an update increases with greater look angles, and low grazing angles may induce navigation errors.

TARGETS OF OPPORTUNITY

T 1. Hostile area altitude . . . . ENTER
                                       ESTIMATED TARGET ALT

T 2. TID DEST switch . . . . ANY POSITION
                                            EXCEPT MAN

T 3. Reference point/target . . . . UNHOOK

T 4. Desired target . . . . VISUALLY
                                       IDENTIFIED BY PILOT

T 5. Target designate switch. . . . . FORWARD
                                            TO CAGE DIAMOND

T 6. Aircraft heading . . . . ALIGN CGTL
                                                WITH TARGET

T 7. Target designate switch. . . . . MOVE UP/DOWN
                                  TO SLEW DIAMOND OVER TARGET

T 8. Target designate switch. . . . . FORWARD
                                       TO DESIGNATE TARGET

T 9. Manual direct steering . . . . AVAILABLE

MAPPING MODE ENTRY

T 1. Map initial point/waypoint . . . . HOOK

T 2. Target coordinates. . . . . STORE/CHECK

T 3. Desired heading . . . . ENTER

T 4. CAP SPD function button . . . . . RUN LENGTH
                                       (even tenths of nmi)

T 5. Target altitude . . . . CHECK

T 6. Map initial point/waypoint . . . . UNHOOK

T 7. IP function button. . . . . HOOK

T 8. CAP SPD function button   DEPRESS

3-40

T 9.	Number of mapping lines required	ENTER
T 10.	CAP ALT function pushbutton	DEPRESS
T 11.	Map line separation distance (S.D.)	ENTER to nearest foot

Note

Map line S.D. must be preceded by a positive symbol using CAP pushbutton N+E if map area is right of flight line, or a negative symbol using CAP pushbutton S-W if map area is left of flight line.

T 12.	IP function button	UNHOOK
T 13.	Map initial point/waypoint	HOOK

UNPLANNED AIR-TO-AIR PHOTOGRAPHY

T 1.	AWG-9	TARPS MODE
T 2.	V/H selector switch	MAN
T 3.	Manual thumbwheels	300 KNOTS, 500 FEET
T 4.	NAV MODE switch	AHRS
T 5.	OWN A/C pushtile	HOOK
T 6.	Altitude	4,100 FEET, ENTER
T 7.	OWN A/C pushtile	UNHOOK
T 8.	Any reference point	HOOK
T 9.	Altitude	4,000 FEET, ENTER
T 10.	Radar altimeter	OFF
T 11.	AWG-9 switch	STBY
T 12.	CPS system switch	RDY
T 13.	PAN switch	CTR
T 14.	Exposure selector switch	AS DIRECTED BY TARGET SCENE
T 15.	Pilot's bomb button	DEPRESS AT 1 SECOND INTERVAL
T 16.	Maneuver aircraft for proper positioning. Once own-aircraft altitude is set to 4,100 feet, do not increase actual barometric altitude by more than 500 feet.	

TARPS ALTITUDE DETERMINATION

Flycatcher 7-002767

T 1.	0000	RADAR ALTIMETER
T 2.	0001	AWG-9 RADAR
T 3.	0002	BAROMETRIC MINUS TARGET ALTITUDE
T 4.	0003	BAROMETRIC MINUS HOSTILE AREA ALTITUDE

TARPS MODE EXIT

RIO

T 1.	CAP CATEGORY switch	NAV
T 2.	CAP function pushbutton no. 5	DESELECT

TARPS DEGRADED MODE PROCEDURES

CAUTION

Prior to initiating corrective actions on malfunctioning sensors, ensure that other sensors are either in OFF or STBY positions.

Serial Frame Camera Failure

T 1.	FILM switch	CYCLE OFF/RUN/OFF
T 2.	FRAME switch	CYCLE OFF/VERT OR FWD
T 3.	FILM switch	RUN
T 4.	FILM switch	OFF
T 5.	V/H	MANUAL
T 6.	Thumbwheels	SET HIGH V/H VALUE
T 7.	FILM switch	RUN

If Not Corrected:

T 8. FILM switch . . . . OFF

T 9. FRAME switch . . . . OFF

Mount Failure

T 1. FRAME switch . . . . CYCLE TO OPPOSITE POSITION

If Not Corrected:

T 2. FRAME switch . . . . OFF

CAUTION

- Initiate corrective action only one time.
- If mount light does not go off, secure sensor and wait 5 minutes to try again.

Panoramic Camera Failure

T 1. FILM switch . . . . CYCLE OFF/RUN

T 2. FILM switch . . . . OFF

T 3. PAN switch . . . . CYCLE OFF/CTR

T 4. FILM switch . . . . RUN

If Not Corrected:

T 5. FILM switch . . . . OFF

T 6. PAN selector . . . . LEFT or RIGHT

T 7. FILM switch . . . . RUN

If Not Corrected:

T 8. FILM switch . . . . OFF

T 9. PAN selector . . . . OFF

CAUTION

Do not initiate BIT.

Manual V/H Failure

T 1. Thumbwheels . . . . 350 KNOTS/200 FEET

T 2. V/H switch . . . . TEST

T 3. MAN V/H light out . . . . GOOD TEST

T 4. MAN V/H light on . . THUMBWHEEL FAILURE

IRLS Failures

COOL DOWN MALFUNCTION (IR NR LIGHT ILLUMINATED)

T 1. IRLS switch . . . . OFF

T 2. IRLS switch . . . . STBY

After cooldown is complete, IR NR light will be out (25 seconds) then illuminate for the remainder of BIT (80 seconds total). If IR NR light remains on for more than 17 minutes, IR cooling system is malfunctioning; turn the system off.

OTHER IR LS MALFUNCTIONS (IR LS LIGHT ILLUMINATED)

T 1. FILM switch . . . . CYCLE OFF/RUN/OFF

T 2. IRLS switch . . . . OFF

If IR LS light remains illuminated, assume IR door malfunction and continue with checklist.

T 3. IRLS switch . . . . STBY

Observe IR NR light illuminated for cooldown period (max. 17 minutes), then out for 25 seconds, then on for remainder of AUTO BIT TEST (80 seconds).

Note

Note time IRLS light illuminates during BIT.

T 4. IRLS switch . . . . NFOV

T 5. FILM switch . . . . CYCLE RUN/OFF

NAVAIR 01-F14AAA-1

T	6.	V/H switch. . . . . MAN
T	7.	Thumbwheels . . . . .350 KNOTS/350 FEET
T	8.	FILM switch. RUN

If Corrected:

T	9.	V/H switch. AUTO

If Not Corrected:

T	10.	FILM switch. . . . . OFF
T	11.	IRLS switch . . . . OFF

Note

For actual combat missions, fail indications may constitute abort criteria.

POSTLANDING—RIO

Note

Before shutdown, run BIT. Note results on BER card.

1. Ejection seat. . . . . SAFE
2. Harnessing . . . . UNSTRAP
3. Radar beacon . . . . OFF
4. IFF. . . . . MODE 4 HOLD, THEN OFF
5. DATA LINK . . . . OFF
6. NAV MODE switch . . . . OFF
7. WCS . . . . OFF
8. LIQ COOLING switch . . . . OFF
9. TACAN function selector . . . . OFF
10. OXYGEN switch. . . . . OFF
11. UHF. . . . . OFF
12. Report . . . . READY FOR SHUTDOWN

After shutdown of both engines.

13. CANOPY handle. . . . . OPEN
    Alert pilot
14. Flightcrew . . . . EGRESS

HOT REFUELING PROCEDURES

Before commencing ground hot refueling operations, a qualified ground crewman shall inspect the exterior of the aircraft for any discrepancies that might be hazardous to refueling or further flight operations. One ground crewman shall remain in a position on the right side of the aircraft within view of both the pilot and refueling crew. Any hazardous condition requires the immediate termination of refueling operations.

After refueling, the flightcrew should refer to appropriate checklists to configure the aircraft for takeoff, depending on intentions.

1. Fire extinguishing equipment. . . . AVAILABLE
2. All emitters . . . . STBY OR OFF
3. Right throttle . . . . OFF
4. Wheels. . . . . CHOCKED
5. Parking brake . . . . .PULL

CAUTION

If heavy braking is used during landing or taxiing followed by application of the parking brake, normal brake operation may not be available following release of the parking brake if the brakes are still hot. Check for normal brake operation after releasing the parking brake and before commencing taxiing.

6. REFUEL PROBE switch. . . . . FUS EXTD/
   ALL EXTD
   (AS DESIRED)
7. WING/EXT TRANS switch . . . . . AS DESIRED

Note

- If external tanks or wings accept fuel in the FUS EXTD position, select ORIDE on WING/EXT TRANS switch.

- If wings or external tanks do not accept fuel in the ALL EXTD position, select FUS EXTD and turn WING/EXT TRANS switch OFF.

8. REFUEL PROBE switch . . . . RET
9. WING/EXT TRANS switch . . . . OFF

Section III
Part 2

3-43

Section III
Part 2

NAVAIR 01-F14AAA-1

DECK-LAUNCHED INTERCEPT PROCEDURES

Note

These procedures assume that a quick reaction, full-mission-capable launch is essential. Prestart procedures and cockpit configuration may vary in accordance with airwing policy and specific EMCON conditions. All CNI equipment, as applicable should be placed in ON or STBY, YAW STAB AUG switch should be selected ON, and the HYD TRANSFER PUMP switch should be in NORMAL before application of electrical power. The LTS, INST, EMERG GEN, MACH LEV, WG SWP, and STICK SW tests on MASTER TEST panel should be conducted and verified during periodic aircraft turn-ups. Compliance with the takeoff checklist is mandatory to ensure proper aircraft configuration before launch.

Pilot Procedures

1. External electrical power . . . . ON
2. Seat . . . . ARM
3. Fire detect . . . . CHECK
4. Left engine . . . . IDLE
5. Right engine . . . . IDLE
6. OBC . . . . SELECT
7. MSL PREP . . . . ON
8. SW COOL . . . . ON
9. OBC . . . . DESELECT
10. Takeoff checklist
11. Ordnance crew . . . . ARM

RIO Procedures

1. NAV MODE . . . . CVA
2. Seat . . . . ARM
3. WCS switch . . . . STBY
4. CAINS/WPT . . . . SELECT
T 5. CPS SYSTEM switch . . . . RDY
T 6. IRLS switch . . . . STBY
7. Takeoff checklist (complete non-OBC functions)
8. NAV MODE . . . . INS AT READY
   (Program restart)
9. Ordnance crew . . . . ARM

Note

- Sparrow tune occurs after CW is enabled and can complete after TX time out.
- With AVC 2102 incorporated PH attack capability is present after launch and Sparrow tune occurs automatically whenever CW is enabled.

HOT SWITCH PROCEDURES

Increased potential hazards exist in hot switch operations when an engine is running with canopy open and front seat unoccupied. To minimize this potential hazard, minimum time should be spent in this condition. Pilot switch should be expedited and crew unstrap should be done with canopy closed. Pilot-to-pilot brief should be accomplished with a pilot in the aircraft.

Note

The RIO will vacate the aircraft first. When he is on the ground, flight deck, or hanger deck, the pilot will exit. This is particularly important during shipboard operations.

1. Parking brake . . . . PULL
2. HYD TRANSFER PUMP switch . . . . NORMAL

THIS PAGE INTENTIONALY LEFT BLANK.

	3.	WCS switch	OFF
T	4.	CPS system switch	OFF
	5.	Left throttle	OFF
	6.	THROTTLE MODE switch	MANUAL
	7.	Throttle friction lever	INCREASE
	8.	Ejection seats	SAFE
	9.	Flightcrew	UNSTRAP
	10.	Cockpit	CHECK FOR FOD
	11.	CANOPY handle	OPEN
	12.	Flightcrews	SWITCH
	13.	Flightcrew	STRAP IN
	14.	Ejection seats	ARMED
	15.	CANOPY handle	CLOSE
	16.	THROTTLE MODE switch	BOOST
	17.	Throttle friction lever	AS DESIRED
	18.	Left engine	START
	19.	WCS switch	STBY

CAUTION

Ensure TARPS maintenance personnel have loaded sensors and cleared aircraft before initiating power to TARPS pod.

T	20.	CPS system switch	RDY

Note

The poststart checklist shall be completed with respect to aircraft configuration and switch positions prior to start.

FIELD CARRIER LANDING PRACTICE (FCLP)

Preflight Inspection

A normal preflight inspection will be conducted with specific attention directed to tire condition, nosestrut extension, angle-of-attack probe conditions, and windshield cleanliness. Check that the hook bypass switch is in the FIELD position.

Takeoff

The takeoff will be individual, using either MIL or a selected zone AB power.

Note

A minimum nozzle position of 2 is required for takeoff to ensure positive afterburner lightoff.

Radio Procedures And Pattern Entry

A radio check with Paddles is advisable before pattern entry to confirm Charlie Time. Approaches to the field for break will be controlled by the tower and then switched to Paddles for FCLP pattern control. At no time will an aircraft remain in the pattern without a UHF receiver. On each succeeding pass, the following voice report will be made at normal meatball acquisition positions:

- Side number
- TOMCAT
- Ball/Clara
- Fuel State
- Type approach (manual, AUTO, DLC, coupled)

Pattern

The pattern should be a race track with the 180° position approximately 1-1/4 miles abeam at 600 feet above field elevation. (See figure 3-5.) The length of the groove should be adjusted to give a wings-level descent on the glide slope of 20 to 25 seconds (approximately 1 mile). For maximum gross weight at touchdown, refer to section 1, part 4, operating limitations. For a 51,800-pound aircraft, an optimum

Section III
Part 2
NAVAIR 01-F14AAA-1

FIELD CARRIER LANDING PRACTICE

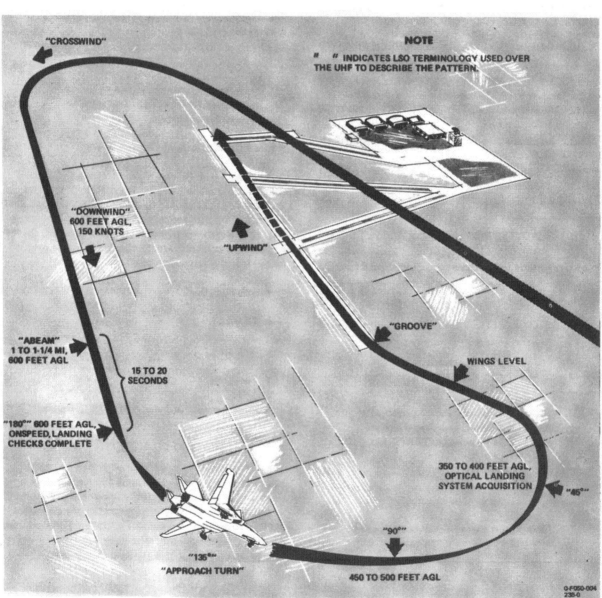

Figure 3-5. Field Carrier Landing Practice

ON SPEED indexer indication results in an airspeed of 128 KIAS. The turn to the downwind leg should be commenced after climbing to pattern altitude (600 feet AGL) utilizing 30° angle of bank and 150 KIAS. Turning from the 180° power should be adjusted to maintain optimum angle of attack. A gradual descent may be commenced at this position with a minimum altitude of 450 feet AGL at the 90° position and 350 feet AGL as a minimum until the pilot is receiving glide slope information. At approximately the 45° position, the meatball appears on the Fresnel lens. A common error is to begin the descent upon first seeing the meatball. Maintain altitude until the meatball is centered on the lens, then adjust power and nose attitude as necessary to start a rate of descent that will keep the meatball centered.

For manual, auto, and DLC approach techniques, refer to Carrier-Based Procedures, part 3, this section.

Night FCLP

All provisions that apply to day FCLP also apply to night FCLP, plus the following items:

- External lights steady - BRIGHT
- Hook bypass switch in the FIELD position

When comfortably situated in the pattern, instruments should be flown as much as possible up to the 45° position.

PART 3 — CARRIER-BASED PROCEDURES

LAUNCH

Applicable aircraft launching bulletins, the CV and LSO NATOPS Manuals and the pertinent CV air operations manual shall be read by all flightcrew members prior to carrier qualification. In addition, the predeployment lecture syllabus contained in Chapter I of the CV NATOPS Manual shall be completed.

BRIEFING

A thorough briefing shall be accomplished by the flight leader prior to launch. This briefing should call particular attention to current BINGO fields, emergency procedures peculiar to carrier operations, operating area NOTAMS, fuel management, and ships NAVAID status. Aircraft configuration, gross weight, expected WOD, and applicable launch trim settings will be verified prior to man-up.

PREFLIGHT

Preflight inspection should be accomplished with particular attention given to nosestrut, main loading gear, tires, hook, and underside of the fuselage. Note carefully the actual wing sweep, the lateral spacing between parked aircraft, and the general direction of engine exhaust. Do not preflight the aircraft topside aft of the bleed air doors if spotted with the tail outboard of the safety nets. In the cockpit, particular attention should be given to the flightcrew displays to ensure that the retaining devices have been installed. Ensure that the wing sweep handle is secure in the oversweep position when applicable. If the wings are not in oversweep, ensure that the emergency wing sweep handle position corresponds with the actual wing position. Leave the emergency WING SWEEP handle guard up, extend the emergency WING SWEEP handle, and pull WING SWEEP DRIVE NO. 1 and WG SWP DR NO. 2 MANUV FLAP circuit breakers (LE1, LE2). Crossbleed starts should not be performed unless the area aft of the aircraft is clear. Tie downs should not be removed and engine starts should not commence unless the auxiliary brake air pressure gage indicates a full charge.

START AND POSTSTART

Start and poststart procedures to be utilized aboard ship are the same as shore-based procedures, except that steps 22 through 37 of the poststart - pilot procedures are omitted. These steps are omitted because aircraft are spotted too close together to allow the wings to be swept forward while tied down. Cranking the left engine prior to starting the right, as outlined in the shore-based procedures will ensure that auxiliary brake pressure is available, and will ensure that backup flight control module is full of hydraulic fluid prior to cycling.

CARRIER ALIGNMENT

There are three methods of aligning the INS on the carrier deck: RF D/L, deck-edge cable, and manual (hand-set) alignment. All three use the same TID alignment display as ground alignment and the STANDBY and/or READY light logic remains the same.

1. RF D/L alignment (CAINS):

   a. WCS switch .... STBY
   b. DATA LINK switch .... ON
   c. CAINS or WAYPOINT .... SELECT
   d. NAV MODE switch .... CVA ALIGN

2. Deck edge cable alignment (SINS):

   a. WCS switch .... STBY
   b. DATA LINK switch .... ON
   c. Switching from RF D/L to cable inputs is automatic upon cable hookup.

3. Hand-set alignment:

   Must be utilized if flashing HS acronym is observed on TID.

   a. CAP .... NAV DATA
   b. OWN A/C .... HOOK/SELECT
   c. LAT .... ENTER
   d. LONG .... ENTER

e. HDG (true) .... ENTER

f. SPD .... ENTER

HS acronym stops flashing.

TAXIING

Shipboard taxi operations differ slightly from the field. Taxiing aboard ship requires higher power settings and must be conducted under positive control of a plane director. Any signal from the plane director above the waist is intended for the pilot and any signal below the waist is intended for deck-handling personnel.

The nosewheel steering system characteristics are excellent and enable extremely tight cornering capability. At full nosewheel steering deflection (70°), the inside mainmount wheel backs down and turn radius will be restricted if the inside brake is locked. For a minimum radius turn, momentarily depress the brake on the inside wheel and then allow the inside wheel to roll freely while controlling the turn rate by braking the outside wheel. For normal turns, symmetric brake applications should be applied to control aircraft forward motion. Forward motion should be initiated before effecting a tight radius turn to reduce power requirements.

Taxi speed should be kept under control at all times, especially on wet decks and approaching the catapult area. Be prepared to use the parking-emergency brake should normal braking fail. The lower ejection handle guard should be down while taxiing. The parking brake is an excellent feature which may be used to prevent leg fatigue during taxi delays.

CATAPULT HOOKUP

Set the VDI and HUD to show level flight at normal strut extension. Proper positioning on the catapult is easily accomplished if the entry is made with only enough power to maintain forward motion and if the plane director's signals are followed explicitly.

WARNING

All functional checks shall be performed before taxiing onto the catapult. Ensure that the takeoff checklist is complete and that the proper trim is set for launch before entering the nosetow approach ramp.

The catapult director will direct the pilot to approach the catapult track, using nosegear steering and brakes. Upon signal from the plane director and when positioned immediately behind the mount of the lead-in track, kneel the aircraft. If the launch bar is to be lowered from the cockpit, upon signal from the plane director, deflect the nosewheel, lower the launch bar, center the nosewheel, and disengage nosewheel steering. If the launch bar is to be lowered by the deck crew, no pilot action is required. After the hold-back bar has been attached to the aircraft and checked by squadron maintenance personnel, the catapult director will direct the aircraft forward until the hold-back bar is snug against the catapult buffer unit. The aircraft will be stopped in position for shuttle tension-up. The HUD and VDI will show 2° to 3° nosedown with the aircraft in the kneeled position.

CAUTION

- If the LAUNCH BAR light illuminates immediately upon selecting the KNEEL position with the NOSE STRUT switch, a malfunction in the system has occurred and the landing gear will not retract following the catapult launch.

- Nosewheel steering is designed to disengage and the NWS ENGA light goes off when deck personnel lower the launch bar on the catapult. The arresting hook must have been cycled on deck and the throttles set at IDLE to enable the system. This feature prevents the pilot from inadvertently damaging the launch bar during control checks after final tensioning.

CATAPULT TRIM REQUIREMENTS

The following requirements are applicable to clean aircraft or any combination of air-to-air store, external tank, gross weight combinations, and launch cg locations between 7.0% and 18.5% MAC.

The following table lists recommended catapult launch longitudinal trim settings based on military power. When using zone 5 afterburner, reduce recommended catapult longitudinal trim setting by 2°. Trim reduction is applicable to all aircraft cg positions and anticipated excess end airspeeds. (See figure 3-6.)

Anticipated End Airspeed Above Minimum (Knots)	Longitudinal Trim (degrees) Trailing Edge Up	
	cg between 7.0% and 16.5% MAC	cg between 16.5% and 18.5% MAC
0 to 9	7	4
10 to 24	5	2
25 to 50	3	0
Note Reduce longitudinal trim settings 2 when using zone 5 afterburner.		

Figure 3-6. Anticipated End Speed

CATAPULT LAUNCH

Aircraft launch gross weight will be cross checked and verified by signal with the flight deck personnel prior to kneel. If the aircraft is to be catapulted with a partial fuel load the pilot should ensure that longitudinal trim settings are adjusted if necessary (figure 3-7). Upon receipt of the "Tension-up and release brakes" signal, release the brakes, ensure the parking brake is off, and advance the throttles to MIL. When a turnup signal is received from the catapult officer, grip the throttles firmly, check engine instruments, ensure that the caution and advisory panel is clear, and the flight officer is ready. When satisfied that the aircraft is functioning properly, salute the catapult officer. Normally, a 3 to 5 second delay will occur before the catapult fires. Optimum launch technique is to maintain a loose grasp on the control stick while allowing it to move aft during the catapult stroke. Initial catapult firing results in a short-term vertical acceleration of 15 to 20g caused by full compression of the stored-energy nosestrut. Firmly restrain the throttles to prevent their aft travel during the catapult stroke.

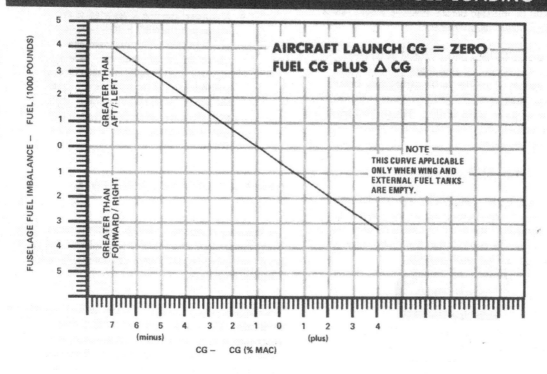

Figure 3-7. Center of Gravity Variation With Fuel Loading

> **CAUTION**
>
> Refer to Section I, Part 4 for external tank limitations.

At shuttle release, the energy stored in the nosestrut is released, rotating the aircraft up to the initial flyaway altitude of approximately 7° to 9° on the VDI and HUD. The control stick will, without pilot input, return to the trimmed position shortly after shuttle release. The pitch trim for launch is designed for hands-off operation. No control programming is required by the pilot to establish the proper attitude for flyaway. During rotation to the flyaway attitude the flightcrew will sense the aircraft's characteristic sink of approximately 5 feet. The flightcrew should be alert for any abnormal launch characteristics and should always be prepared to make control inputs as necessary to ensure a safe flyaway. When a MAX power launch is scheduled, the following signals will be used:

1. Two-finger turnup power advanced to MIL.

2. Catapult officer responds with 5 fingers (open hand held towards pilot).

3. Pilot selects MAX power, checks instruments and positions himself, then salutes the catapult officer.

CATAPULT ABORT PROCEDURES (DAY)

If after turnup on the catapult, the pilot determines that the aircraft is down, the pilot gives the no-go signal by shaking his head from side to side. Never raise the hand into view or make any motion that might be construed as a salute. After the catapult officer observes the pilot's no-go signal, he will cross his forearms over his head, and then give the standard release tension signal. When the catapult is untensioned, the catapult officer will signal the pilot to raise the launch bar. The pilot shall ensure that the throttles are seated in the catapult detent and will raise the launch bar with the LAUNCH BAR ABORT switch.

> **CAUTION**
>
> To avoid damage to the launch bar retract mechanism, do not actuate the LAUNCH BAR ABORT switch with the nosewheel deflected off center.

When the launch bar is clear of the shuttle, the catapult officer will move the shuttle forward of the aircraft launch bar. At this point the aircraft is no longer in danger of being launched. The catapult officer will signal the pilot to lower the launch bar and then step in front of the aircraft and signal the pilot to throttle back.

> **CAUTION**
>
> - If the aircraft is down prior to it being pushed or pulled back for release from the holdback fitting and when directed by the catapult officer, the launch bar shall be raised by the LAUNCH BAR ABORT switch.
>
> - Unkneeling the nosegear while the launch bar is in the catapult track or shuttle will damage the launch bar linkage and bungees. The pilot should unkneel the aircraft only when he is sure that the launch bar is free to rise and upon signal from the catapult officer or taxi director.

Note

The LAUNCH BAR ABORT switch is spring-loaded and must be held in the ABORT position until the catapult officer signals to lower the launch bar.

If the aircraft is down after the go signal is given, transmit the words "SUSPEND, SUSPEND"; however, the flightcrew should be prepared for the catapult stroke and to perform emergency procedures if required.

LANDING

Carrier Landing Pattern (VFR)

The visual flight rules (VFR) carrier landing pattern (figure 3-8) shall be in accordance with the CV NATOPS Manual. The pattern starts with the level break at 800 feet, and 300 to 350 KIAS. The break interval will be approximately one-half of the desired ramp interval time (15 to 17 seconds normal interval). When established wings-level on the downwind leg, descend to and fly the pattern at 600 feet MSL. Engage DLC upon completion of flap extension. Slow to on-speed AOA, engage the APC, compare airspeed and AOA for proper indications, check for proper functioning of the DLC, and complete the landing checklist prior to reaching the 180° position (ANTI SKID SPOILER BK switch OFF). The 180° turn is commenced 1 to 1-1/2 nmi abeam the LSO platform to arrive at the 45° position at approximately 450 feet MSL. The nominal bank angle throughout the turn should be 25°. Glide slope meatball

3-51

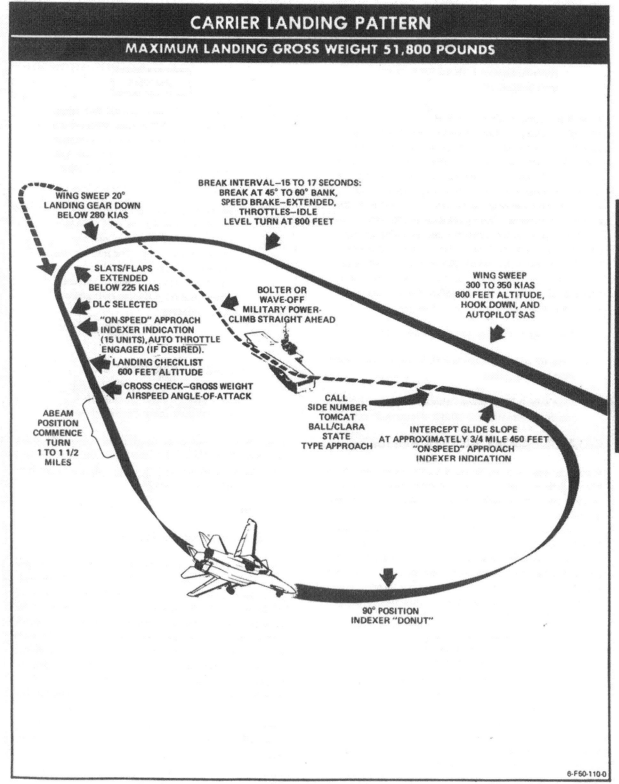

Figure 3-8. Carrier Landing Pattern

acquisition will occur at approximately 3/4 mile. Do not descend below 350 feet prior to acquiring the ball. On rollout to final, slightly overshoot the ships wake.

WARNING

- The LSO and tower must be informed if the landing is to be made in any wing or flap configuration other than 20° wing sweep, flaps and slats down, to ensure wind-over-deck requirements are met.

- Do not attempt shipboard landing with inoperative roll SAS and store asymmetry greater than 170,000 inches per pound due to lateral pilot-induced oscillation in the approach unless field divert is not possible. (Example: weapon rail at station 6 and AIM-54 missile at station 8 equals 170,000 inches per pound.)

Note

With the hook down, airspeed in excess of 320 KIAS may cause the hook transition light to illuminate.

Manual Approach Technique

The aircraft exhibits significant vertical response in gusty wind conditions.

Relatively low engine rpm and slow engine response dictate that throttle adjustments be small, and that sufficient time be allowed for the effects of thrust changes to become apparent. It is especially important that large power reductions be avoided in close, as large descent rates could develop, which cannot be countered. Once established on the glide slope, keep the scan going, cross-checking meatball, lineup, and angle of attack. Make positive corrections immediately but smoothly. Do not accept a low glide slope condition for any period of time. Be alert for foul deck or technique waveoff indications and be prepared to provide immediate response. With rough seas and pitching decks, some erratic meatball movements may be encountered. If this is the case, average out the bouncing ball to maintain a smooth and safe rate of descent. In no case overcorrect if the ball moves to a high indication. To avoid being "cocked up", arrest a "come down in close" with power and nose attitude. Attempts to arrest high sink rates with nose attitude alone will probably result in landing damage to ventral fins and afterburner. Upon touchdown, add full power (MIL) in anticipation of a bolter.

Approach Power Compensator Technique

Practice is required to develop the proper control habits necessary to use APC. For the APC to perform satisfactorily, smooth attitude control is again essential. Large abrupt attitude changes result in excessive thrust changes.

As the initial turn from the 180° position is made, the aircraft will momentarily indicate up to a two-unit slow condition. The APC will cause power adjustments to correct back to within one unit of on-speed throughout the remainder of the turn. Upon roll out on glide slope the pilot must override the tendency of the nose to come up by maintaining slight forward stick. The aircraft will indicate a 1- to 2-unit fast angle of attack, which should slow to on-speed within 5 seconds. In gusty wind conditions all available cues should again be utilized to minimize glide slope deviations and attendant attitude change requirements. Close-in corrections are very critical. If a large attitude correction for a high, in-close situation develops, the recommended procedure is to stop meatball motion and not attempt to recenter it. A low, in-close condition is difficult to correct with APC and often results in an over-the-top bolter. It may be necessary to disengage or manually override APC in order to safely recover from a low in-close situation. Throughout the approach, the pilot should keep his hand on the throttles in the event APC disengages inadvertently.

In aircraft BUNO 160909 and subsequent and aircraft incorporating AFC 554, a smooth throttle transition from AUTO to BOOST mode can be achieved by depressing the CAGE button on the outboard throttle grip.

Note

- Approaches that result in steep bank angles (greater than 30°) and/or short groove lengths should be avoided because of the large thrust reductions required and the short periods of time available to make proper corrections.

- Disengagement of APC by any means other than THROTTLE MODE switch, will result in illumination of the AUTO THROT light for 10 seconds.

Direct Lift Control Technique

Direct lift control (DLC) should be engaged for all approaches.

DLC is selected upon completion of the flap extension cycle. When the aircraft is established at approach AOA, the APC is engaged and a quick check of the DLC thumbwheel controller is made to ensure an operable DLC.

Note

Automatic pitch compensation for DLC activation is optimized for approach angle-of-attack speeds. Activation of DLC at higher airspeeds will result in inducing noticeable changes in pitch attitude.

Upon meatball acquisition, fly a normal APC approach until established on glidepath. Once established within one ball of on-glide slope, the pilot observes any deviations and returns to glidepath with DLC while making a slight corrective attitude adjustment. This technique is employed to the in-close position at which DLC only is used for the ball control while maintaining an essentially constant attitude. At the in-close position, if the meatball shows more than 1/2 ball low, DLC up commands may be coordinated with slight noseup corrections.

DLC may be employed by vernier or bang-bang control, depending upon the extent of correction required. The proficient pilot can observe slight glidepath deviations and fly the aircraft through proportional DLC control to effect smooth corrections. Selection of MIL will automatically disengage DLC.

WARNING

Selection of DLC during the flap extension cycle can generate excessive pitch rates. DLC is to be selected only upon completion of the flap cycle. DLC must be deselected prior to flap retraction to avoid an excessive pitch trim change with automatic DLC stowage during the flap retraction cycle.

Waveoff Technique

A waveoff will be initiated immediately upon a signal or voice call from the LSO. In addition, anytime the meatball is lost in-close, in the groove, the pilot will initiate his own waveoff. MIL should generally be used for waveoffs. Tests have shown that slow afterburner lighting and staging times render application of MAX power insignificant in arresting normal sink rates at initiation of waveoff. However, the pilot should not hesitate to use MAX power if conditions warrant. Static ground tests indicate that there is no significant thrust loss due to nozzle opening prior to AB light off. Power addition should be accompanied by a slight aircraft rotation, maintaining 15 units AOA to minimize sink. Do not overrotate the aircraft in-close. Overrotation on a close-in waveoff significantly increases the possibility of an in-flight engagement. Normally, waveoffs will be taken straight ahead, especially when close in. When using APC, waveoff technique is the same as for manual approaches except that a force of approximately 8 pounds is required to disengage the throttle torque switches. Disengagement of the APC by overriding the throttle forces results in the throttle MODE switch automatically returning in the BOOST position and illuminates the AUTO THROT light on the pilot's DDI panel. A time delay relay holds the AUTO THROT light on for 10 seconds following APC disengagement.

CAUTION

If a force in excess of 14 pounds is applied to break the throttles out of the auto mode, the throttle MODE switch will return to the BOOST position but the throttle mode will revert to manual. The switch must be cycled to MAN and back to BOOST to regain the BOOST mode.

BOLTER TECHNIQUE

The bolter maneuver is effected by selecting MIL and slight aft control stick until the desired flyaway attitude is established.

CAUTION

The use of excessive backstick on a bolter may cause the tail surface to stall, delaying aircraft rotation and causing the aircraft to settle off the angle.

BINGO FUEL

Fuel reserves should be programmed depending on distance of the field from the CV, aircraft configuration, and enroute weather. This bingo fuel quantity should be set before takeoff.

ARRESTED LANDING AND EXIT FROM THE LANDING AREA

As the aircraft touches down, advance throttles to MIL. Upon completion of landing rollout, reduce power to IDLE. Raise the hook and flaps and select wing sweep BOMB while allowing the aircraft to roll aft. Apply brakes on signal. Flaps retraction requires approximately 7 seconds. When the flaps are fully retracted the wings will sweep aft. Engage nosewheel steering and taxi forward on the come-ahead signal. If the wings sweep aft to 55°, auxiliary and main flap retraction has been verified and full aft wing sweep may be selected using the emergency handle. The RIO should monitor wing sweep position while taxiing. Oversweep should be selected prior to final spot and shutdown. The engines should remain running until the cut signal is given by the plane director. If at any time during this phase of operations a brake failure occurs, pull the parking-emergency brake. If the aircraft continues to roll, drop the hook, advise the tower, and signal for chocks to be installed (use nosewheel steering to ensure that the aircraft remains on the deck). Do not unstrap, dearm the ejection seat, or leave the cockpit until tiedowns have been installed.

CARRIER-CONTROLLED APPROACHES

Should these procedures conflict with the applicable CV Air Operations Manual, the latter shall govern. Detailed pilot-controller voice procedures must be established in accordance with each ships CCA doctrine.

Figure 3-9 shows a typical carrier-controlled approach. Typical automatic carrier landing approaches are shown in section VI, part 1.

Note

Pending completion of system development, ACLS Mode I/IA approaches are not authorized. ACLS mode II and III, and AN/SPN-41 or AN/ARA-63A ILS approaches are authorized.

Holding Phase

Five minutes before penetration, defogging shall be actuated and maximum comfortable interior temperature will be maintained to prevent possible fogging or icing on the windshield and canopy.

Note

Fuel dump is accomplished by gravity flow and is ineffective during the penetration descent. Fuel dump, if required, should be planned accordingly for the level leg.

1. Before descent, check shoulder harness handle locked, set lights as directed by existing weather, and lower arresting hook.

2. Accomplish final changes to radio and IFF upon departing Marshall or earlier. After these changes are made, the pilot should make no further changes except under emergency conditions.

3. When commencing penetration, initiate a standard descent: 250 KIAS, 4000 feet per minute, speed brakes as required.

WARNING

If a gear and/or flaps down penetration is required, ensure that the wings are programmed forward of 22° prior to lowering flaps. If flaps are lowered with swings swept aft of 22°, auxiliary flap extension will be inhibited resulting in rapid nosedown pitching rates.

4. Radar and barometric altimeters shall be cross-checked continuously when below 5,000 feet.

Section III
Part 3

NAVAIR 01-F14AAA-1

CARRIER-CONTROLLED APPROACH (TYPICAL)
MAXIMUM LANDING GROSS WEIGHT 51,800 POUNDS

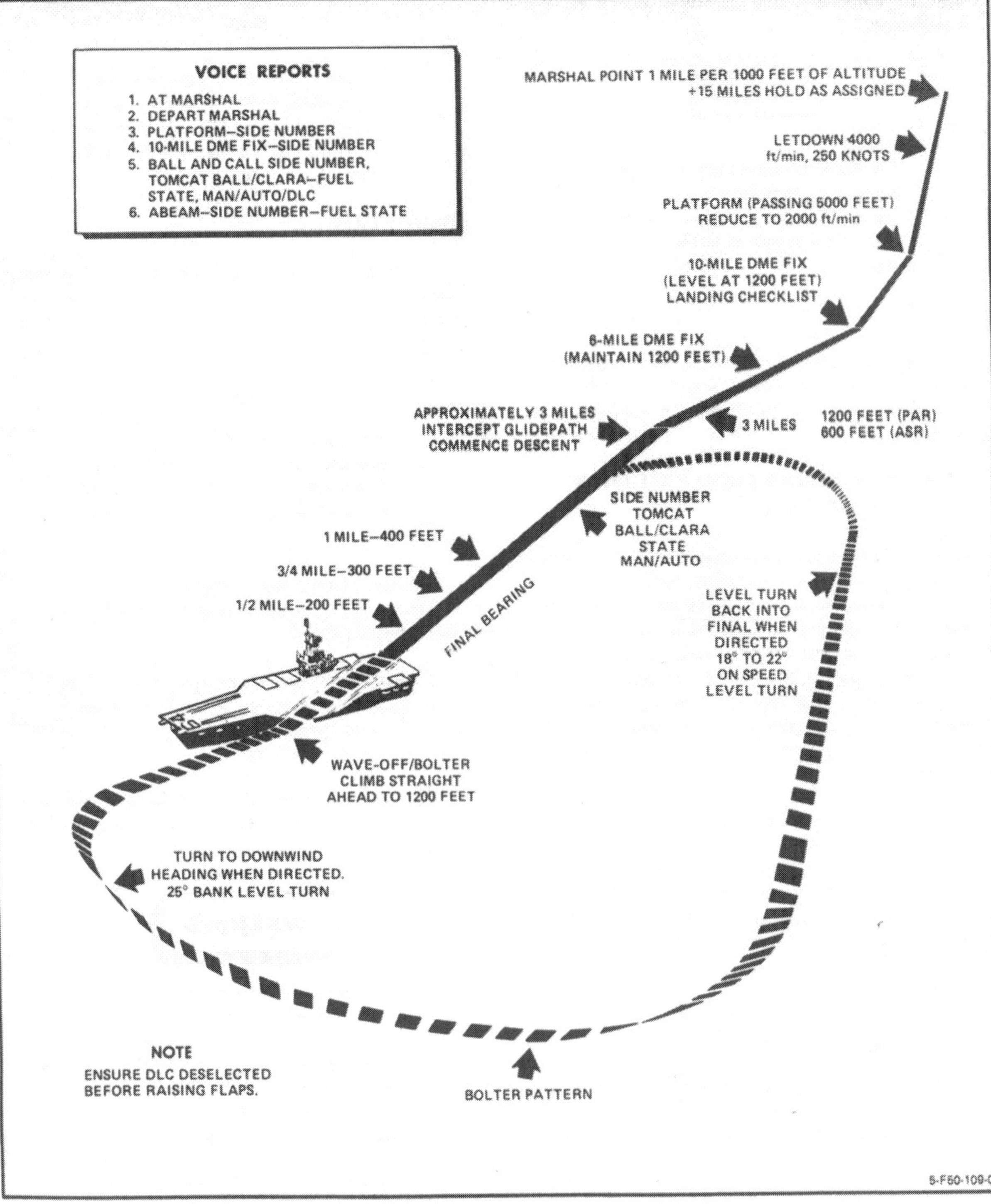

VOICE REPORTS
1. AT MARSHAL
2. DEPART MARSHAL
3. PLATFORM—SIDE NUMBER
4. 10-MILE DME FIX—SIDE NUMBER
5. BALL AND CALL SIDE NUMBER, TOMCAT BALL/CLARA—FUEL STATE, MAN/AUTO/DLC
6. ABEAM—SIDE NUMBER—FUEL STATE

MARSHAL POINT 1 MILE PER 1000 FEET OF ALTITUDE +15 MILES HOLD AS ASSIGNED

LETDOWN 4000 ft/min, 250 KNOTS

PLATFORM (PASSING 5000 FEET) REDUCE TO 2000 ft/min

10-MILE DME FIX (LEVEL AT 1200 FEET) LANDING CHECKLIST

6-MILE DME FIX (MAINTAIN 1200 FEET)

APPROXIMATELY 3 MILES INTERCEPT GLIDEPATH COMMENCE DESCENT

3 MILES

1200 FEET (PAR)
600 FEET (ASR)

SIDE NUMBER
TOMCAT
BALL/CLARA
STATE
MAN/AUTO

1 MILE—400 FEET
3/4 MILE—300 FEET
1/2 MILE—200 FEET

FINAL BEARING

LEVEL TURN BACK INTO FINAL WHEN DIRECTED 18° TO 22° ON SPEED LEVEL TURN

WAVE-OFF/BOLTER CLIMB STRAIGHT AHEAD TO 1200 FEET

TURN TO DOWNWIND HEADING WHEN DIRECTED. 25° BANK LEVEL TURN

NOTE
ENSURE DLC DESELECTED BEFORE RAISING FLAPS.

BOLTER PATTERN

5-F50-109-0

Figure 3-9. Carrier-Controlled Approach (Typical)

Platform

At 20 miles passing through 5,000 feet, aircraft descent shall be slowed to 2,000 feet per minute. At this point, a mandatory unacknowledged voice report will be broadcast by each pilot. The aircraft side number will be given and "platform" will be reported. Continue descent to 1,200 feet.

Ten-Mile DME Fix

1. At 10 miles, report side number and 10-mile gate.

2. Commence transition to landing configuration, unless otherwise directed by CCA, maintaining 1,200 feet.

3. Gear and flaps shall be down by 6 miles.

4. Complete the landing checklist. Check anti-ice, lights, and rain removal, as required.

Six-Mile DME Fix

For a precision approach radar (PAR) approach, maintain 1,200 feet at approach speed until intercepting the guidepath at 2-3/4 miles, unless otherwise directed.

For an air surveillance radar (ASR) approach, a gradual descent of 600 feet per minute shall be commenced departing the 6-mile DME fix. Altitude of 600 feet is maintained until the final controller calls "commence descent". At 600 feet, the aircraft intercepts the center of the glide slope at 1-1/2 miles on a 3.5° slope. Commence descent of 500 to 700 feet per minute, using the following check-points:

1. 1 mile – 500 feet

2. 3/4 mile – 400 feet

3. 1/2 mile – 300 feet

Meatball Contact

When transitioning to a visual approach, report call, side number, TOMCAT, meatball or CLARA (no meatball), fuel state, and type pass. The LSO will acknowledge and instructions from the final controller will cease. Pilots are cautioned against premature contact reports and transition to visual glide slope during night recoveries when visibility permits sighting the ship beyond 2 to 3 miles. The height and dimension of the lens or mirror optical beam at 1-1/4 miles is over 200 feet and the true center cannot be distinguished. This, coupled with the relatively short length of the runway lights, will give the pilot the illusion of being high when, in fact, the aircraft may be well below optimum glide slope. An additional advantage of delaying the meatball report (even though the ball is in sight) is that the final controller will continue line-up instructions that can greatly assist the pilot in establishing satisfactory line-up. Use the vertical velocity indicator to set up a rate of descent of 500 to 700 feet per minute. The AN/ARA-63A is an excellent aid during the approach and should be used whenever possible.

WAVEOFF AND BOLTER

In the event of a waveoff or bolter, climb straight ahead to 1,200 feet and maintain 150 KIAS. When directed by CCA, initiate a level turn to the downwind leg reporting abeam with fuel state. (If no instructions are received within 2 minutes or 4 miles DME, assume communications failure and initiate the downwind turn to the reciprocal of final bearing reporting abeam with fuel state. If no acknowledgment is received, start a turn at 4 miles or 2 minutes to intercept final bearing.) A 25° bank angle at 150 KIAS on the upwind turn establishes the aircraft at the desired 1-1/2 miles abeam on the downwind leg.

Slow to proper approach speed when approaching the abeam position. CATCC clears the aircraft to turn inbound to intercept final bearing. A level, on-speed approach turn of 18° to 22° bank angle from the normal downwind position allows the aircraft to properly intercept final bearings at 1-1/4 to 1-1/2 miles aft of the ship. Traffic spacing ahead may require that the aircraft continue on downwind leg well past the normal abeam position before being directed to turn to final bearing. No attempt should be made to establish visual contact with the ship when executing a CCA until the final approach turn has been executed.

Note

The Radar Beacon (AN/APN-154) should be turned off as soon as practicable after landing to avoid causing interference with AN/SPN-42 control of other aircraft in the pattern.

NIGHT FLYING

Night carrier operations will have much slower tempo than daylight operations and it is the pilot's responsibility to maintain this tempo. Normal day carrier operations shall be used except as modified below.

Briefing

Before initial night flight operations, all pilots should receive an additional briefing from the following persons:

  a. Flight deck officer
  b. Catapult officer
  c. Arresting gear officer
  d. LSO
  e. CATCC

Individual flight briefings will include all applicable items outlined above, with particular emphasis on weather and bingo fuel.

Preflight

In addition to normal cockpit preflight, ensure that external light switches are properly positioned for poststart light check. Install night filters on applicable cockpit displays.

Poststart

Adjust cockpit light to desired brightness. When ready for taxi, indicate with appropriate signal.

Taxi

Night deck-handling operations are of necessity slower than those used during the day. When a doubt arises as to the meaning of a signal from a taxi director, stop.

CATAPULT HOOKUP (NIGHT)

Maneuvering the aircraft hookup at night is identical to that used in day operation. However, it is difficult to determine your speed or degree of motion over the deck. The pilot must rely upon, and follow closely, the plane director's signals.

Catapult Launch

On turnup signal from the catapult officer, ensure throttles in MIL and check all instruments. When ready for launch, place external light master switch ON (bright and steady). After launch, establish an 8° to 10° pitch angle on the VDIG, cross-checking the pressure instruments to ensure a positive rate of climb. Retract the landing gear. An altitude of 500 feet is considered to be minimum altitude for retraction of flaps. When well established in a climb, switch lights to flashing or as applicable for an instrument climb-out. The standby indicator should be used in the event of a VDIG malfunction.

WARNING

- If wings are swept back manually, close attention should be paid to maintaining positive rates of climb. The loss of lift incurred by premature wing sweep aft can result in significantly decreased rates of climb, with very little change in pitch attitude and trim requirements.

- Failure of the CSDC will result in the loss of all primary attitude references on the VDIG in which case the pilot must use the standby gyro and turn and bank indicator for attitude reference.

CATAPULT ABORT PROCEDURES (NIGHT)

The pilot's no-go signal for night launches will be to not turn on his exterior lights, and to transmit on land or launch his aircraft side number, the catapult he is on, and the words "SUSPEND, SUSPEND". After the catapult is untensioned, the catapult officer will signal to raise the launch bar. The pilot shall ensure that the throttles are seated in the catapult detent or throttle friction is full forward before raising the launch bar with the LAUNCH BAR ABORT switch. When the launch bar is clear of the shuttle, the catapult officer will move the shuttle forward of the aircraft launch bar. At this point the aircraft is no longer in danger of being launched. The catapult officer will signal the pilot to lower the launch bar and then step in front of the aircraft and signal the pilot to throttle back.

CAUTION

- If the aircraft is down prior to it being pushed or pulled back for release from the holdback fitting and when directed by the catapult launching officer, the launch bar shall be raised by the LAUNCH BAR ABORT switch.

- Unkneeling the nosegear while the launch bar is in the catapult track or shuttle will damage the launch bar linkage and bungees. The pilot should unkneel the aircraft only when he is sure that the launch bar is free to rise and upon signal from the catapult officer or taxi director.

Note

The LAUNCH BAR ABORT switch is spring-loaded and must be held in the ABORT position until the catapult officer signals to lower the launch bar. If the aircraft is down after the go signal is given, transmit the words "SUSPEND, SUSPEND"; however, the flightcrew should be prepared for the catapult stroke and to perform emergency procedures if required.

ARRESTED LANDING AND EXIT FROM LANDING AREA (NIGHT)

During approach all lights shall be on bright and steady. At the end of arrestment rollout, turn off external lights and follow the director's signals while effecting the normal aircraft clean-up procedures.

PART 4 — SPECIAL PROCEDURES

IN-FLIGHT REFUELING PROCEDURES

Note

Before commencing in-flight refueling operations, each flightcrew member shall become familiar with the NATOPS Air Refueling Manual and the in-flight refueling portion, Section I, Part 2 of this manual.

In-flight Refueling Controls

Regardless of fuel management panel switch positioning, at low fuel states the initial resupply of fuel is discharged into the left and right wing box tanks. Thereafter, distribution of the fuel to the forward, aft, wing, and external drop tanks is controlled by the fuel management switch position. The split refueling system to the left and right engine feed group provides for a relatively balanced center of gravity condition during refueling. Selective refueling of the fuselage or all fuel tanks is provided on the REFUEL PROBE switch with the probe extended. In the FUS/EXTD position, normal fuel transfer and feed is unaltered. This position is used for practice plug-ins, fuselage only refueling, or return flight with a damaged air-refueling probe. The ALL/EXTD position shuts off wing and external drop tank transfer to permit the refueling of all tanks. The REFUELING PROBE switch circuit uses essential dc no. 2 power to control operation of the probe actuator through redundant-extend solenoids and a single-retract solenoid.

In-flight Refueling Checklist

The in-flight refueling checklist shall be completed before plug-in.

1. WCS switch . . . . STBY

2. Arming switches . . . . SAFE

3. DUMP switch . . . . OFF

4. AIR SOURCE pushbutton . . . . L ENG

5. REFUEL PROBE switch . . . . AS DESIRED
   (TRANSITION LIGHT OFF)

6. Wings . . . . AS DESIRED

7. Visors . . . . RECOMMENDED DOWN

To prevent fuel fumes from entering the cockpit through the ECS due to possible fuel spills during in-flight refueling, select AIR SOURCE pushbutton L ENG.

In-flight Refueling Techniques

Note

The following procedures, as applied to tanker operation, refer to single-drogue tanker only.

Refueling altitudes and airspeeds are dictated by receiver and/or tanker characteristics and operational needs, consistent with the tankers performance and refueling capabilities. This covers a practical spectrum from the deck to 35,000 feet, 170 to 300 KIAS and wing sweep angles of 20° to 68°. Optimum airspeed and wing sweep position is 240 KIAS and approximately 40° wing sweep. This configuration increases aircraft angle of attack enough to lower the receivers vertical tails below the tankers jet wash and decreases bow wave effect.

APPROACH

Once cleared to commence an approach and with refueling checklists completed, assume a position 5 to 10 feet in trail of the drogue with the refueling probe in line in both the horizontal and vertical reference planes. Trim the aircraft in this stabilized approach position and ensure that the tanker's (amber) ready light is illuminated before attempting an approach. Select a reference point on the tanker as a primary alignment guide during the approach phase; secondarily, rely on peripheral version of the drogue and hose and supplementary remarks by the RIO. Increase power to establish an optimum 3 to 5 knots closure rate on the drogue. It must be emphasized that an excessive closure rate will cause a violent hose whip following contact and/or increase the danger of structural damage to the aircraft in the event of misalignment; too slow a closure rate results in the pilot fencing with the drogue as it oscillates in close proximity to the aircraft's nose. During the final phase of the approach, the drogue has a tendency to move slightly upward and to the right as it passes the nose of the receiver aircraft, due to the aircraft-drogue airstream interaction. Small corrections in the approach phase are acceptable. However, if alignment is off in the final phase it is best to immediately return to the initial approach position and commence another approach, compensating for previous misalignment by adjusting the reference point selected on the tanker. Small lateral corrections with a "shoulder probe" are made with the rudder, and vertical corrections with the horizontal stabilizer. Avoid any corrections about the longitudinal axis since they cause probe displacement in both the lateral and vertical reference planes.

MISSED APPROACH

If the receiver probe passes forward of the drogue basket without making contact, a missed approach should be initiated immediately. Also, if the probe impinges on the canopy-lined rim of the basket and tips it, a missed approach should be initiated. Realization of this situation can be readily ascertained through the RIO. A missed approach is executed by reducing power and backing to the rear at an opening rate commensurate with the optimum 3 to 5 knots closure rate made on an approach. By continuing an approach past the basket, a pilot might hook his probe over the hose and/or permit the drogue to contact the receiver aircraft fuselage. Either of the two aforementioned hazards require more skill to calmly unravel the hose and drogue without causing further damage than to make another approach. If the initial approach position is well in line with the drogue, the chance of hooking the hose is diminished when last minute corrections are kept to a minimum. After executing a missed approach, analyze previous misalignment problems and apply positive corrections to preclude a hazardous tendency to blindly stab at the drogue.

CONTACT

When the receiver probe engages the basket it will seat itself into the drogue coupling and a slight ripple will be evident in the refueling hose. Here again, the RIO can readily inform the pilot by calling "contact". The tanker's drogue and hose must be pushed forward 3 to 5 feet by the receiver probe before fuel transfer can be effected. This advanced position is evident by the tanker's amber ready light going out and the green fuel transfer light coming on. While plugged-in, merely fly a close-tail-chase formation on the tanker. Although this tucked-in condition restricts the tanker's maneuverability, gradual changes involving heading, altitude and/or airspeed may be made. A sharp lookout doctrine must be maintained due to the precise flying imposed on both the tanker and receiver pilots. In this respect the tanker can be assisted by other aircraft in the formation. The receiver RIO can also assist in maintaining a visual lookout since the receiver radar is in the STBY position.

DISENGAGEMENT

Disengagement from a successful contact is accomplished by the reducing power and backing out at a 3- to 5-knot separation rate. Care should be taken to maintain the same relative alignment on the tanker as upon engagement. The receiver probe will separate from the drogue coupling when the hose reaches full extension.

When clear of the drogue:

1. REFUEL PROBE switch . . . . RET

2. Probe transition light . . . . OFF

3. AIR SOURCE pushbutton . . . . BOTH ENG

4. Wing sweep mode switch . . . . AUTO

Resume normal flight operations.

FORMATION FLIGHT

Parade Formation

The following formation description is a recommended guideline for the F-14 basic parade position. Line-of-bearing is determined by placing the upper leading edge of the lead aircraft's intake on the red explosive seat triangle below the RIO's cockpit. Sufficient stepdown is provided if the pilot can see a part of the opposite engine nacelle under the near nacelle.

The triangulation for wingtip clearance/overlap is as follows:

- WINGSWEEP 20°, CLEAN AND DIRTY CONFIGURATION

   Use above line-of-bearing and stepdown indicators, and line up the leading edges of the lead aircraft's ventral fins.

   This position should provide the wingman with approximately 10 feet of stepdown and approximately 8.5 feet of wing overlap.

- WINGSWEEP 68°:

    Use above line-of-bearing, increase stepdown to show approximately one foot of the opposite engine nacelle under the near nacelle, and allow approximately one foot of the forward edge of the opposite ventral fin to show in front of the near ventral fin.

    This position should provide the wingman with approximately 10 feet of stepdown and no wingtip overlap.

BANNER TOWING

Ground Procedures

The following procedures are provided for guidance. Local course rules may dictate modification of these steps:

1. When tower clearance onto the duty runway has been received, the tow aircraft taxis to a position as directed by the tow hookup crew. The tow pilot holds this position until released by the tow hookup crew. The escort pilot maintains his position on the taxiway at the approach end of the runway.

2. When signaled to do so by the tow hookup crew, the tow pilot proceeds to taxi down the runway.

3. Upon receipt of a visual taxi signal from the tow hookup crew to slow down, the escort pilot relays this signal to the tow pilot via UHF radio.

4. Upon receipt of a visual taxi signal from the tow hookup crew to stop, the escort pilot relays this signal to the tow pilot via UHF radio.

5. Upon receipt of a signal from the tow hookup crew that the tow hookup is complete, the escort pilot requests the tow pilot to take up slack.

6. The tow pilot proceeds to taxi down the runway.

7. When the banner moves forward onto the runway the escort pilot transmits, "tow aircraft hold-good banner", and taxis onto the runway abeam the banner for takeoff.

8. When ready, the tow pilot transmits, "Tower, Lizard 616 for banner takeoff, escort to follow a good banner".

9. After the banner becomes airborne, the escort pilot commences his takeoff roll.

Flight Procedures

Flight tests have demonstrated no significant degradation of F-14A performance and handling characteristics when towing a banner.

CAUTION

Angle of bank should be limited to 30° or less to preclude contact between the tow cable and afterburner nozzle.

Refer to Section I, Part 4 for banner towing restrictions.

TAKEOFF

Normal takeoff procedures, including rotation speeds and techniques, are suitable for takeoff with the banner. Except when operating in high ambient temperatures or at high density altitude airfields, takeoff in MIL or zone 1-3 AB is recommended to reduce rotation attitude. However, takeoff can be made with up to MAX AB without incurring heat damage to the tow cable.

CAUTION

- Takeoff ground roll with banner can be estimated by adding a factor of 10% to basic airplane takeoff performance. If aircraft lift-off will not occur prior to crossing the long field arresting gear, the gear must be removed to preclude the banner being torn off.

- If the crosswind component is in excess of 10 knots, the takeoff roll should be made on the upwind side of the runway to prevent the banner from striking the runway lights on the downwind side of the runway.

Note

Adequate clearance exists to prevent contact between the tow cable and speedbrakes during ground operation. If takeoff is aborted, basic emergency procedures are applicable. The tow cable will be released when the tailhook is lowered.

After lift-off, continue rotation to 15° (MRT) or 25° (MAX AB), while raising the landing gear. Do not exceed 17 units AOA. Climb out at 180 to 200 KIAS until the flaps are up, then continue climb at 200 to 220 KIAS.

Note

Tow airspeeds in excess of 220 KIAS will result in excessive banner fraying.

CRUISE/PATTERN

No special pilot techniques are required when towing a banner. Enroute cruising speed of 180 to 220 KIAS will provide adequate energy for mild maneuvering while minimizing banner fray. If a low pattern airspeed is desired, extend flaps/slats if necessary to maintain AOA at or below 12 units. The tow aircraft must call all turns to allow the chase aircraft to position himself on the outside of the turn.

If the banner is shot off or falls off in flight, the remaining cable should be dropped in the gunnery area or in a confirmed clear area. After the cable is released, a chase aircraft should join to verify that the cable has been dropped.

WARNING

Without the banner, any remaining cable will flail unpredictably. The chase should approach the tow aircraft from abeam, avoiding a cone-shaped area defined by the tow's 4 to 8 o'clock positions.

DESCENT

Airspeeds of 160 to 220 KIAS should be used for descent. Flaps and slats may be utilized to increase the rate of descent as desired.

CAUTION

Speedbrakes should not be used while towing, since limited clearance exists between the cable and speedbrakes during extension and retraction in flight.

BANNER DROP

The tow aircraft should extend its flaps and reduce airspeed (140 to 160 KIAS, 12 units AOA maximum) for the drop. The banner should be dropped in wings level flight at a minimum aircraft altitude of 1,000 feet AGL. The chase aircraft should ensure adequate clearance exists between the banner and ground obstacles during approach to the drop zone and provide calls to assist in line-up. Release is normally called by the tower when the banner is over the center of the drop zone. Release is accomplished by lowering the tailhook. Because of the low release altitude, crosswind will have no appreciable effect on the banner impact point (the banner will hit down range of the release point). Following banner release, the tailhook should be raised.

BANNER RELEASE FAILURE

If the arresting hook fails to extend, the banner cannot be released. In this case, the following procedure is recommended:

1. In the gunnery range (or other cleared area) descend to low altitude, extend the flaps, slow to 140 to 160 KIAS, 12 units AOA maximum and descend to 100 to 200 feet AGL. This will drag the banner off on the ground (or water). Have the escort pilot confirm that the banner breaks off on ground collision, and determine the length of the remaining tow cable.

WARNING

The escort pilot must remain well clear of the remaining cable. The last 25% of the remaining cable will flail unpredictably.

2. If 100 feet or greater remaining tow cable length is confirmed by the escort pilot, plan to touchdown 1,000 to 1,500 feet long, runway length permitting.

CAUTION

Every effort must be made by the tow pilot not to drag the remaining tow cable across lines, fences, or other obstacles due to property damage that will result.

Note

The long touchdown should be carefully planned because long field arrestment is impossible.

IN-FLIGHT COMPASS EVALUATION

The tactical software provides an accurate comparison of the inertial navigation system (INS) and the attitude heading reference set (AHRS) to the nearest tenth of a degree (±6 minutes), by comparing calculated magnetic variation (vC) to the RIO-entered manual magnetic variation (vM). The difference between the two (vC-vM) should not exceed +2° or -2°.

This procedure for in-flight compass evaluation produces best results at shore-based installations. Errors in the ship's navigation system are transferred to the aricraft navigation system during CVA align and can produce invalid results. Postflight closeout groundspeed should be less than 6 knots; otherwise, results will be invalidated. Closeout groundspeed of 6 knots or more indicates a bad IMU, a bad CSDC, or bad alignment. The results of the evaluation should be ignored until the INS problem has been resolved and the evaluation performed again.

Note

Before performing an in-flight evaluation, the aircraft should not be flown in a manner that would tumble the AHRS gyros.

The following steps should be performed and data recorded on a form similar to figure 3-10.

Annual Compass Compensation In-Flight Evaluation

1. Flightcrew start engine.
2. WCS Switch — ST BY
3. Conduct INS alignment to fine align complete.
4. Select INS navigation mode.
5. Verify that AHRS caution light is not illuminated. The AHRS light may illuminate momentarily during a power interrupt, or, depending on the type of gyro, it may illuminate for as long as 3 minutes while the displacement gyro is erecting.
6. Enter magnetic variation (vM) to the nearest tenth of a degree.
7. Verify that the value entered above is displayed on the TID as vM.
8. During routine flight, fly a steady heading for 30 seconds. Update vMs, if required. On the compass controller panel, set the mode switch to SLAVED, press and hold the HDG pushbutton until the SYNC indicator nulls.
9. Hold the heading as steady as possible and record TID vC, vM, true heading, magnetic heading, and BDHI and standby compass headings.
10. Repeat steps 8 and 9 for various headings during the flight, preferably two headings in each quadrant.
11. At completion of the flight, calculate vC-vM and the difference between the TID magnetic heading and the standby compass heading.
12. For the AHRS to meet the annual compensation requirements, vC-vM should not exceed ±2°.
13. The standby magnetic compass should be within ±5° of the magnetic heading to meet requirements.

SAMPLE

COMPASS EVALUATION

AIRCRAFT	DATE

1. Align IMU to fine align complete.
2. Enter magnetic variation to the nearest tenth.
3. Compass mode switch — slaved.
4. Establish eight different headings (two in each quadrant). Press HDG pushbutton each time until synch nulls, then release. Record the following:

	HDG 1	HDG 2	HDG 3	HDG 4
MAG HDG (TID)				
vC (TID)				
STBY COMPASS				
BDHI				
TRUE HDG (TID)				
vM (TID)				

	HDG 5	HDG 6	HDG 7	HDG 8
MAG HDG (TID)				
vC (TID)				
STBY COMPASS				
BDHI				
TRUE HDG (TID)				
vM (TID)				

Closeout groundspeed _____

Range to homebase _____ (land based)

NOTE

Do not update the INS during entire flight (land based).

FLIGHT CREW

SAMPLE

Figure 3-10. Compass Evaluation

Section III
Part 5

NAVAIR 01-F14AAA-1

PART 5 — FUNCTIONAL CHECKFLIGHT PROCEDURES

FUNCTIONAL CHECKFLIGHTS

Functional checkflights will be performed when directed by, and in accordance with OPNAVINST 4790.2 series and the directions of NAVAIRSYSCOM TYPE Commanders, or other appropriate authority. Functional checkflight requirements and applicable minimums are described below. Functional checkflight checklists are promulgated separately.

CHECKFLIGHT PROCEDURES

To complete the required checks in the most efficient and logical order, a flight profile has been established for each checkflight condition and identified by the letter corresponding to the purpose for which the checkflight is being flown (A, B, or C, as shown in figure 3-11). The applicable letter identifying the profile prefixes each check in the Functional Checkflight Checklist, NAVAIR 01-F14AAA-1F. The post-maintenance checkflight procedures are specific, supplementary checks to be performed in conjunction with NATOPS normal procedures (Section III, Part 2). Checkflight personnel will familiarize themselves with these requirements before each flight. Thorough professional checkflights are a vital part of the squadron maintenance effort. Check crews perform a valuable service to the maintenance department by carrying out this function. The quality of service provided by check crews reflects directly in the quality of maintenance and, reciprocally, enhancement of flight operations. To avoid degradation of this valuable service, always adopt the attitude that thoroughness, professionalism, and safety are primary considerations for all checkflights. A daily inspection is required before each checkflight.

Note

Shipboard constraints can preclude completion of some items on the applicable flight profile checklist.

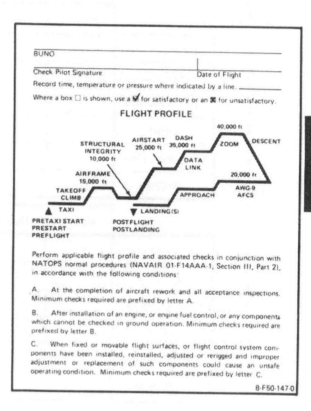

Figure 3-11. Flight Profile

3-66

NAVAIR 01-F14AAA-1

Section III
Part 5

FUNCTIONAL CHECKFLIGHT PROCEDURES (PILOT)

PROFILE

PRESTART

ABC

1. Fuel quantity and distribution CHECK .... ☐
   Check for proper fuel quantities in each system. Left tape 6200 pounds maximum, right tape 6600 pounds maximum, wings approximately 2000 pounds each, and the external tanks approximately 1800 pounds. Check total.

	Left	Right
FEED		
FUS		
WING		
EXT		
TOTAL		

A

2. ICS – CHECK
   a. Normal .... ☐
   b. Backup .... ☐
   c. Emerg .... ☐

A

3. Refuel PROBE – CHECK

   a. Extend (with handpump) .... ☐

   b. Retract (with handpump) .... ☐

START

ABC

1. Idle engine instrument readings .... ☐

	Left	Right	Nominal
NOZ position			5 (open)
OIL (PSI)			40 to 50 (30 min)
RPM (%)			63 to 74
TIT (°C)			500 to 600
FF (PPH)			900 to 1200

PROFILE

POSTSTART

ABC

1. VMCU operation – CHECK .... ☐
   Voltage drop obtained when emergency generator is turned off will activate VMCU. This will result in instantaneous illumination of following lights for 1 to 2 seconds:

   PITCH STAB 1 and 2
   ROLL STAB 1 and 2
   YAW STAB OP and OUT
   HZ TAIL AUTH
   RUDDER AUTH
   SPOILERS
   AUTO PILOT
   MACH TRIM
   R INLET

   When normal voltage is regained, VMCU is deactivated; all lights will go out except RUDDER AUTH

ABC

2. MASTER TEST switch – WG SWP .... ☐

   Wingsweep mode switch must be in AUTO

A

3. UHF .... ☐
   Check complete operation of throttle UHF switch – FWD, AFT, BOTH, ICS.

AB

4. OBC (Autopilot ENGAGE) .... ☐

   a. After ramps are extended – SELECT RAMPS TO STOW .... ☐

   b. Verify RAMP lights go off and INLET lights illuminate .... ☐

3-67

Section III
Part 5

PROFILE

	c. When OBC complete
	(1) Reset both AICS and check that INLET RAMPS are in AUTO ☐
	(2) Reinitiate complete normal OBC ☐
	d. Speed brake switch
	(1) EXT-RET ☐
	(2) Verify stabilizers shift 1° nose down (clean) or 3° nose down (weapons rails) on extension and opposite on retraction (ITS) ☐

ABC 5. ECS check

   a. AIR SOURCE pushbuttons should be depressed; L ENG, R ENG, OFF, then back to BOTH ENG. There will be slight reduction in flow with single engine air source selected. There should be no excessive interruptions in airflow when source changes are made ☐

   b. Cockpit temperature control (TEMP mode selector switch) should be checked in both MAN and AUTO modes to ensure proper temperature control ☐

ABC 6. Flaps down ☐

   a. Verify stabilizer shifts 3° nose up (ITS) ☐

ABC 7. Flight controls check

   a. Trim - FULL NOSEDOWN (9° TRAILING EDGE DOWN) ☐

PROFILE

   b. Stick full aft - CHECK FOR FREE MOVEMENT ☐

   c. Trim - FULL NOSEUP (LESS THAN 20° TEU) (17 to 19 SECONDS) ☐

   d. Stick full forward - CHECK FOR FREE MOVEMENT ☐

   e. Yaw trim - 7° LEFT TO 7° RIGHT (12 TO 14 SECONDS) ☐

   f. Trim - FULL LEFT ☐

   g. Stick full left - CHECK POWER APPROACH SPOILER GEARING AND UNIFORM SPOILER EXTENSION ☐

   h. Trim - FULL RIGHT (16 TO 18 SECONDS), CHECK UNIFORM SPOILER EXTENSION AND 6° DIFFERENTIAL TAIL SPLIT ☐

ABC 8. Flight controls - CYCLE ☐
   Observe following:

   a. Longitudinal - 33° TEU TO 12° TED HORIZONTAL TAIL ☐

   b. Lateral - 24° TOTAL DIFFERENTIAL TAIL ☐

   c. Depress autopilot emergency disengage paddle - 14° TOTAL DIFFERENTIAL TAIL ☐

   d. Directional – 30° RUDDER ☐

   e. Longitudinal – LATERAL COMBINED – 35° UP TO 15° DN ☐

   f. Spoilers – 55° ☐

NAVAIR 01-F14AAA-1

Section III
Part 5

PROFILE		
ABC	9. Spoiler checks	
	a. DLC operation – CHECK	
	(1) Engage DLC, verify stabilizer shifts 6° down and SPOILER indicators show extended	☐
	(2) Full up DLC, verify stabilizer returns to trim position and SPOILER indicators show drooped	☐
	(3) Full down DLC, verify stabilizer shifts 8° below trimmed position and SPOILER indicators show extended	☐

Note

If either right SPOILER position indicator shows one position higher than actual spoiler position (that is, DN vice drooped and extended vice DN), a spoiler zero degree switch has failed. Spoiler symmetry protection circuitry is inoperative and spoiler ground roll braking inflight is possible.

	b. SPOILER BK – SELECT	☐
	c. SPOILER FLR ORIDE operation – CHECK (AFC 573 incorporated)	
	(1) INBD and OUTBD SPOILER FLR ORIDE switches – ORIDE	☐

PROFILE		
ABC	(2) MASTER TEST switch – STICK SW . . . . Verify all spoilers remain extended (55°).	☐
	(3) INBD SPOILER FLR ORIDE SWITCH – NORM . . . . Verify INBD spoilers fail down and SPOILER light illuminates.	☐
	(4) OUTBD SPOILER FLR ORIDE switch – NORM . . . . Verify OUTBD spoilers fail down.	☐

Note

With SPOILER FLR ORIDE switches in ORIDE, spoiler symmetry protection circuitry is disabled.

	d. MASTER TEST switch – STICK SW	☐
	(1) Verify GO light illuminates with one inch lateral stick inputs left and right	☐
	(2) MASTER TEST switch – OFF	☐
	(3) MASTER RESET – DEPRESS . . . . Verify SPOILER light goes out and all spoilers extend to 55°.	☐
	e. SPOILER BK/THROTTLE interlocks – CHECK	☐
A	10. Radar altimeter - TEST	☐
A	11. Displays - CHECK	☐
A	12. TACAN - BIT	☐

3-69

Section III
Part 5

PROFILE

| A | 13. | ARA-63A - BIT .... ☐ |
| A | 14. | Gun sight - CHECK (MANUAL MODE) .... ☐ |

    a. Select GUN and AIR TO AIR

    b. Superimpose gunsight reticle over ADL. ELEV LEAD window should read 0 mils.

    c. Set +46 mils.

    d. Uncage gunsight.

    e. Verify HUD display

Horizontal displacement with respect to ADL is 2 mils left for diamond and 1 mil left for movable reticle.

| AB | 15. | Autopilot emergency disengage paddle switch — HOLD DEPRESSED |

(IAFC 623 and IAYC 2342 incorporated)

    a. Verify throttles in manual mode .... ☐

PROFILE

| AB | TAXI |

1. MASTER TEST switch - MACH LEV .... ☐

	Left	Right	Limits	Diff
RPM (%)			81 to 86	4

    a. GO/NO GO light .... ☐

If NO GO light is illuminated after successful engine RPM check, aircraft is still up.

| AB | 2. Engine runup - CHECK AT MIL .... ☐ |

	Left	Right	Limits
NOZ position			0 (closed)
OIL (PSI)			40 to 50
RPM (%)			99 to 100 nominal (105 maximum)
TIT (°C)			1080 to 1185
FF (PPH)			7000 to 8000

    a. Throttles - MIL .... ☐

    b. THROTTLE MODE switch - MAN .... ☐

    c. Throttles - IDLE .... ☐

    d. THROTTLE MODE switch - BOOST .... ☐

    e. Afterburner - CHECK LIGHTOFF .... ☐

NAVAIR 01-F14AAA-1

Section III
Part 5

PROFILE		PROFILE	
	TAKEOFF AND CLIMB		g. CABIN PRESS switch - NORM ☐
A	1. Landing gear - RETRACT ☐ (9 TO 15 SECONDS NOMINAL)		h. RAM AIR switch - INCR (35 TO 50 SECONDS) ☐
A	2. Servo and radar altimeters - CHECK BELOW 5000 FEET ☐		As ram air door opens (up to 50 seconds to open fully) there will be barely perceptible increase in cockpit airflow. Slight cockpit pressurization can be noticed on cabin pressure altimeter.
A	3. REFUEL PROBE switch - EXTD-RET ☐		
	15,000-FOOT CHECKS		
ABC	1. Fuel transfer ☐		i. AIR SOURCE pushbutton - RAM ☐
AB	2. ECS check (250 KIAS) ECS check should be performed at altitude above 8,000 feet so cabin pressurization can be checked, but low enough to prevent large cockpit pressure changes when cockpit air is secured.		With RAM selected, 400° manifold is repressurized, which maintains canopy seal, air bags, and antenna waveguides pressurization. As canopy seal reinflates, cockpit pressurization available from ram air will be much more apparent.
	a. Cabin altitude approximately 8,000 feet ☐		j. AIR SOURCE pushbutton - BOTH ENG ☐
	b. Air distribution - CANOPY DEFOG-CABIN AIR ☐		k. WCS switch - XMT ☐ (Coordinate with RIO.)
	c. WSHLD DEFOG switch - MIN-NORM-MAX ☐	AC	3. Stabilize at 300 KIAS
	d. WCS switch - STBY ☐ (Coordinate with RIO.)		L R
	e. AIR SOURCE pushbutton - OFF ☐		a. Full stick roll - SAS OFF-ON ☐ ☐
	Cockpit pressurization will quickly bleed off with cabin pressure altimeter moving upward toward aircraft altitude.		Note full extension of down wing spoilers. SAS on roll acceleration will be slightly higher due to increased horizontal tail authority
			L R
	f. CABIN PRESS switch - DUMP ☐		b. Rudder pulse - SAS OFF-ON ☐ ☐
	Cockpit will completely depressurize.		Yaw excursions should cease immediately upon engagement of YAW STAB.

Section III
Part 5

PROFILE

AC

c. Pitch pulse - SAS OFF-ON .... ☐

   At slow speeds, pitch SAS OFF is hardly noticeable. Lack of pitch SAS becomes more apparent as speed increases.

4. Wing sweep and maneuver devices (Check at 0.5 IMN.)

   a. Maneuver devices - EXTD .... ☐

   b. WING SWEEP MODE switch - AFT .... ☐

      (1) Manually sweep wings fully aft and check that wings stop at 50° .... ☐

   c. Maneuver flaps partial up with thumbwheel to ensure that devices will retract .... ☐

   d. WING SWEEP MODE switch - BOMB .... ☐

      (1) Verify maneuver devices automatically retract and then wings sweep to 55° .... ☐

   e. WING SWEEP MODE switch - MAN FULL AFT .... ☐

   f. WING SWEEP MODE switch - AUTO .... ☐

   g. EMERGENCY WING SWEEP handle - CYCLE 22°, -68°, -22° .... ☐

      (1) Verify spider detent engaged, emergency WING SWEEP warning light off .... ☐

PROFILE

ABC

   (2) MASTER RESET pushbutton - DEPRESS, check WING SWEEP ADVISORY LIGHT OFF .... ☐

   h. Maneuver devices - EXTD .... ☐

   i. Accelerate to greater than 0.75 IMN and check maneuver devices automatically retracted .... ☐
      (maneuver devices start auto retract at 0.67 IMN.)

   j. Decelerate to less than 0.68 IMN and check maneuver devices remain retracted .... ☐

   k. WING SWEEP MODE switch - AUTO .... ☐

5. High AOA Mach lever and AUTO MAN devices (approximately 15,000 feet) - CHECK .... ☐

   a. Throttles - IDLE .... ☐

   b. Maneuver devices extend at 10.5 units AOA .... ☐

   c. RPM increases at 16 ±1 units AOA (greater than 80%) .... ☐

   CAUTION

   If Mach levers fail high AOA check, do not perform approaches to stalls.

   d. Maneuver devices retract at 8 units AOA .... ☐

6. Approaches to stalls and approach configuration check (approximately 15,000 feet)

   a. Clean stall .... ☐

      (1) Stabilize in level flight at 15 units AOA, 90% speed brakes out.

      (2) Ensure maneuver devices retracted maintaining power setting.

3-72

PROFILE		
	(3)	Slowly decelerate to buffet onset. Note AOA (approximately 16.5 to 17.0 units). Set MIL PWR.
	(4)	Continue deceleration to 28 units AOA. Note any abrupt or significant roll-off tendencies.
	b.	Approach configuration check
	(1)	Perform landing checklist ☐
	(2)	DLC - ENGAGE ☐ (Check for excessive lateral trim requirements.)
	(3)	Automatic throttle and DLC - CHECK ☐
		(a) Response to longitudinal stick ☐
		(b) Response to turn entry, steady rollout ☐
		(c) Response to DLC thumbwheel inputs ☐
		(d) Response in HOT/NORM/COLD ☐
		(e) AUTO THROT light
		1 Manual override ☐
		2 CAGE/SEAM button ☐
	c.	Dirty stall ☐ Slowly decelerate in level flight to 16.5 to 17.0. Set MIL PWR, continue deceleration to maximum of 20 units AOA. Note any abrupt or significant roll-off tendencies.
	d.	Attempt speed brake extension at MIL power ☐
ABC	7.	Structural integrity check (0.9 IMN at 10,000 ft)
	a.	High-speed dash - MIL THRUST ☐
	b.	High-g turn ☐
	c.	Anti-g valve operation ☐
	d.	Accelerometer - CHECK ☐

WINDMILL AIRSTARTS (25,000 FEET)

AB	1.	WCS switch - OFF ☐ (Coordinate with RIO.)

PROFILE		
AB	2.	Stabilize at 295 KIAS, right engine at IDLE ☐
AB	3.	Right throttle - OFF ☐
	a.	Perform normal airstart - AIR SOURCE pushbutton - BOTH ENG, AIRSPEED - 295 KIAS, RPM STABILIZED, 15% MINIMUM ☐
AB	4.	Stabilize at 310 KIAS, left engine at IDLE ☐
AB	5.	Left throttle - OFF ☐
	a.	Perform normal airstart - AIR SOURCE pushbutton - BOTH ENG, AIRSPEED 310 KIAS, RPM STABILIZED, 15% MINIMUM ☐
AB	6.	AIRSTART switch - OFF ☐
AB	7.	WCS switch - STBY ☐ (Coordinate with RIO.)

CAUTION

Windmill air start sub-idle stall can be evidenced by low engine rpm, a corresponding lack of engine response to throttle advancement, and a slowly increasing TIT. The sub-idle stall can be cleared by cycling the throttle to OFF and immediately returning it to IDLE.

CLIMB TO 35,000 FEET

AB	1. Fuel management - CHECK ☐			

	Left	Right
FEED		
FUS		
WING		
EXT		
TOTAL		

AB	2.	ECS check
	a.	Automatic cabin temperature control ☐
	b.	Manual cabin temperature control ☐

Section III
Part 5

NAVAIR 01-F14AAA-1

PROFILE

AB

   c. Cabin altitude schedule .... ☐
      (approximately 14,000 feet
      at 35,000 feet)

3. Engine instruments -
   CHECK .... ☐
   (engine MIL power at
   0.9 IMN)

	Left	Right	Limits
OIL (PSI)			40 to 50
RPM (%)			105 maximum
TIT (°C)			1040 to 1185

HIGH-SPEED DASH (35,000 FEET)

AB

1. Jam throttles - MAX
   AFTERBURNER .... ☐

   a. Engine instruments -
      MONITOR .... ☐

	Left	Right	Limits
NOZ position			4-1/2 to 5 (open)
OIL (PSI)			40 to 50
RPM (%)			105 maximum
TIT (°C)			1040 to 1185

   b. Mach trim compensation
      CHECK .... ☐

   c. Compare pitot-static
      instruments .... ☐
      (pilot and RIO)

   d. Wing sweep -
      PROGRAM .... ☐

PROFILE

AB

   e. Glove vane - AUTO
      PROGRAM................. ☐
      (full out at 1.45 IMN)

2. Engine mach lever - CHECK .... ☐

CAUTION

- Monitor RPM decay while retarding throttles to IDLE to ensure proper mach lever operation. Discontinue mach lever check if RPM decays to less than 92% above 1.5 IMN. Place throttles to MIL to decelerate.
- Engine surge to 106% is permitted while deselecting afterburner.

   a. Both throttles to MIL until
      nozzles closed, then IDLE
      above 1.5 IMN .... ☐

	Left	Right	Nominal	Diff
RPM (%)			92 to 98	4

ZOOM (40,000 FEET)

AB

1. Pitch up to FL 400 .... ☐

2. Cabin pressurization and
   ECS - CHECK................. ☐
   (approximately 17,000 feet
   at 40,000 feet)

20,000-FOOT CHECKS

A

1. AFCS check

   a. Attitude hold

      (1) Autopilot - ENGAGE,
          VERIFY NO
          TRANSIENT............. ☐

      (2) Check for
          smooth operation
          in CSS .... ☐

PROFILE

   b. Heading hold

     (1) Heading hold - ENGAGE . . . . ☐

     (2) Left and right pedal sideslip - CHECK RETURN TO REFERENCE HEADING . . . . ☐

     (3) CSS left or right to 5° roll - AIRCRAFT SHOULD RETURN TO 0° ANGLE OF BANK . . . ☐

   c. Altitude hold

     (1) Altitude hold - SELECT, verify A/P REF LIGHT ILLUMINATES . . . . ☐

     (2) A/P REF - NWS pushbutton - DEPRESS, verify that A/P REF LIGHT GOES OFF . . . . ☐

     (3) Check for altitude control

        • ±30 feet in level flight . . . . ☐

        • ±60 feet in 30° angle of bank . . . . ☐

     (4) Autopilot emergency disengage paddle - OFF . . . . ☐

     (5) PITCH and ROLL STAB AUG switches - ON . . . . ☐

A

2. Air-to-air check

   a. Weapon selector switch - PH - (IFT) . . . . ☐

PROFILE

   b. Attack steering - LAR-Vc . . . . ☐

   c. Collision steering . . . . ☐

   d. Attack steering - LAR-Vc . . . . ☐

     (1) SP . . . . ☐

     (2) SW . . . . ☐

   e. Target intercept

     (1) VSL high . . . . ☐

     (2) VSL low . . . . ☐

     (3) MRL . . . . ☐

     (4) PLM . . . . ☐

     (5) PAL . . . . ☐

     (6) Select gun . . . . ☐

     (7) Observe proper HUD display . . . . ☐

   f. Gun sight - CHECK (MANUAL MODE) . . . . ☐

     (1) Select GUN and AIR TO AIR

     (2) Uncage gunsight

     (3) Fly level coordinated turn pulling enough g's to place the corner of the diamond on a line with the horizontal line of the ADL

        (SHOWN FOR LEFT TURN)

     (4) Results should be 3g turn in 45 ± 6 seconds with diamond displaced 15 mils horizontally

Section III
Part 5

NAVAIR 01-F14AAA-1

PROFILE

A

3. Air-to-ground check

    a. Select A/G .... ☐

        (1) Dive angle greater than 10° .... ☐

        (2) Check symbology; when in range (gun - 6000 feet) diamond disappears .... ☐

A

4. Fuel dump check

    a. Speed brake switch - EXT .... ☐

    b. DUMP switch - DUMP .... ☐
       (Observe no fuel dump.)

    c. Speed brake switch - RET .... ☐
       (Observe fuel dump.)

    d. DUMP switch - OFF .... ☐

PROFILE

APPROACH

ABC

1. Landing checklist complete .... ☐

2. ACLS/ARA-63 - CHECK .... ☐

3. Airspeed and AOA (15 units AOA) - CHECK. .... ☐

    a. AOA, INDEXER, HUD .... ☐

    • Gross weight _____ pounds

    • Airspeed _____ KIAS
      (119 KIAS plus 3 KIAS/2000 pounds over 42,000 pounds gross weight)

ABC

4. Walkaround inspection - COMPLETE. .... ☐

3-76

NAVAIR 01-F14AAA-1   Section III
Part 5

FUNCTIONAL CHECKFLIGHT PROCEDURES (RIO)

PROFILE	
A	**PRESTART**
	1. ICS - CHECK ☐
	a. Normal ☐
	b. Backup ☐
	c. Emerg ☐
ABC	2. IND LT - TEST ☐
	POSTSTART
ABC	1. Computer address panel - INSERT HOME BASE AND OWN A/C ☐

Time	Latitude	Longitude

A	2. ECMD check ☐
	a. Verify ECM display ☐
	b. Verify navigation display ☐
	c. Verify data block and ADF indication ☐
A	3. ECM switch - ON-RESET ☐
	a. ALR-45/ALR-50-BIT ☐
	b. ALQ-100-BIT-STBY ☐
ABC	4. Upon completion of OBC, record acronyms on TID.

[blank box]

A	5. Servo altimeter - SET AND CHECK, ERROR ☐

When the local barometric pressure is set, the altimeter should agree within 75 feet at field elevation in both modes, and the primary or standby readings should agree within 75 feet. In addition, the allowable difference between primary mode readings of altimeters is 75 feet at all altitudes.

PROFILE	
	TAXI
A	1. Compass - CROSS-CHECK HEADING ☐
ABC	2. WCS BIT 1 through 4 - INITIATE AND RECORD ON BER CARD ☐
A	3. INS - CHECK ☐ (at takeoff end of runway)

Groundspeed	Time

ABC	4. WCS switch - STBY ☐
	TAKEOFF AND CLIMB
A	1. Airspeed - CHECK ☐ (200 KIAS)

Pilot	RIO
KIAS	TAS

A	2. Altimeter - CHECK ☐

INS	SERVO

A	3. TACAN and INS position - CROSS-CHECK ☐
A	4. INS navigation and radar mapping - CHECK ☐

3-77

PROFILE

    a. Radar map - CHECK
       ALL RANGE SCALES .... ☐

15,000-FOOT CHECKS

AB 1. Structural integrity check

    a. Anti-g valve operation .... ☐

    b. Set WCS switch to
       STBY before PILOT'S
       ECS check .... ☐

25,000-FOOT CHECKS

AB 1. WCS switch - OFF .... ☐
       (Prior to airstarts)

AB 2. WCS switch - STBY .... ☐
       (Airstarts complete)

CLIMB TO 35,000 FEET

A   1. D/L - CHECK .... ☐

A   2. Select assigned frequency
       and ADDRESS .... ☐

A   3. Receive D/L messages

    a. Steering symbols .... ☐

    b. TID target data .... ☐

    c. DDI lights .... ☐

DESCENT

A   1. Air-to-air check

    a. Radar mode - CHECK .... ☐

       PULSE .... ☐
       PD SRCH .... ☐
       RWS .... ☐
       TWS AUTO .... ☐
       TWS MAN .... ☐
       HRWS .... ☐

PROFILE

    b. MLC switch - OUT-
       AUTO-IN (PD SRCH) .... ☐

    c. Navigation - CHECK.......... ☐

       (1) TACAN fix - RECORD ..... ☐
           (Do not initiate update.)

Δ Latitude	Δ Longitude

       (2) Radar fix - RECORD...... ☐
           (Do not initiate update.)

Δ Latitude	Δ Longitude

       (3) Visual fix - RECORD..... ☐
           (Do not initiate update.)

Δ Latitude	Δ Longitude

    d. IFT check

       (1) Pilot select PH,
           missile preparation,
           and TNG .... ☐

       (2) RIO verify PH6
           after PREP timeout ...... ☐

    e. WCS checks against
       suitable airborne
       targets .... ☐

       (1) Intercept targets and
           set ASPECT switch
           to TAIL .... ☐

           (a) Check operation in

               PD SRCH .... ☐

NAVAIR 01-F14AAA-1

Section III
Part 5

PROFILE		
	RWS	☐
	TWS MAN	☐
	(b) Initiate PH attack and verify WCS MODE pushbuttons are set for	
	TWS AUTO	☐
	PDSTT	☐
	(c) Observe transition to	
	PULSE STT	☐
	(d) Return to PULSE SRCH	☐
	(e) Close to visual range and verify DDD display	☐
	(2) VSL mode - HI-LO LOCK-ON	☐
	(3) MRL mode - CHECK LOCK-ON	☐
	f. IFF - CHECK MODES 1, 2, 3, and 3C	☐
A	2. Air-to-ground check	
	a. Select A/G	☐
	b. A/G gun - OFF	☐
	c. WPN TYPE - OFF	☐
	d. Attack mode - CMPTR PILOT	☐
ABC	3. Perform WCS BIT sequence 1 through 4 and record on BER card (BITS 5-8 as required)	☐
A	4. NAV MODE switch - AHRS	☐

PROFILE		
	a. Pilot check VDI display, maneuver aircraft, and fast erect	☐
	b. Radar antenna scan - CHECK	☐
ABC	5. Verify pilot's OBC disabled	☐
ABC	6. Perform airborne OBC	☐
	Record acronyms	

APPROACH

A	1. Airspeed - CHECK	☐
	a. Compare with pilot's airspeed at 15 units AOA	☐
	b. Error _____ KIAS	

LANDING

ABC	1. WCS switch - STBY	☐

IN CHOCKS

ABC	1. INS and visual - CHECK AND UPDATE IN CHOCKS	☐
	a. Record closeout error on flight lines	☐

Δ Latitude	Δ Longitude	Δ Time	Ground-speed

3-79

Section III
Part 5

NAVAIR 01-F14AAA-1

PROFILE

ABC

   b. Initiate
      fix enable .... ☐

   c. Observe aircraft
      symbol shift
      on TID .... ☐

2. Call up OBC
   maintenance
   display .... ☐

PROFILE

ABC

Record acronyms

POSTFLIGHT

1. Walkaround inspection -
   COMPLETE .... ☐

SECTION IV — FLIGHT CHARACTERISTICS

TABLE OF CONTENTS

Primary Flight Controls ... 4-1	Cruise and Combat Configuration ... 4-8
Secondary Flight Controls ... 4-2	Engine Operating Characteristics ... 4-11
Asymmetric Thrust Flight Characteristics ... 4-3	Takeoff and Landing Configuration ... 4-13
High Angle-of-Attack Flight ... 4-6	Degraded Approach Configuration ... 4-15

PRIMARY FLIGHT CONTROLS

Primary flight controls are devices that change the flight path of the aircraft. They consist of the differential horizontal tail for pitch and roll control, the spoilers for supplementary roll control, and the rudders for yaw control.

Pitch Control

The horizontal tail provides excellent pitch control from under 100 KIAS to over Mach 2. Its effectiveness gives the aircraft several capabilities not enjoyed by other fighters, including the ability to reach or exceed limit load factors over much of the subsonic and all of the supersonic envelope, low takeoff rotation speeds; and it is also an excellent drag device below 100 KIAS on landing rollout. The only major disadvantages are that the pilot can generate high enough pitch rates (particularly in the nosedown direction) to cause coupling under certain conditions, and that a pitch attitude sufficient to scrape tailpipes can be attained on landing rollout.

Roll Control

Differential deflection of the horizontal tail surfaces provides primary roll control throughout the flight envelope and is the only roll control when the wings are swept beyond 57° (disabling the spoilers). Differential tail generates high roll rates, but it also causes adverse yaw that gets worse as the angle of attack is increased. The differential tail imposes high torsional load on the fuselage at high g's but these are alleviated by spoilers and roll SAS under normal conditions.

Spoilers augment the differential tail for roll control at wing-sweep angles forward of 57°. They are very effective at low to medium angles of attack for roll control, and reduce the adverse yaw end torsional loads caused by the differential tail. The spoilers are also the primary mechanism for direct lift control and for spoiler braking.

Yaw Control

Twin rudders furnish yaw control. Through strong dihedral effect, good roll control is also available from rudders at medium and high angles of attack. Rudder power is sufficient to provide adequate control under all asymmetric store loading conditions.

Stability Augmentation (SAS)

SAS is provided in the pitch, roll, and yaw axes. Pitch SAS improves damping of longitudinal, short period dynamic response, but the aircraft can be operated safely throughout its envelope without it. Without pitch SAS, the nose will feel somewhat more sensitive at high speed.

Roll SAS does a number of things for the aircraft. Roll acceleration is augmented during the initial lateral stick input. The SAS reduces differential tail deflection to limit maximum roll rate to less than 180° per second for structural considerations and to prevent roll coupling in the transonic speed range. An undesirable byproduct of the roll rate limiting is an oscillatory roll motion encountered in full deflection rolls at medium subsonic speeds and higher.

Because roll SAS provides structural protection, flight above indicated Mach 0.93 is prohibited without roll SAS. Should tactical considerations necessitate violating this restriction, restrict rolls to less than full lateral stick deflection and to not more than 180° of bank angle change at one time. This minimizes the possibility of aircraft damage. Below indicated Mach 0.93, lack of roll SAS slows initial roll acceleration. High angle-of-attack handling qualities are improved below indicated

Mach 0.93 because unwanted SAS induced differential tail inputs are eliminated.

YAW SAS is the most critical of the stability augmentation functions. Directional dynamic response (yaw oscillations) is poorly dampened without it. In regions of reduced directional stability above 24 units AOA or supersonic, the SAS dampens out yaw rates that might otherwise cause loss of control, engine stall, or structural damage, depending on the situation. Below the SAS OFF limit speed of Mach 0.93, with yaw SAS OFF, normal maneuvering can be accomplished if extra care is taken to control yaw excursions with rudder.

Trim

Trim rates in pitch, roll, and yaw are slow. Lateral control authority and roll rates at slow speeds will be reduced by almost one-half with full stick deflection in the direction of full lateral trim due to decreased spoiler deflection. When maximum lateral control authority is required, the trim in the direction of stick displacement should be reduced. Runaway trim in any axis is controllable. During field landing the aircraft can be recovered safely, if runaway trim is experienced. However, carrier approaches with full runaway pitch trim may be difficult.

SECONDARY FLIGHT CONTROLS

Secondary flight controls affect the flightpath of the aircraft although they have other primary purposes, such as increasing lift or drag. Secondary flight controls of the aircraft include main, auxiliary, and maneuver flaps, speed brakes, DLC, landing gear, variable-sweep wing, and throttles.

Flap, Slats, And DLC

Trim changes during extension and retraction of high-lift devices are significant. During extension of flaps/slats at 220 KCAS, an initial push force of approximately 5 pounds is required followed by a pull force of up to 15 pounds. Engagement of DLC at approach speeds causes a slight nose-down pitch that requires an approximate 3 to 4 pound pull force to counter. Forces during retraction of the high lift devices are generally opposite and approximately the same magnitude. The force required during retraction of flaps/slats may be less objectionable, as the flaps are generally raised at a slower airspeed and therefore require less opposing force.

CAUTION

During simultaneous retraction of flaps/slats/DLC, up to 30 pounds push force may be required to maintain pitch attitude.

Speed Brake

The speed brakes provide some deceleration capability throughout the flight envelope. However, the most effective means to slow the aircraft is to reduce thrust while applying g. Extension/retraction of the speed brakes results in a pitch trim change that varies with flight conditions. In general, this change is not objectionable except at higher airspeeds where the rapidity of the change (1.5 seconds for full extension) may cause a bobble in gunsight tracking and lead to a minor case of PIO.

Static Stability

The static longitudinal stability of the aircraft can be roughly divided into four regimes. At slow speeds where the wings are not sweeping, static longitudinal stability is slightly positive. At speeds where the wings are automatically sweeping, static stability becomes neutral to slightly negative.

Note

Positive static stability in this area can be regained by fixing the wings at a constant angle.

In the transonic region, from Mach 0.8 to 1.5, static longitudinal stability is essentially neutral. There are, however, some minor reversals in the stick force gradient. This is noticeable at approximately Mach 0.95 and again at approximately Mach 1.4 when the glove vanes begin extending. Above Mach 1.5, the stick force gradient becomes neutral.

Trimmability

Ease of trimming the aircraft to level flight can be broken down into two areas. At airspeeds slower than those using automatic wing sweep programming, the aircraft is relatively easy to trim to level flight because it has positive static stability. At airspeeds where the wings automatically move with a change in airspeeds, it becomes very difficult to achieve a hands-off trim. Because of the change in aircraft pitching moment caused by movement of the wings, the nose tends to pitch further down with each increase in speed or further up with each decrease in speed.

Changes in thrust settings normally require a trim change, particularly in the approach configurations. A reduction in power causes a slight nosedown pitch.

Note

Trim actuator rates in all axes are relatively slow.

Roll Performance

The roll performance of the aircraft is generally satisfactory, particularly at higher airspeeds. At lower speeds, however, the great mass and inertia of the aircraft restrict its roll rate to considerably less than that of smaller, more nimble tactical aircraft (A-4, T-38). Roll rate at low airspeeds and high angle of attack are definite tactical limitations.

Roll response to control inputs is good with two exceptions. At higher airspeeds, the roll rate limiting feature of the SAS causes marked variations in acceleration, which feels to the pilot as a roll oscillation. Natural pilot compensation for this characteristic may lead to a form of lateral PIO during maximum roll rate maneuvers at high airspeeds. At most airspeeds, roll response to small inputs is overly sensitive due primarily to low breakout forces and a small nonlinearity caused by an abrupt increase in roll rate when the stick is displayed laterally just enough to break the spoilers out. This can cause bank-angle overshoots during maneuvering flight.

Note

Although it differs from aircraft to aircraft, most of the aircraft tend to roll faster to the right than to the left.

Maneuvering Stick Force

Maneuvering stick force, or stick force per g of the aircraft is relatively good throughout most of the flight envelope. It is essentially linear; that is, an increase in force commands an increase in g, and is approximately 4 pounds per g. The force per g is essentially the same throughout the flight envelope and changes very little with altitude, airspeed, loading, or cg position. One exception is a low-speed, high angle-of-attack maneuver, where the force per g can be as high as 10 pounds per g.

Although not directly related, stick displacement during maneuvering may affect the pilot similarly to force per g. Stick displacement required during maneuvering in the aircraft are relatively large and may be uncomfortable to some pilots. While the stick forces are not high, the stick must be placed relatively close to the pilot's torso to attain a given g. This allows the pilot less leverage with his arm and becomes more tiring. This is especially true at lower airspeeds and higher angles of attack.

ASYMMETRIC THRUST FLIGHT CHARACTERISTICS

With one engine inoperative, flight characteristics are considerably affected by the thrust asymmetry generated by the operating engine. The distance of the engines from the aircraft centerline induce flight control requirements and flying qualities not present in center-line thrust aircraft. Flight control requirements are a function of the thrust setting on the operating engine. The thrust required to maintain flight, and therefore the magnitude of the thrust asymmetry, is a function of the following:

GROSS WEIGHT. Heavier gross weights require higher thrust settings to maintain level flight and therefore larger control deflections.

CONFIGURATION. Aircraft configuration greatly varies the amount of thrust required. At cruise configuration airspeeds, control requirements will be significantly reduced compared to landing configurations, which will require significantly higher thrust settings.

AIRSPEED. Minimum thrust is required, and therefore minimum asymmetric moment will be present at maximum endurance airspeeds. Higher or lower airspeeds will require higher power settings and increased thrust asymmetry. At airspeeds above maximum endurance, the greater asymmetry will be largely offset by the additional control power available due to the higher airspeed. Minimum control speed is reached at the point when maximum rudder deflection is no longer sufficient to maintain directional control.

ALTITUDE. Net thrust is strongly dependent on altitude. For a constant throttle setting the asymmetric thrust is considerably higher at sea level than at higher altitudes.

BANK ANGLE. Bank angle increases induced drag and therefore requires higher thrust settings to maintain level flight. The higher thrust setting demands increased rudder deflection in a turn compared to that required in level flight at the same airspeed. Turn direction into or away from the failed engine, significantly affects rudder requirements. In straight line flight, some amount of rudder deflection will be required to offset the aircraft yaw moment from asymmetric thrust at zero bank angle. If bank angle into the dead engine is introduced, an increased amount of rudder is required to counter the additional sideforce component of gravity. Conversely, a bank angle away from the dead engine will introduce a sideforce component of gravity countering the thrust asymmetry and therefore reducing the rudder requirement.

CRUISE. Sufficient controllability exists in cruise configuration to control thrust asymmetry up to maximum afterburner in most cases; however, as aircraft speed approaches zero, flight control effectiveness also approaches zero and thrust asymmetry will generate a yawing motion much like a pin wheel regardless of flight control inputs. The most effective pilot action under these conditions is to eliminate the asymmetric thrust by retarding throttles to idle while simultaneously reducing angle of attack and countering yaw with rudder. In all other flight conditions airplane response to engine stall is generally mild and is characterized by slow build up in yaw rate followed by slowly increasing roll off in the same direction as yaw. Usually, the aircrew will only notice the roll as it completely masks the yaw rate. Rudder is the primary control to offset yawing moment from asymmetric thrust. Higher airspeeds provide more rudder effectively and increase pilot ability to control yaw due to asymmetric thrust.

WARNING

Departure resistance is significantly degraded with the addition of external stores for both asymmetric and symmetric thrust maneuvering flight. The use of lateral stick to offset roll due to yaw from asymmetric thrust will generate adverse yaw and aggravate the yaw due to asymmetric thrust. The result may be a yawing, rolling department.

For example, if engine stall occurs in an aircraft configured with two each tanks, Phoenix, Sparrows and Sidewinders in maximum afterburner during a steady state 4 "g" turn at 0.9 IMN/30,000 feet, a bank angle change of about $100°$ will be generated in 5 seconds if no pilot action is taken. The low yaw rate (less than $15°$/second) generated by this engine stall will be masked by roll. This yaw rate and roll can be countered easily with rudder alone or by retarding the throttles to IDLE. If, however, full lateral stick (Roll SAS ON) is applied to offset the roll, the bank angle will exceed $200°$ in 5 seconds. The result will be a violent high speed departure. Buildup of yaw rate after an engine stall may be completely masked from the pilot by roll if he does not recognize that stall has occurred and that airplane motion is the result of that stall. Therefore, when any roll off or yaw rate occurs during maneuvering flight with maximum thrust, the pilot should reduce AOA counter with rudder, and avoid the use of lateral stick alone. Additionally, if an engine stall is suspected both throttles should be retarded to IDLE.

TAKEOFF CONFIGURATION. Aircraft controllability during asymmetric thrust takeoff emergencies is influenced primarily by rudder position and airspeed. Rudder control power to offset yawing moment due to thrust asymmetry is a function of airspeed. Figure 4-1 refers. Rudder is the primary control for countering yaw due to asymmetric thrust since lateral stick inputs alone will induce adverse yaw in an already critical flight regime. At the first indication of an engine failure, the pilot should not hesitate to apply up to full rudder to counter roll and yaw.

Failure to limit pitch attitude will place the aircraft in a regime of reduced directional stability, rudder control and rate of climb. The aircraft may be uncontrollable at angles of attack above 20 units. Smooth rotation to $10°$ pitch attitude (approximately 14 units) will provide a good initial fly away attitude, ensure single engine acceleration and generate adequate rate of climb performance.

LANDING CONFIGURATION. While sufficient controllability exists under most flight conditions in cruise configurations, flight in the landing configuration must be approached with caution. For the asymmetric thrust emergency situation, reducing gross weight prior to landing, in addition to improving wave off performance, will reduce landing distance. Flying a higher airspeed approach (14 units AOA) provides additional rudder control power to counter thrust asymmetry.

The role of the RIO is critical in this regime. He should closely monitor airspeed, bank angle, altitude and angle of attack throughout the approach. Steep angle of bank turns, particularly into the dead engine, reduce climb performance, and may also result in rudder requirements exceeding available control deflection causing loss of control. If this situation is encountered at low altitude and

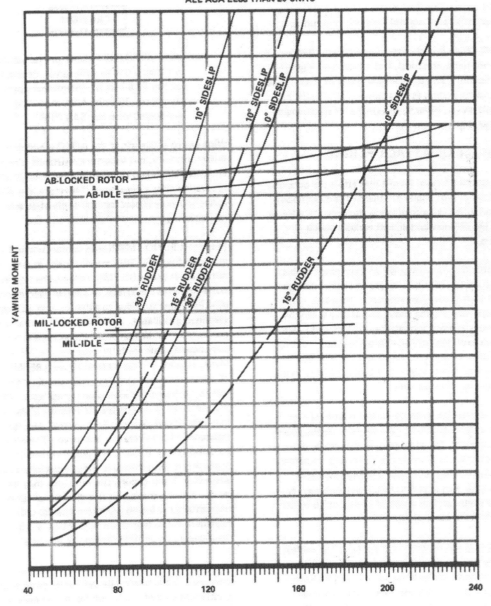

Figure 4-1. Rudder Effectiveness

available rudder control is not sufficient, the pilot may have only the option of reducing thrust on the operating engine to regain control, which may not be feasible at extremely low altitude. By performing turns away from the failed engine, both thrust and rudder control requirements will be reduced. Turns in the landing configuration should be accomplished using bank angles of 20° or less and angle of attack of 12 units or less.

Zone 5 A/B single engine waveoffs, especially at slow approach speeds (light gross weights), require the use of rudder before or simultaneously with throttle advance. Late rudder application that allows sideslip buildup may result in loss of aircraft control even with full rudder.

The aircraft design is such that no one system (flight control, pneumatic, electrical, etc) depends on a specific engine. Therefore, loss of an engine does not result in loss of any complete system as long as the HYD TRANSFER PUMP is operative. Refer to section XI for single-engine performance data.

HIGH ANGLE-OF-ATTACK FLIGHT

There are several physical factors that affect the behavior of the aircraft at high angles of attack. The most important are horizontal tail effectiveness, fuselage lift, adverse yaw from the differential tail, and variable-sweep wing effects.

The very effective horizontal tail of the aircraft enables the pilot to easily attain high angles of attack and high pitch rates even at very low airspeeds. In general this is an advantage, but in some cases, particularly when the stick is moved forward too rapidly, the high pitch rates can combine with yaw or roll motion to cause coupling and a resultant oscillation. This could cause a departure from controlled flight, aggravate a departure in progress, or even cause an inverted spin.

Fuselage lift characteristics are largely responsible for the lack of a definable stall in the aircraft. As the angle of attack increases, the wings stall, but the wide, flat fuselage continues to produce lift, so there is no sudden stall indication. There is, however, a large increase in induced drag (the byproduct of lift), which causes very high descent rates at high angles of attack; up to 9,000 feet per minute in some cases.

The differential tail provides very effective roll control throughout most of the flight envelope, but as the angle of attack increases, differential tail deflection produces increasing amounts of adverse yaw. When the angle of attack gets high enough, the differential tail produces sufficient adverse yaw to cause the aircraft to yaw, then (through dihedral effect) to roll opposite the direction of lateral stick input. The angle of attack at which this occurs varies from about 25 units AOA at slow speed to as low as 17 units AOA at Mach 0.9. Because the differential tail is such a powerful producer of adverse yaw, lateral stick deflection should be minimized at high angles of attack. Generous use of the rudders provides adequate roll control at high angles of attack and helps correct the adverse yaw produced by differential tail deflection. A summary of these characteristics is shown in figure 4-2.

> **CAUTION**
>
> With roll SAS ON, departure resistance is reduced due to adverse yaw generated by roll SAS inputs. Intentional operation above 17 units AOA should be conducted with roll SAS OFF.

Wing sweep angles aft of the AUTO schedule reduce buffet intensity, but departure resistance is reduced and more altitude is required to dive pullout when recovering after a departure. Therefore, the AUTO sweep schedule is best for high angle-of-attack maneuvering.

In aircraft BUNO 159825 and subsequent and aircraft incorporating AFC 364, maneuver-slat extension delays buffet onset below 0.7 IMN, reduces the intensity of the buffet, reduces the effects of adverse yaw, improves dihedral effect (roll due to sideslip), and increases the sustained g available. Above 0.7 IMN, buffer onset occurs prior to the maneuver-slat-extension threshold, but once the slats are fully extended, buffet is reduced. Maneuver slats will not extend above 0.85 IMN because of the wing sweep interlocks. Longitudinal trim change due to maneuver slat extension is negligible. Although the maneuver slats increase the severity of the wing rock between 20 and 28 units AOA, overall departure resistance of the aircraft is improved. The wing rock may be damped with rudders, but greater difficulty may be encountered with maneuver slats, especially at low airspeeds. If this occurs, the wing rock may be damped by neutralizing the lateral and directional controls and momentarily reducing AOA to below 20 units. Since maneuver-slat extension and retraction is fully automatic, no changes in high AOA flying techniques are required.

In aircraft BUNO 159637 and earlier without AFC 364, maneuver-flap extension delays buffer onset approximately one unit of AOA, but has no noticeable effect on high angle-of-attack handling characteristics. Speed brake position has no effect except in takeoff or landing

NAVAIR 01-F14AAA-1

HIGH ANGLE-OF-ATTACK SUMMARY
ROLL SAS ON OR OFF (NO ARI)

AIRCRAFT CONFIGURATION:
ALL EXTERNAL STORE CONFIGURATIONS

DATE: 1 MAY 1977
DATA BASIS: FLIGHT TEST

NOTE

- FULL RUDDER DEFLECTION SHOULD BE USED IF REQUIRED TO COUNTERACT ADVERSE YAW
- ROLL SAS ON INCREASES THE SEVERITY OF DEPARTURES.
- ENGINE STALLS MAY OCCUR ABOVE 25 UNITS AOA DURING THROTTLE TRANSIENTS OR STEADY STATE OPERATION LESS THAN 85% RPM.

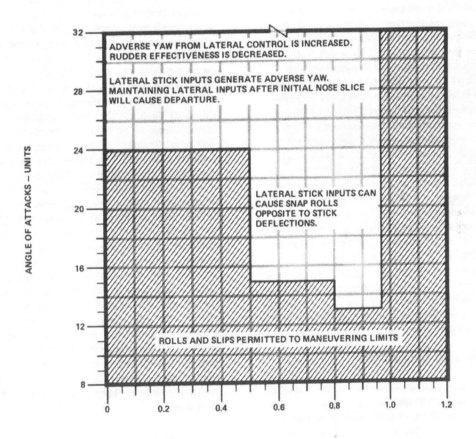

Figure 4-2. High Angle-of-Attack Summary

configurations, where extending the speed brakes improves directional stability.

External store loadings that result in aft cg locations reduce directional and longitudinal stability, resulting in increasing yaw sensitivity and a tendency for longitudinal stick force lightening above 24 units AOA. With external fuel tanks, initial aircraft response to lateral and directional control inputs are slower. The net result is a degradation in bank angle control and/or departure resistance.

CRUISE AND COMBAT CONFIGURATION

Normal Stall Approaches

In a 1-g deceleration, light buffet starts at 13 to 14 units AOA and increases to a moderate level by about 15 units. After 15 units AOA, buffet intensity varies only slightly with increasing angle-of-attack and provides no usable indication of the aircraft angle-of-attack. At 20 to 28 units AOA, reduced directional stability is apparent, and even small control inputs will cause yaw oscillations that, if unchecked, can produce a mild wing rock (±10° to 15°). Aft cg locations aggravate the yaw excursion slightly. Above 25 units AOA, lateral stick deflection causes yaw opposite to the stick deflection. The stick should be centered laterally above 25 units AOA, and the rudders used to maintain balanced flight. Rudders are effective in controlling yaw and bank angle at all angles of attack. Large rudder or lateral stick inputs produce an increase in angle of attack as sideslip increases, particularly at aft cg locations. If the deceleration is continued to full aft stick deflection, angle of attack is equivalent to about 42° to 45°, although the cockpit indicator pegs at 30 units AOA. Pitch attitude is between 10° to 20° above the horizon (no external stores) and 10° to 15° below the horizon (maximum external load). Some longitudinal porpoising may occur at full aft stick. Maneuver slats delay buffet onset to 20 to 21 units AOA and reduce buffet intensity throughout the buffet region. Wing rock at 20 to 28 units AOA will be more severe (±25°) and more difficult to damp.

Accelerated Stall Approaches

Aircraft behavior does not differ significantly from that observed in normal stall approaches. Lateral stick inputs should be avoided above 25 units AOA below Mach 0.5 and above 17 units at Mach 0.5 to 1.0. During accelerated stall approaches at higher g and higher speeds, the rates of adverse yaw and subsequent roll away from the direction of lateral stick deflection increase. Recovery from accelerated high angle-of-attack flight is the same as for normal stall.

Departure From Controlled Flight

A departure from controlled flight in cruise and combat configuration is caused by excessive sideslip and/or yaw rate. Such a departure can occur with lateral and/or directional control inputs or large thrust asymmetry due to a stalled engine at high AOA. At subsonic airspeed, the AOA at which departure occurs decreases as Mach number increases. No departures have been encountered during supersonic flight testing.

Departures usually result from the adverse yaw generated by excessive or prolonged lateral stick deflection in maneuvering flight. These departures can be minimized by using rudder for roll control at high AOA. Above 0.7 IMN, however, abrupt rudder inputs alone may generate sufficient sideslip to induce violent departures. A single engine stall at high AOA produces a powerful yawing moment that may lead to a departure if rapid action to reduce AOA is not undertaken. The departures are characterized by a rapid increase in yaw rate and lateral acceleration either away from the lateral stick deflection, or in the direction of the rudder input or the stalled engine. The yaw rate may develop into a series of uncommanded rolls. The departure may occur as low as 25 units AOA below 0.5 IMN to 17 units AOA at higher speeds. The severity of the departure increases with increasing Mach number. Indicated angle-of-attack remains pegged at 30 units AOA throughout the maneuver. The pilot's natural tendency is to oppose the roll with lateral stick. This aggravates the departure, especially if the roll SAS is engaged, and should not be done. During ACM and all other high AOA maneuvering, the aircraft should be flown with generous use of rudders, leading with or using rudder simultaneously with lateral stick deflection. Use maximum rudder, if necessary, to reduce or eliminate adverse yaw. The aircraft is significantly less prone to depart if flown with the wing sweep control in the AUTO position. Manually sweeping the wings aft of the programmed sweep angle significantly reduces the departure resistance of the aircraft. Manually selecting ROLL STAB AUG switch OFF increases departure resistance and reduces the severity of departures once they occur. While maneuver slats increase the departure resistance of the aircraft, neither maneuver flaps nor speed brakes have any noticeable effect on flight characteristics once the aircraft does depart.

Accelerated departures are initially characterized by a rapid increase in lateral acceleration, but may become violently oscillatory about all three axes. Tests have shown aircraft displacement rates in excess of 120° per second in roll and 70° per second in yaw, and positive load factors that almost double the g at entry. Pitch rates oscillate up to ± 30° per second and lateral acceleration oscillates up to ±0.6 g. These oscillations may cause pilot disorientation, and proper recovery controls may not be obvious. If this occurs, the proper response would be to neutralize rudders and lateral stick and apply forward longitudinal stick. Recovery indications should become apparent within two turns.

Note

The most important action of any departure recovery is reducing the AOA. This is enhanced by timely application of forward stick and reducing or countering, to the extent feasible, any yawing moment on the aircraft.

Yawing moments may be generated by:

- Pilot commanded control system inputs (either intentional or unintentional)
- Uncommanded roll SAS inputs
- Thrust asymmetry

The first two may be minimized by verifying that lateral stick and rudders have been neutralized. Military power should be applied to both engines. This action will provide maximum compressor stall margin; yawing moment resulting from a stalled engine can be neutralized with opposite rudder except at very slow airspeeds. The flightcrew should lock shoulder harnesses as soon as possible into a departure to maintain body position and delay incapacitation.

Note

Refer to Vertical Stall Approaches for engine procedures in extremely nose high, decaying airspeed situations.

Flat Spin

A flat spin is recognized by the flat aircraft attitude (approximately 10° nose down with no pitch or roll oscillations), steadily increasing yaw rate, and longitudinal acceleration (eyeballs-out g). It may develop within two to three turns following a departure if yaw is allowed to accelerate without rapid, positive steps to effect recovery. Yaw rate can result from departure induced by aerodynamic controls, a large thrust asymmetry, or a combination of both. The aircraft may first enter an erect oscillatory spiral as airspeed rapidly decreases. Frequent hesitations in yaw and roll may occur as yaw rate increases. The turn needle is the only valid indication of spin direction. It always indicates turn direction correctly, whether erect or inverted. AOA will peg at 30 units and airspeed will oscillate between 0 and 100 knots. The aircraft may also depart by entering a coupled roll where yaw rate may build up without being noticed to the point that when roll stops, yaw rate is sufficient to sustain a flat spin. A large thrust asymmetry; that is, engine stall at high angle of attack or low airspeed, can also produce sufficient yaw rate to drive the aircraft into a flat spin. In all instances, recovery should be accomplished by prompt application of departure recovery procedures to reduce angle of attack and control yaw rate.

Regardless of the method of entry, once the flat spin has developed, the flat aircraft attitude, (10° nose down) steadily increasing yaw rate, and buildup of longitudinal g forces not accompanied by roll and/or pitch rates will be apparent to the flightcrew. AOA will be pegged at 30 units, yaw rate will be fast (as high as 180° per second) and altitude loss will be approximately 700 feet per turn. Longitudinal acceleration can reach as high as 6.5g. Time between aircraft departure and flightcrew recognition of a fully developed flat spin depends upon the nature of the entry (accelerated departure, low speed stalled engine, etc.). The time between recognition of a flat spin and buildup of incapacitating longitudinal g forces is dependent upon aircraft loading, thrust asymmetry, flight control position during spin entry and spin, locked or unlocked harness, tightness of the lap restraints, and flightcrew physical condition and stature. Test data indicate that following recognition of a flat spin, the pilot may be able to maintain antispin controls for 15 to 20 seconds (approximately 7 to 10 turns) but may severely jeopardize his ability to eject due to the incapacitation that occurs as the g forces build. Successful F-14 flat spin recover procedures have not been demonstrated; therefore, once the aircraft is confirmed to be in a flat spin, the flightcrew should jettison the canopy and eject, this decision should not be delayed once the flat spin is recognized.

It is important to understand that longitudinal g forces can be present in accelerated departures from controlled flight and ejection initiated solely because of longitudinal g forces is premature.

To preclude premature ejection from a recoverable aircraft, verify that the aircraft is not rolling or oscillating in pitch or is not in a coupled departure. If any of these characteristics are evident, then a flat spin has not developed and departure recovery procedures should be continued. (Refer to Section V, Part 3.)

Vertical Stall Approaches

If the aircraft is allowed to decelerate to zero airspeed in a vertical or near vertical attitude, it will slide backwards momentarily, then pitch down to a near vertical dive. The rate of nosedown pitch is about 20° per second at all wing sweeps up to 50°. Aft of 50°, the pitch rate increases until, at 68° wing sweep, the pitch rate is about 30° per second. The aircraft may pass through the vertical to near level flight attitude, yaw in one direction, and then return to a vertical dive attitude. This may occur more than once. This tendency is more pronounced at aft wing sweeps, but can usually be controlled with longitudinal control inputs. Some pitchovers may be accompanies by yaw and/or roll, which will dampen without pilot action as the aircraft accelerates. Rudder and lateral stick are also effective in damping oscillations once the aircraft is nose-low and accelerating. The aircraft is very responsive to longitudinal stick inputs at all angles-of-attack at speeds above 100 KIAS.

VERTICAL RECOVERY

In a nose-high, rapidly decreasing airspeed condition:

- At less than 130 KIAS or above 40,000 feet altitude, if in afterburner, leave throttles fixed until recovery is complete.

- Below 40,000 feet altitude at greater than 130 KIAS, if in afterburner, reduce throttles to MIL as soon as possible and while forward airspeed still exists.

- Any altitude, throttles are below MIL, advance them to MIL as soon as possible and while forward airspeed still exists.

1. 100 KIAS use longitudinal stick to pitch the nose down. At extreme nose-high attitudes, aft stick facilitates recovery time and will avoid prolonged engine operation with zero oil pressure.

2. Below 100 KIAS, release controls and wait for aircraft to pitch nosedown. This prevents depleting hydraulic pressure if both engines fail, and provides the quickest recovery.

3. If roll and/or yaw develop, wait until aircraft is nose-down and accelerating before correcting with rudder or lateral stick.

4. Use longitudinal control as necessary to keep nose down and accelerate.

5. Above 100 KIAS, pull out, using 17 units AOA.

6. Recovery to level flight from point of pitchover can normally be completed in less than 10,000 feet.

Inverted Stall Approaches

As in normal stall approaches, there is no clearly defined inverted stall. A moderate rate application of full forward stick in inverted flight results in a negative angle of attack of about -30°.

CAUTION

Dynamic forward stick inputs of moderate rate may exceed the negative g limit of -2.5 g.

Indicated AOA will show zero beyond about -5° true angle of attack. Lateral stick provides effective roll control and rudders are very powerful in generating yaw and roll at all attainable negative angles of attack. Dihedral effect is negative. Therefore, a right rudder input produces right yaw, but left roll. This feels natural to the pilot in inverted flight, and enables raising a wing with opposite rudder when inverted. Oil pressure will indicate zero and the OIL PRESS caution light and MASTER CAUTION light, will illuminate while at negative angles of attack.

WARNING

Zero or negative g flight in excess of 10 seconds in afterburner or 20 seconds in military power or less depletes fuel feed tanks (cells 3 and 4), causing flameout of both engines.

INVERTED STALL RECOVERY

Recovery from an inverted stall is performed by applying full aft stick, while neutralizing lateral stick, to return to positive g flight. Recovery from negative g conditions will usually occur immediately. Return to level flight can then be performed from the resultant nosedown attitude by rolling erect with rudder and/or lateral stick and pulling out at 17 units AOA.

Inverted Spin

An inverted spin may be encountered if the aircraft unloads while there is a yaw rate present. In flight tests, the inverted spin has been caused by holding full forward stick while inverted, applying full rudder and holding this combination through 360° of roll. Prospin controls need not be held to maintain the aircraft in a spin. The inverted spin is primarily identified from cockpit instruments by less-than zero g and an angle of attack of zero. Since the inverted spin is quite disorienting, spin direction must be determined by observing the turn need deflection. Altitude loss during the inverted spin is 800 to 1800 feet per turn and time per spin turn is 3 to 6 seconds. Nose attitude in the inverted spin is approximately 25° below the horizon. The engines will encounter compressor stalls regardless of thrust setting during the inverted spin. In order to have the greatest possible engine stall margin, military thrust should be maintained until a compressor stall occurs. Upon encountering an engine compressor stall, both throttles should be moved smoothly to IDLE to reduce asymmetric thrust and to avoid engine overtemperature. If TIT remains high and RPM low, shut down the affected engine. If both engines have stalled and overtemped, shut down only one engine to have hydraulic pressure available for recovery and subsequent flight path control. After recovery the shutdown engine can then be restarted and the remaining stalled engine shut down. Warning of possible inverted spin usually occurs sufficiently in advance for the aircrew to take corrective action. Warning is usually very noticeable in the form of a nose down pitch (negative g) with a yawing and possible rolling motion that is quite uncomfortable to the aircrew. In the fully-developed inverted spin, rudder opposite yaw/turn needle is the strongest antispin control. Aft stick is a strong antispin control during the incipient spin phase and a weak antispin control in the fully developed inverted spin. In the absense of asymmetric thrust, the antispin control inputs will recover a fully-developed inverted spin within one turn. Although lateral stick opposite yaw is an antispin control, it is not included in the recovery procedures because at full aft stick only 4° differential tail split is possible. Spin tunnel tests indicate this small lateral control has negligible effect on inverted spin recovery characteristics. In addition, opposite rudder recovers the spin very quickly and lateral control present right after spin recovery could induce a post-recovery departure. (Refer to Section V, Part 3.)

ENGINE OPERATING CHARACTERISTICS

High Angle of Attack (AOA) and Tactical Maneuvering

The engines generally operate satisfactorily at extremely high positive or negative AOA but are prone to compressor stalls if high AOA is combined with:

- Sideslip
- High altitude and low airspeed
- Throttle settings below 90% RPM
- Throttle movements
- Jet wash

Each of these conditions reduces the engine's compressor stall margin, which is already reduced when operating at high AOA. Excessive sideslip, even at low AOA, may result in compressor stalls. High AOA combined with sideslip may result in compressor stalls anywhere in the aircraft flight envelope. The greatest stall margin (least likelihood of stall) is provided when the engine is stabilized at MIL power. When operating at 20 units AOA or greater, throttles should remain fixed at MIL power or stabilized afterburner (AB). There is no way to predict when a compressor stall will occur, but there is a high likelihood of inducing one by changing throttle settings at high AOA. In particular, AOA should always be reduced below 20 units prior to selecting or deselecting AB. Engine stalls may occur at any power setting anywhere in the aircraft flight envelope, but they are most prevalent in the regions shown in figure 4-3.

Subsonic compressor stalls may be indicated by audible banging or chugging, or they may be inaudible, in which case high TIT and low RPM are the first indications. At high AOA and slow airspeed, rapid yaw acceleration or nose slice may be the first indication. The higher the thrust setting at stall, the more likely the stall will be audible, and the more likely the engine will suffer turbine damage. The applicable emergency procedure is described in Section V, Emergency Procedures. Particular emphasis must be placed on reducing AOA, smoothly retarding both throttles to IDLE, rapidly determining the stalled engine(s), and, if the stall does not clear, shutting the stalled engine down.

WARNING

When a compressor stall occurs at high AOA and/or extreme nose high attitudes, ensure that yaw rate and AOA are under control and thrust asymmetry is minimized before attempting to analyze engine instruments.

Since a stalled engine produces negligible thrust, single engine compressor stalls at high power settings result in large asymmetric thrust moments from the good engine. Should this occur at slow airspeed or high AOA, there is a danger of rapid yaw rate buildup and possible flat spin entry if corrective action is not taken immediately. Corrective action consists of unloading the aircraft, reducing

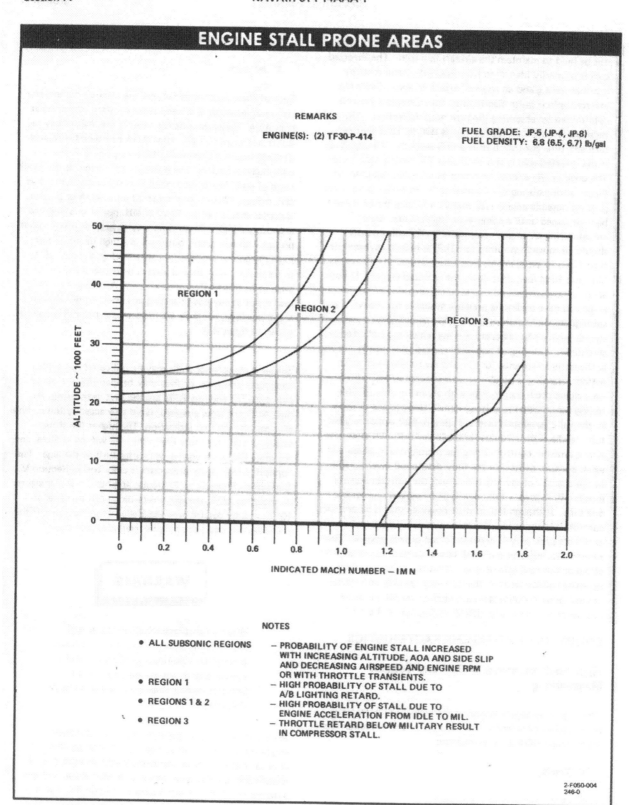

Figure 4-3. Engine Stall Prone Areas

the thrust asymmetry, and countering yaw rate. How these are accomplished depends on the flight conditions; in the worst case, an extreme nose high attitude and airspeed rapidly decreasing below 100 KIAS, the pilot will be facing several seconds where flight controls will be ineffective due to the lack of airspeed. The pilot should immediately retard both throttles smoothly to IDLE, thereby minimizing thrust asymmetry, while maintaining neutral controls. If performed expeditiously, these actions should prevent yaw rate buildup and allow the aircraft nose to fall through and regain flying speed. After throttles are reduced, the pilot should lock his harness in anticipation of a possible departure.

WARNING

Compressor stalls encountered while operating at high AOA must be dealt with immediately, since flight control effectiveness may be insufficient to counter the yaw rate generated by asymmetric thrust.

Note

Although reducing throttles to IDLE may induce a compressor stall on the good engine, this is preferable to entering a flat spin.

Once the pilot has minimized thrust asymmetry and ensured that yaw rate/AOA are under control, he should immediately check engine instruments to determine if any further action is required. If TIT remains above 1000°C at IDLE power and/or engine response is not normal, the stalled engine must be shut down to avoid serious turbine damage, and the sooner this is accomplished, the better.

CAUTION

Engine turbine damage may result from prolonged stalls (greater than 20 seconds) with TIT stabilized as low as 1,000°C, or from short duration stalls (less than 5 seconds) with TIT stabilized above 1,300°C.

If both engines are stalled after retarding throttles to IDLE, at least one engine must be secured immediately to prevent turbine damage and provide maximum potential for an airstart. If possible, secure the engine that did not initiate the event (the second engine to stall). The cause of the first engine stall may not be known at this point; however, it is highly probable that the second stall will have been induced during the throttle transient to IDLE. Leaving one engine in hung stall minimizes the likelihood of total loss of hydraulic and electrical power (emergency generator). The method of relight to be used (windmill or spooldown) depends on the situation. If sufficient altitude, airspeed, and hydraulic pressure exist, a windmill airstart should be attempted. Should flight conditions (such as low altitude) require a rapid airstart, perform a spooldown airstart as soon as TIT drops below 400°C. See Section V for a more detailed discussion of dual engine stall and airstart procedures.

Missile Gas Ingestion

Refer to NAVAIR 01-F14AAA-1A.

TAKEOFF AND LANDING CONFIGURATION

Normal Stalls

During deceleration in a level, 1-g stall approach, light buffet starts at about 19 units AOA. Buffet does not significantly change thereafter as the angle of attack is increased and provides no usable stall warning. A reduction in stick force is felt between 24 and 28 units AOA. At 25 units AOA divergent wing rock and yaw excursions define the stall. Yaw angle may reach 25°, and bank angle 90° within 6 seconds if the angle of attack is not lowered. Above 24 units AOA, lateral stick inputs produce considerable adverse yaw. Large sideslip angles, whether produced by rudder or lateral stick, are accompanied by a nose rise, which may require forward stick application to prevent the angle of attack from increasing to 28 units. Extending the speed brakes slightly aggravates the stick force lightening at 24 units AOA but improves directional stability significantly, reducing the wing rock and yaw tendency at 25 units AOA. Stall approaches should not be continued beyond the first indication of wing rock. When wing rock occurs, the nose should be lowered and no attempt should be made to counter the wing rock with lateral stick or rudder.

Stalls with the landing gear extended and flaps up are similar to those with flaps extended. Buffet starts at 16 to 18 units AOA and wing rock at 26 units AOA.

Figure 4-4 shows stall speeds for standard day temperature at sea level with slats/flaps extended and gear down.

Note

In aircraft BUNO 159825 and subsequent and those aircraft incorporating AFC 400, a rudder pedal shaker is installed as a warning to the pilot of excessive angle of attack during approach. With the flaps 25° or greater and angle of attack greater than approximately 20 units, the pedal shaker is automatically activated.

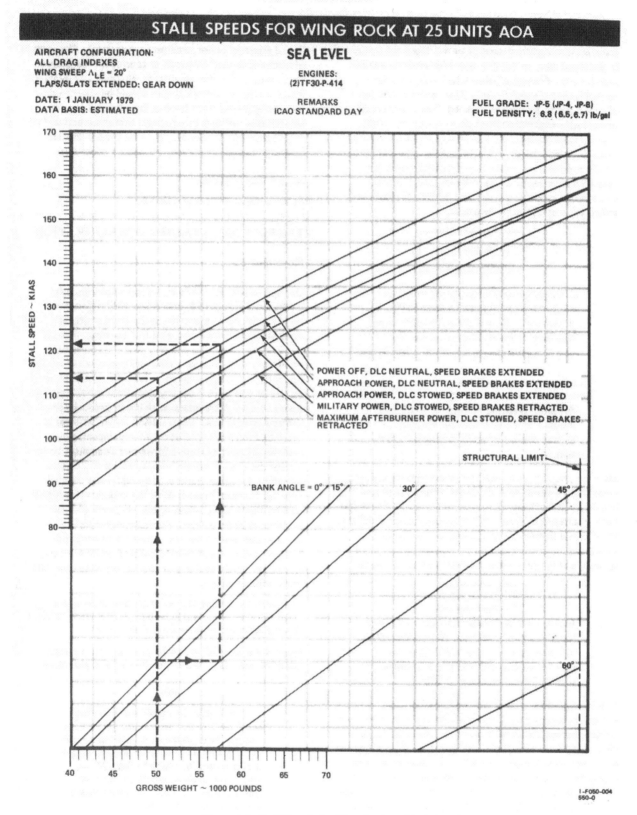

Figure 4-4. Stall Speeds for Wing Rock at 25 Units AOA

STALL RECOVERY

1. Relax aft stick force and ease stick forward, if necessary, to decrease angle-of-attack to less than 16 units.

2. Maintain 15 to 16 units, 2 g or less, and military thrust during roundout to level flight.

Note

If in afterburner, remain there.

3. Recovery to level flight requires about 1,000 feet of altitude.

DEGRADED APPROACH CONFIGURATION

No Flaps, No Slats, and Wings at 20°

VFR carrier approaches flown without flaps/slats available should be initiated at 180° position 1-1/2 miles from the ship. The approach is flown at 15 units AOA (14 units AOA if AFB 198 has not been complied with). Airframe buffet will occur at 16 to 17 units AOA with wing drop (5 to 10°) and/or an increase in sink rate occurring at 16.5 to 17.5 units AOA. If AFB 198 has not been complied with, airframe buffet and wing drop may occur at 0.5 to 1 unit lower AOA. Spoiler effectiveness is slightly degraded because of the absence of the aerodynamic slot formed when the flaps are extended. Precise airspeed control is essential for a no-flap/no-slat approach. Fast or high/fast approaches result if timely throttle adjustments are not made throughout the approach. The pilot must waveoff approaches that result in large throttle reductions (to near idle) in close.

CAUTION

Nose altitude control is more sensitive during a no-flap approach, and care must be exercised not to overcontrol nose corrections in close. Cocked-up, high-sink landing can result in damage to ventral fins and/or afterburners. Avoid slow approaches. Wing drop and increased sink rate may occur at 16 to 17 units AOA.

Single Engine

A straight-in-approach should be flown for single engine approaches. Any maneuvering required prior to rolling out on final approach should be accomplished using 12 units AOA or less, at bank angles of 20° or less. Turns into the failed engine should be avoided whenever practicable. Once established on final approach, optimum angle of attack is 14 units. Rudder trim, augmented as necessary by additional rudder pedal deflection should be used to counter thrust asymmetry. Airspeed control is more difficult than for two-engine approaches, therefore, there may be a tendency to overcontrol power. Speed brakes should remain retracted during actual single engine approaches. DLC should not be selected since it will be either unavailable with the left engine secured or subject to frequent, inadvertent disengagements if military power is required during left engine only operation. If afterburner is used, the pilot should be prepared to use up to full rudder to counter the large yaw moment produced. Rudder should be simultaneously applied with afterburner. Lateral stick should not be used alone to counter yawing moment. However, if lateral stick is used, it should be applied coincidental with or after rudder input. A maximum of 14 units AOA should be used on all waveoffs since rotation to higher angles of attack does not significantly reduce sink rates, but does increase the likelihood of an in-flight engagement. Single engine approaches, waveoffs, and bolters may be satisfactorily performed up to the gross weight limits of two-engine operations. (See Asymmetric Thrust Flight Characteristics.)

Outboard Spoiler Module Failure

When the wings are forward of 57°, loss of outboard spoilers results in a decrease in roll authority and in lateral control effectiveness. Such loss causes no significant degradation in approach handling characteristics and is generally only apparent when large bank angles changes are commanded, such as during roll into and out of the approach turn. If the outboard spoiler module fails when the flaps and slats are down, the spoilers may float up and lock at some position above neutral. This may be accompanied by trim changes in all three axes, which can be trimmed out. Approach speed will increase slightly if a spoiler float occurs. If the failure occurs when the flaps are up, spoiler float is minimized.

WARNING

In the event of outboard spoiler module failure, do not engage DLC or ACLS.

SAS Off

Approach characteristics are not significantly degraded with partial or total SAS failure. The aircraft is slightly more sensitive to longitudinal control inputs if pitch SAS is lost. Lateral-directional response to turbulence increases if yaw SAS is lost. Roll SAS failure results in slightly increased roll sensitivity.

Note

Pitch SAS loss may result in loss of outboard spoilers. Roll SAS loss may result in loss of inboard spoilers.

Aft Wing Sweep Landings

Optimum angle of attack for aft wing sweep approaches is 15 units. Airspeeds for various configurations are shown in figure 4-5. As aft wing sweep increases, trimmed stick position moves aft and the aircraft becomes more sensitive to pitch and roll inputs. Both vertical and lateral gusts response decrease during landing at aft wing sweep angle. Aft of 57° wing sweep, roll control is achieved by use of differential tail alone and increasing yawing movements are apparent for lateral control inputs (adverse yaw). At aft wing sweeps, approach power compensator performance is degraded and it should not be used.

The 180° position should be flown wider than normal in proportion to the increase in airspeed; however, for wing-sweep angles greater than 57°, straight-in approaches should be used to minimize effects of adverse yaw. Main flaps and slats or maneuver flaps should be used (if available) to decrease approach airspeeds.

WARNING

Sweeping wings aft with main flaps and slats extended results in significant nose-down trim changes. Conversely, sweeping wings forward causes significant noseup trim changes.

If speed brakes are not available, thrust requirements on glide slope are decreased and judicious throttle management is more critical.

WARNING

If maneuver flaps are used, the pilot must ensure that the maneuver flap thumbwheel is not actuated when at optimum angle of attack.

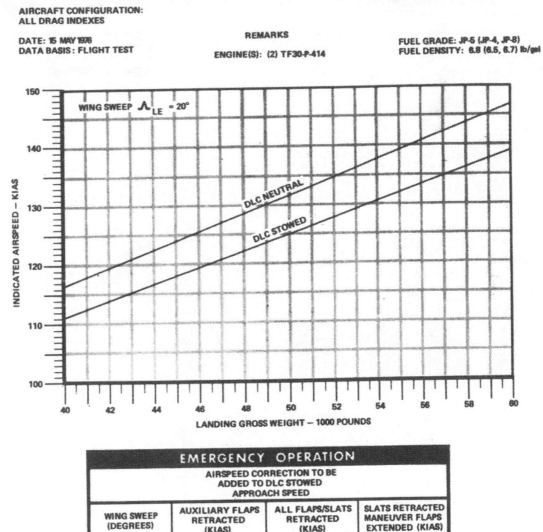

Figure 4-5. Landing Approach Airspeed (15 units AOA)

THIS PAGE INTENTIONALY LEFT BLANK.

ISBN #978-1-935327-71-4   1-935327-71-2

Made in the USA
Middletown, DE
15 October 2023

40812441R00194